C. F. Lewis

A MANUAL

OF THE PRINCIPLES AND PRACTICE OF

ROAD-MAKING:

COMPRISING THE

LOCATION, CONSTRUCTION, AND IMPROVEMENT

OF

ROADS,

(COMMON, MACADAM, PAVED, PLANK, ETC.,)

AND

RAIL-ROADS.

BY W. M. GILLESPIE, LL.D., C.E.

TENTH EDITION, WITH LARGE ADDENDA.

EDITED BY
CADY STALEY, A.M., C.E.

"Every judicious improvement in the establishment of Roads and Bridges increases the value of land, enhances the price of commodities, and augments the public wealth."
DE WITT CLINTON.

A. S. BARNES & COMPANY,
NEW YORK AND CHICAGO.
1874.

PREFACE.

The common roads of the United States are inferior to those of any other civilized country. Their faults are those of direction, of slopes, of shape, of surface, and generally of deficiency in all the attributes of good roads. Some of these defects are indeed the unavoidable results of the scantiness of capital and of labor in a new country, but most of them arise from an ignorance either of the true principles of road-making, or of the advantages of putting these principles into practice. They may therefore be removed by a more general diffusion of scientific instruction upon this subject, and to assist in bringing about this consummation is the object of the present volume. In it the author has endeavored to combine, in a systematic and symmetrical form, the results of an engineering experience in all parts of the United States, and of an examination of the great roads of Europe, with a careful digestion of all accessible authorities, an important portion of the matter having never before appeared in English. He has striven to reconcile the many contradictory theories and practices of road-making; to select from them those which are most in accordance with the teachings of science; to present as clearly and precisely as possible the leading features of those approved, laying particular stress on such as are most often violated or neglected; and to harmonize the successful but empirical practice of the English engineers with the theoretical but elegant deductions of the French.

Before the construction of a road is commenced, its makers should well determine "*What it ought to be*," in the vital points of direction, slopes, shape, surface and cost. This is therefore the first topic discussed in this volume. The next is the "*Location*" of the road, or the choice of the ground over which it should pass, that it may fulfil the desired conditions. In this chapter are given methods of performing all the necessary measurements of distances, directions and heights, without the use of any instruments but such as any mechanic can make, and any farmer use. The "*Construction*" of the road is next explained in its details of Excavation, Embankment, Bridges, Culverts, &c. At this stage of progress our road-makers too generally stop short, but the road should not be considered complete till "*The Improvement of its surface*" has been carried to as high a degree of perfection as the funds of the work will permit. Under this head are examined earth, gravel, McAdam, paved, plank and other roads. "*Rail-roads*," and their motive powers, are treated of in the next chapter. The "*Management of town roads*" is last taken up, the evils of the present system of Road-tax are shown, and a better system is suggested. In the "*Appendix*" are minute and practical examples of the calculations of Excavation and Embankment.

To enable this volume the better to attain its aim of being doubly useful, as a popular guide for the farmer in improving the roads in his neighborhood, and as a College Text book, introductory to the general study of Civil Engineering, the mathematical investigations and professional details have been printed in smaller type, so as to be readily passed over by the unscientific reader.

Note.—The additions in this edition of the Manual of Roads and Rail-roads, are from the notes of the author's lectures to the Civil Engineering classes in Union College. C. S.

AUTHORITIES REFERRED TO.

Alexander. Amer. Ed. of Simms on Levelling, Baltimore, 1837.
Annales des Ponts et Chaussées, Paris.
Anselin. Experiences sur la main-d'œuvre des differens travaux, Paris.
Babbage. Economy of Machinery and Manufactures, London, 1831.
Berthault-Ducraux. De l'Art d'entretenir les Routes, Paris, 1837.
Bloodgood. Treatise on Roads, Albany, 1838.
Chevalier. Les voies de communication aux Etats Unis, Paris, 1843.
Civil Engineers' and Architects' Journal, London.
Cresy. Encyclopedia of Civil Engineering, London, 1847.
Davies. Elements of Surveying, &c., New York, 1845.
Delaistre. La Science des Ingenieurs, Paris, 1825.
Dupin. Applications de Géometrie, Paris, 1822.
" Travaux civils de la Grande Bretagne, Paris, 1824.
Eaton. Surveying and Engineering, Troy.
Edgeworth. Construction of Roads and Carriages.
Flachat & Bonnet. Manuel et Code des Routes et Chaussées, Paris.
Frome. Trigonometrical Surveying, London, 1840.
Gayffier. Manuel des Ponts et Chaussées, Paris, 1844.
Gerstner. Memoire sur les grandes routes, Paris, 1827.
Grieg. Strictures on Road-police, London.
Griffith. On Roads, London.
Hughes Making and Repairing Roads, London.
Journal de l'Ecole Polytechnique, Paris.
Journal of the Franklin Institute, Philadelphia.
Jullien. Manuel de l'Ingenieur Civil, Paris, 1845.
Laws of Excavation and Embankment on Railways, London, 1840.
Lecount. Treatise on Railways, London, 1839.
Macneill. Tables for calculating Cubic quantities of Earthwork, London
Mahan. Course of Civil Engineering, New York, 1846.
Marlette. Manuel de l'Agent-voyer, Paris, 1842.
McAdam. System of Road-Making, London, 1825

Millington Civil Engineering, Philadelphia, 1839.
Morin. Aide-Memoire de Mécanique, Paris, 1843.
Mosely. Mech principles of Engineering and Architecture, London, 1843
Navier. Travaux d'entretien des Routes, Paris, 1835.
" Application de la Mécanique aux constructions, Paris.
Nimmo. On Roads of Ireland, &c.
Parnell. Treatise on Roads, London, 1838.
Paterson. Practical Treatise on Public Roads, &c., Montrose, 1820.
Penfold. On Making and Repairing Roads, London, 1835.
Poncelet. Mécanique Industrielle, Paris, 1841.
Potter. Applications of Science to the Arts, New York, 1847.
Railroad Journal, New York, 1832–1847.
Renwick. Practical Mechanics, New York, 1840.
Reports of U. S. Commissioner of Patents, Washington.
Reports of U. S. Engineer Corps, Washington.
Reports to Parliament on Holyhead roads, &c., London.
Ritchie. Railways, London, 1846.
Road Act of New York, Rochester, 1845.
Roads and Railroads, London, 1839.
Sganzin. Course of Civil Engineering, Boston, 1837.
" Cours de Construction par Reibell, Paris, 1842.
Simms. Telford's rules for making and repairing roads, London
" Public Works of Great Britain, London.
" Sectio-Planography, London.
Stevenson. Civil Engineering of North America, London, 1838
Telford. Reports on Holyhead roads, London.
Tredgold. On Railroads, London, 1835.
Wood. On Railroads, Philadelphia, 1832.

ANALYTICAL TABLE OF CONTENTS.

APPENDIX A.

A

MANUAL OF ROAD-MAKING.

INTRODUCTION.

The Roads of a country are accurate and certain tests of the degree of its civilization. Their construction is one of the first indications of the emergence of a people from the savage state; and their improvement keeps pace with the advances of the nation in numbers, wealth, industry, and science—of all which it is at once an element and an evidence.

Roads are the veins and arteries of the body politic, for through them flow the agricultural productions and the commercial supplies which are the life-blood of the state. Upon the sufficiency of their number, the propriety of their directions, and the unobstructedness of their courses, depend the ease and the rapidity with which the more distant portions of the system receive the nutriment which is essential to their life, health, and vigor, and without a copious supply of which the extremities must languish and die.

But roads belong to that unappreciated class of blessings, of which the value and importance are not fully felt because of the very greatness of their advantages, which are so manifold and indispensable, as to have rendered their extent almost universal and their origin forgotten. Perhaps we will better appreciate them, if we endeavor to imagine what would be our condition if none had ever been constructed.

Suppose, then, that a traveller had occasion to go from Boston to Albany, and that no road between the two places was yet in existence. In the first place, how would he find his way? Even if he knew that his general direction should be towards the setting sun, the sun would be often hidden by day, and the stars by night; and, there being no roads, there would be no engineers and no surveyor's compass. The moss upon the north side of the trees might be in some degree a guide to him, if he were skilled in woodcraft; but he would at last become so bewildered, that, like lost hunters on the prairies, he would begin to believe that the sun rose in the west, set in the east, and was due north at mid-day.

Allowing, however, that he was fortunate enough to retain the true direction, would he be able to follow it? In the forest he must force for himself a passage through the tangled underwood, and make long circuits around the fallen trees, which no axe-men have as yet cleared away. Through the swamp he must struggle amid the slippery and deceitful mud, for no road-maker has yet built the causeway. Over the mountain he must clamber only to again descend, for topographical science has not taught him how much he would gain by winding around its base. The rocky walls of precipices he must arduously climb, and perilously descend, for no engineer has as yet blasted a passage through them. Meeting a deep river, or even a mere mountain torrent, if he cannot ford or swim it, he must seek its head with many miles of added travel, to be doubled again by his return to his original direction. All this while, too, he can subsist only by precarious hunting; for, there being no roads, there would be no inns, and he can scarcely carry himself along, much less a store of provisions.

Look now at the contrast, and at the ease, speed, and comfort with which the modern traveller flies from place to place upon that best of all roads, a railroad.

But the increase of personal comfort is only a petty item in estimating the importance of roads, even in despite of Dr. Johnson's exclamation, that life has no greater pleasure than being whirled over a good road in a post-chaise. More important is the consideration, that, in the absence of such facilities, the richest productions of nature waste on the spot of their growth. The luxuriant crops of our western prairies are sometimes left to decay on the ground, because there are no rapid and easy means of conveying them to a market. The rich mines in the northern part of the state of New York are comparatively valueless, because the roads among the mountains are so few and so bad, that the expense of the transportation of the metal would exceed its value. So, too, in Spain, it has been known after a succession of abundant harvests, that the wheat has actually been allowed to rot, because it would not repay the cost of carriage.* In that country, for similar reasons, sheep are killed for their fleece only, and the flesh is abandoned; as is likewise the case with cattle in Brazil, slaughtered merely for their hides.

Such are the effects of the almost total want of roads. Among those which do exist, the difference, as to ease, rapidity, and economy of transportation, caused by the various degrees of skill and labor bestowed upon them, is much greater than is usually imagined, particularly by farmers, whom they most concern.

---

* Edinburgh Review, lxv. 448.

2

One important difference lies in the *grades* or longitudinal slopes of a road. Suppose that a road rises a hundred feet in the distance of two thousand feet. Its ascending slope is then one in twenty, and (as will be hereafter proven) one-twentieth of the whole load drawn over it in one direction, must be actually *lifted up* this entire height of one hundred feet. But upon such a slope a horse can draw only *one half* as much as he can upon a level road, and two horses will be needed on such a road to do the usual work of one. If the road be intrusted to the care of a skilful engineer, and be made level by going round hills instead of over them, or in any other way, there will be a saving of one half of the former expense of carriage on it.

Another great difference in roads lies in the nature of their *surfaces*; one being hard and smooth, and another soft and uneven. On a well-made road of broken stone, a horse can draw *three times* as much as he can upon a gravel road. By making, then, such a road as the former (according to the instructions in Chapter IV.) in the place of the latter, the expenses of transportation will be reduced to one-third of their former amount, so that two-thirds will be completely saved, and two out of three of all the horses formerly employed can then be dispensed with.* If such an improvement can be made for a sum of money, the interest of which will be less than the total amount of the annual saving of labor, it will be true economy to make it, however great the original outlay; for the de

* In the absence of such an improvement, when the Spanish government required a supply of grain to be transferred from Old Castile to Madrid, 30,000 horses and mules were necessary for the transportation of 480 tons of wheat. Upon a broken-stone road of the best sort, *one-hundredth* of that number could easily have done the work.

cision of all such questions depends on considerations of comparative profit. This part of the subject will be more minutely examined at the end of Chapter I., in considering "*What roads ought to be as to their cost.*"

The profits of such improvements are not confined to the proprietors of a road, (whether towns, or companies re munerated for these expenditures by tolls) but are shared by all who avail themselves of the increased facilities; consumers and producers, as well as road-owners. If wheat be worth in a city a dollar per bushel, and if it cost 25 cents to transport it thither from a certain farming district, it will there necessarily command only 75 cents. If now by improved roads the cost of carriage is reduced to 10 cents, the surplus 15 cents on each bushel is so much absolute gain to the community, balanced only by the cost of improving the road. Supposing that a toll of 5 cents will pay a fair dividend on this, there remains 10 cents per bushel to be divided between the producer and the con sumer, enabling the former to sell his wheat at a higher price than before, while at the same time the latter obtains it at a less cost.

Agriculture is thus directly, and likewise indirectly, de pendent in a great degree upon good roads for its success and rewards. *Directly*, as we have just seen, these roads carry the productions of the fields to the markets, and bring to them in return their bulky and weighty materials of fertilization, at a cost of labor which grows less and less as the roads become better. *Indirectly*, the cities and towns, whose dense population and manufacturing indus- try make them the best markets for farming produce, are enabled to grow and to extend themselves indefinitely by roads alone, which supply the place of rivers, to the banks of which these great towns would otherwise be necessarily confined.* While therefore, it would be an inexcusable waste of money to construct a costly road to connect two small towns which had little intercourse, it would be equally wasteful, and is a much more frequent short-sightedness of economy, to leave unimproved and almost in a state of nature, the communications between a great city and the interior regions from which its daily sustenance is drawn, and into which its own manufactures are conveyed.

Some of the advantages thus to be attained, have been well summed up in a report of a committee of the House of Commons :

" By the improvement of our roads, every branch of our agricultural, commercial, and manufacturing industry would be materially benefited. Every article brought to market would be diminished in price ; and the number of horses would be so much reduced that, by these and other retrenchments, the expense of FIVE MILLIONS [*pounds sterling*] would be ANNUALLY saved to the public. The expense of repairing roads, and the wear and tear of carriages and horses, would be essentially diminished ; and thousands of acres, the produce of which is now wasted in feeding unnecessary horses, would be devoted to the production of food for man. In short, the public and private advantages which would result from effecting that great object, the improvement of our highways and turnpike roads, are incalculable ; though, from their being spread over a wide surface, and available in various ways, such advantages will not be so apparent as those derived from other sources of improvement, of a more restricted and less general nature."

---

* McCulloch, Dictionary of Commerce.

The changes in the condition of a country which such improvements effect, are of the highest importance. There is as much truth as blundering in the famous couplet written by an enthusiastic admirer of the roads which Marshal Wade opened through the Scottish Highlands :

> " Oh, had you only seen these roads before they were made,
> You would lift up your eyes and bless Marshal Wade !"

His military road is said to have done more for the civilization of the Highlands than the preceding efforts of all the British monarchs. But the later roads under the more scientific direction of Telford, produced a change in the state of the people which is probably unparalleled in the history of any country for the same space of time. Large crops of wheat now cover former wastes ; farmers, houses and herds of cattle are now seen where was previously a desert ; estates have increased sevenfold in value and annual returns ; and the country has been advanced at least one hundred years. In Ireland similar effects have been produced, and the face of the country in some districts has been completely renovated. The enlarged labors of the public works, now undertaken in that country by the government, though commenced only for temporary relief, will not fail to produce great permanent benefits.

The moral results of such improvements are equally admirable. Telford testifies that in the Highlands they greatly changed for the better the habits of the great working class. Thus, too, when Oberlin wished to improve the spiritual condition of his rude flock, he began by bettering their physical state, and led out his whole people to open a road of communication between their secluded valley and the great world without. The wonderful moral and intellectual amelioration which ensued was an unmistakeable tribute to the civilizing and elevating influence of good roads.

Among the most remarkable consequences of the improvement of roads, is the rapidly increasing proportion in which their benefits extend and radiate in every direction, as impartially and benignantly as the similarly diverging rays of the sun. Around every town or market-place we may conceive a number of concentric circles to be drawn, enclosing areas from any part of which certain kinds of produce may be profitably taken to the town; while from any point beyond each circumference, the expense of the carriage of the particular article would exceed its value. Thus the inner circle, at the centre of which is the town, may show the limit in every direction from beyond which perishable vegetables, or articles very bulky or heavy in proportion to their value, cannot be profitably brought to market; the next larger circle may show the limit of fruits; and so on. If now the roads are improved in any way, so as in any degree to lessen the expense of carriage, the radius of each circle is correspondingly increased, and the area of each is enlarged as the *square* of this ratio of increase. Thus, if the improvement enables a horse to draw twice as much or to travel twice as fast as he did before, each of the limiting circles is expanded outward to twice its former radius, and embraces *four* times its former area. If the rate of improvement be threefold, the increase of area is *ninefold*; and so on All the produce, industry, and wealth, which by these improvements finds, for the first time, a market, is as it were a new creation.*

The number of passengers is governed by similar laws;

* Dr. Anderson.

and the increased facilities of a better road attract them from inferior ones, as the digging of a new and deep well often drains the water from all the shallow ones in its neighborhood. The distance to the right and left of the new road, from which it will attract passengers, admits of a mathematical investigation, which will be found at the end of Chapter I.; and the deductions of theory are amply corroborated by the observations of experience, and more than realized in the improvement of every old road and the opening of every new one; for not only is the former travel attracted from great distances in every direction, but a very considerable amount is created.

Supposing that by these improvements the average speed over a whole country be only doubled, the whole population of the country (to borrow a metaphor from an accomplished writer) would have advanced in mass, and placed their chairs twice as near to the fireside of their metropolis, and twice as near to each other. If the speed were again doubled, the process would be repeated; and so on. As distances were thus gradually annihilated, the whole surface of the country would be, as it were, contracted and condensed, till it was only one immense city; and yet, by one of the modern miracles of science wedded to art, every man's field would be found not only where it always was, but as large as ever it was, and even far larger, estimating its size by the increased profits of its productions. The more perfect the roads, the more rapidly would this result be attained, and therefore most quickly of all by railroads.

But however great the advantages of *railroads*, as to mere speed, and however precious to the hurrying traveller their triumphs over time and space, COMMON ROADS will always be incomparably more valuable to the community at large. The distinguishing characteristic of a modern rail road, as compared with a "tram road," and that to which its peculiar power is chiefly due, is the projecting flanges of the wheels of its carriages, by which they are retained upon the rails. But this peculiarity, in an equal degree, lessens its advantages to the agricultural population; since the vehicles which are adapted to travel on railroads can not be used on the common roads leading to them, nor in the ordinary labors of the farm; while on all other im proved roads the same wagons, horses, and men, employed at one season in cultivating the ground, can also be profitably employed, in their otherwise idle moments, in conveying the produce to a market. For these reasons, even if a railroad came to every man's door, he could more economi cally use a good common road; but since, on the contrary, the expense of the construction of railroads must al ways restrict them to important lines of communication, (where, indeed, their value can scarcely be estimated too highly) in every other situation, the greatest good of the greatest number, and the most universal benefits with the fewest accompanying evils, will be most effectually secured, by improving (in accordance with the principles to be presently set forth) the people s highways—the common roads of the country.

In this analytical examination of the subject of ROAD-MAKING, it will be considered under the following general heads:

1. *What Roads ought to be.*
2. *Their Location.*
3. *Their Construction.*
4. *Improvement of their Surface*

CHAPTER I.

WHAT ROADS OUGHT TO BE.

"The art of Road-making must essentially depend for its success on its being exercised in conformity with certain general principles; and their justness should be rendered so clear and self-evident as not to admit of any controversy."

SIR HENRY PARNELL.

RAPIDITY, safety, and economy of carriage are the objects of roads. They should therefore be so located and constructed as to enable burdens, of goods and of passengers, to be transported from one place to another, in the least possible time, with the least possible labor, and, consequently, with the least possible expense.

To attain these important ends, a road should fulfil certain conditions, which the nature of the country over which it passes, and other circumstances, may render impossible to unite and reconcile in one combination; but to the union of which we should endeavor to approximate as nearly as possible in forming an actual road upon the model of this ideally perfect one. We will therefore investigate—

WHAT ROADS OUGHT TO BE,

1. *As to their direction.*
2. *As to their slopes.*
3. *As to their cross-section*
4. *As to their surface.*
5. *As to their cost.*

1. WHAT ROADS OUGHT TO BE, AS TO THEIR DIRECTION.

IMPORTANCE OF STRAIGHTNESS.

Every road, other things being equal, should be *perfectly straight*, so that its length, and, therefore, the time and labor expended in travelling upon it, should be the least possible; *i. e.*, its *alignemens*, or directions, departing from one extremity of it, should constantly tend towards the other

Any unnecessary excess of length causes a constant threefold waste; firstly, of the interest of the capital expended in making that unnecessary portion; secondly, of the ever-recurring expense of repairing it; and, thirdly, of the time and labor employed in travelling over it. It will therefore be good economy to expend, in making topographical examinations for the purpose of shortening the road, any amount less than not only that sum which the distance thus saved would have cost, but, in addition, that principal which corresponds to the annual cost of the repairs and of the labor of draught which would have been wasted upon this unnecessary length.

ADVANTAGES OF CURVING.

The importance of making the road as *level* as possible will be explained in the next section, and as a road can in few cases be at the same time straight and level, these two requirements will often conflict. In such cases, *straightness should always be sacrificed to obtain a level, or to make the road less steep.* This is one of the most important principles to be observed in laying out or improving a road, and it is the one most often violated.

A *straight* road over an uneven and hilly country may, at first view, when merely seen upon the map, be pro nounced to be a *bad* road; for the straightness must have been obtained either by submitting to steep slopes in ascending the hills and descending into the valleys, or these natural obstacles must have been overcome by incurring a great and unnecessary expense in making deep cuttings and fillings.

A good road should wind around these hills instead of running over them, and this it may often do without at all increasing its length. For if a hemisphere (such as half a bullet) be placed so as to rest upon its plane base, the halves of great circles which join two opposite points of this base are all equal, whether they pass horizontally or vertically. Or let an egg be laid upon a table, and it will be seen that if a level line be traced upon it from one end to the other, it will be no longer than the line traced between the same points, but passing over the top. Precisely so may the curving road around a hill be often no longer than the straight one over it; for the latter road is straight only with reference to the vertical plane which passes through it, and is curved with reference to a horizontal plane; while the former level road, though curved as to the vertical plane, is straight as to a horizontal one. Both lines thus curve, and we call the latter one straight in preference, only because its vertical curvature is less apparent to our eyes.

The difference in length between a straight road and one which is slightly curved is very small. If a road between two places ten miles apart were made to curve so that the eye could nowhere see farther than a quarter of a mile of it at once, its length would exceed that of a perfectly straight road between the same points by only about one hundred and fifty yards.*

* Sganzin, p. 89.

But even if the level and curved road were very much longer than the straight and steep one, it would almost always be better to adopt the former; for on it a horse could safely and rapidly draw his full load, while on the other he could carry only part of his load up the hill, and must diminish his speed in descending it. As a general rule, the horizontal length of a road may be advantageously increased, to avoid an ascent, by at least twenty times the perpendicular height which is to be thus saved; that is, to escape a hill a hundred feet high, it would be proper for the road to make such a circuit as would increase its length two thousand feet.* The mathematical axiom that "a straight line is the shortest distance between two points," is thus seen to be an unsafe guide in road-making, and less appropriate than the paradoxical proverb that "the longest way around is the shortest way home.'

The gently curving road, besides its substantial advantages, is also much more pleasant to the traveller upon it; for he is not fatigued by the tedious prospect of a long straight stretch of road to be traversed, and is met at each curve by a constantly varied view.

It cannot be too strongly impressed upon a road-maker, that straightness is *not* the highest characteristic of a good road. As says Coleridge—

> "*Straight* forward goes
> The lightning's flash, and *straight* the fearful path
> Of the cannon-ball."

But in striking contrast he adds—

> "The ROAD the human being travels,
> That on which blessing comes and goes, doth follow
> The river's course, the valley's playful *windings*,
> *Curves* round the cornfield and the hill of vines."†

---

* This proportion depends on the degree of friction assumed, a subject to be investigated in a following section.

† The Piccolomini. i. 4

The passion for straightness is the great fault in the location of most roads in this country, which too often remind us how

> "The king of France, with forty thousand men,
> Marched *up a hill*, and then—marched *down again*;"

so generally do they clamber over hills which they could so much more easily have gone around; as if their makers, like Marshal Wade, had "formed the heroic determination of pursuing straight lines, and of defying nature and wheel-carriages both, at one valiant effort of courage and of science."

One reason of this is, that the houses of the first settlers were usually placed on hill-tops, (to escape the poisonous miasmata of the undrained swamps, and to detect the approach of the hostile savages) and that the first roads necessarily ran from house to house. Our error consists in continuing to follow these primitive roads with our great thoroughfares. These original paths were also traversed only by men, and therefore very properly followed the shortest though steepest route. Tracks for pack-horses came next, and a considerable degree of steepness is admissible in them also. Wheeled carriages were finally introduced and brought into use upon the same tracks, though too steep for true economy of labor with them—the standard of slope being very different for foot, horse, and carriage roads Before sufficient attention was paid to the subject, the lands on either side of the road had been fenced off and appropriated by individuals, and thus the random tracks became the legal highways.

The evil is now perpetuated by the unwillingness of farmers to allow a road to run through their farms in a winding line. They attach more importance to the square ness of their fields than to the improvement of the lines of their roads—not being aware how much more labor is wasted by them in travelling over these steep roads, than there would be in cultivating an awkward corner of a field.

This feeling is seen carried to excess in some of the new states of the West, in which the roads now run along "section-lines," and as these sections are all squares, with sides directed towards the cardinal points of the compass, a person wishing to cross the country in any other direction than North, South, East or West, must do so in rectangular zigzags.

PLEASURE DRIVES.

In roads designed solely for pleasure drives, such as those laid out by landscape gardeners in parks, cemeteries, &c., curvature is the rule, and straightness only the exception. In them the object is to wind as much as possible, in Hogarth's "line of grace," so as to obtain the greatest development of length which the area of the ground will permit, but at the same time never to appear to turn for the mere sake of curving. Some reason for the windings must always be suggested, such as a clump of trees, a rise of ground, a good point of view, or any object which may conceal the artifice employed. The visiter must be deceived into the belief that he is travelling over a large area, while he is truly only retracing his steps and constantly doubling upon his track; but he must do it unconsciously, or at least without knowing the precise manner in which the pleasant deception is effected. *Ars est celare artem.*

The map on the opposite page, representing the roads and paths in Greenwood Cemetery, will somewhat illustrate this principle

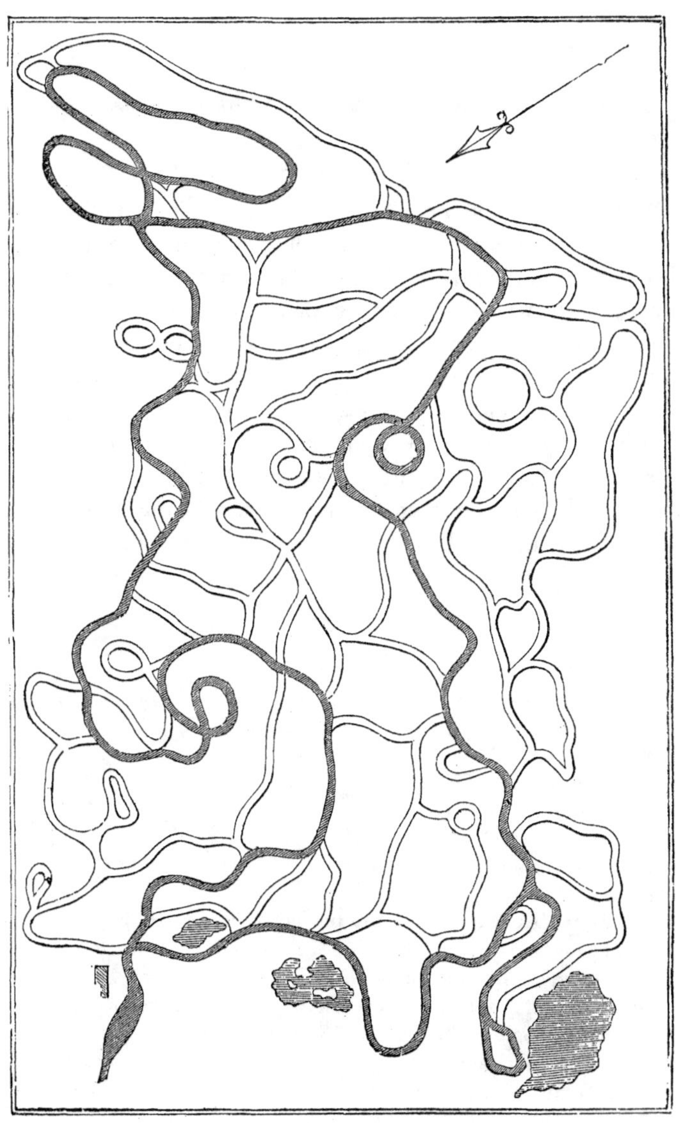

2. WHAT ROADS OUGHT TO BE AS TO THEIR SLOPES.

LOSS OF POWER ON INCLINATIONS.

Every road should be *perfectly level.* If it be not, a large portion of the strength of the horses which travel it will be expended in raising the load up the ascent. When a weight is drawn up an inclined plane, the resistance of the force of gravity, or the weight to be overcome, is such a part of the whole weight, as the height of the plane is of its length If, then, a road rises one foot in every twenty of its length, a horse drawing up it a load of one ton is compelled to actually lift up one-twentieth of the whole weight, *i. e.,* one hundred pounds, through the whole height of the ascent, besides overcoming the friction of the entire load.

Fig. 2.

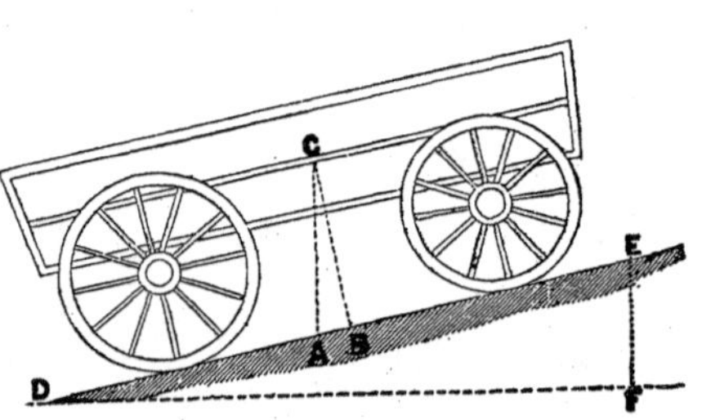

Let DE represent the inclined surface of a road upon which rests a wagon, the centre of gravity of which is supposed to be at C. Draw CA perpendicular to the horizon, and CB perpendicular to the surface of the hill. Let CA represent the force of gravity, or the weight of the wagon and its load. It is equivalent, in magnitude and direction, to its two rectangular component forces, CB and BA. CB will then represent the force with which the wagon presses on the surface of the road, and AB the resisting force of gravity *i. e.*, the force (independent of friction) which resists the ascent of the wagon, or which tends to drag it down hill.

To find the amount of this force, from the two similar triangles, ABC and DEF, we get the proportion

$$CA : AB :: DE : EF.$$

Representing the *length* of the plane by l, its *height* by h, and the *weight* of the wagon and load by W, this proportion becomes

$$W : AB :: l : h,$$

whence $AB = W\frac{h}{l}$; that is, the resistance of gravity due to the inclination, is equal to the whole weight, multiplied by the height of the plane and divided by its length. If the inclination be one in twenty, then this resistance is equal to $\frac{1}{20}$ W.

In this investigation, we have neglected three trifling sources of error: arising from part of the weight being thrown from the front axles to the hind ones, in consequence of the inclination of the traces; from the diminution of the pressure of the weight, owing to its standing on an inclined surface; and from the hind wheels bearing more than half of the pressure, in consequence of the line of gravity falling nearer them.

The results of experiments fully confirm the deductions of theory as to the great increase of draught upon inclinations. The following table exhibits the force required (according to Sir Henry Parnell) to draw a stage coach over parts of the same road, having different degrees of inclination:

Inclination.	FORCE OF DRAUGHT REQUIRED.		
	At 6 miles per hour.	At 8 miles per hour.	At 10 miles per hour.
1 in 20	268	296	318
1 in 26	213	219	225
1 in 30	165	196	200
1 in 40	160	166	172
1 in 600	111	120	128

Putting into a different form the results of these and other experiments, we establish the following *data:*

Calling the load which a horse can draw on a level, 1.00

on a rise of	1 in 100	a horse	can draw	only	.90*
"	1 in 50	"	"	"	.81*
"	1 in 44	"	"	"	.75‡
"	1 in 40	"	"	"	.72†
"	1 in 30	"	"	"	.64†
"	1 in 26	"	"	"	.54†
"	1 in 24	"	"	"	.50‡
"	1 in 20	"	"	"	.40†
"	1 in 10	"	"	"	.25*

In round numbers, upon a slope of 1 in 44, or 120 feet to the mile, a horse can draw only three-quarters as much as he can upon a level; on a slope of 1 in 24, or 220 feet to the mile, he can draw only half as much; and on a slope of 1 in 10, or 528 feet to the mile, only one quarter as much.

This ratio will, however, vary greatly with the nature and condition of the road; for, although the actual resistance of gravity is always *absolutely* the same upon the same inclination, whether the road be rough or smooth, yet it is *relatively* less upon a rough road, and does not form so large a proportional share of the whole resistance.

Thus, if the friction upon a road were such as to require, upon a level, a force of draught equal to $\frac{1}{40}$ of the load, the total force required upon an ascent of 1 in 20, would be $\frac{1}{40}+\frac{1}{20}=\frac{3}{40}$. Here, then, the resistance of gravity is two-thirds of the whole.

If the road be less perfect in its surface, so that its friction

---

* *Gayffier.* Experiments on a French road.

† *Parnell.* Experiments on an English road at average of the three velocities.

‡ Interpolations.

$=\frac{1}{20}$, the total force upon the ascent will be $\frac{1}{20}+\frac{1}{20}$; and here, then, the resistance of gravity is one-half of the whole.

If the friction increase to $\frac{1}{10}$, the total resistance is $\frac{1}{10}+\frac{1}{20}=\frac{3}{20}$; and here, gravity is only one-third of the whole

We thus see that on a rough road, with great friction, any inclination forms a much smaller part of the resistance than does the same inclination on a smooth road, on which it is much more severely felt, and proportionally more injurious; as the gaps and imperfections which would not sensibly impair the value of a common knife, would render a fine razor completely useless.

The loss of power on inclinations is indeed even greater than these considerations show; for, besides the increase of draught caused by gravity, the power of the horse to overcome it is much diminished upon an ascent, and in even a greater ratio than that of man, owing to its anatomical formation and its great weight. Though a horse, on a level, is as strong as five men, yet on a steep hill it is less strong than three; for three men, carrying each 100 lbs., will ascend faster than a horse with 300 lbs.*

Inclinations being always thus injurious, are particularly so, where a single steep slope occurs on a long line of road which is comparatively level. It is, in that case, especially important to avoid or to lessen this slope, since the load carried over the whole road, even the level portions of it, must be reduced to what can be carried up the ascent. Thus, if a long slope of 1 in 24 occurs on a level road, as a horse can draw up it only one half of his full load, he can carry over the level parts of the road only half as much as he could and should draw thereon.

This evil is sometimes partially remedied by putting on a full load and adding extra horses at the foot of the steep

* *Emerson.* Mechanics.

slope. Oxen are thus employed to assist carriages up the high hills, on the summits of which, for safety in time of war, the Etruscans built their cities of Perugia, Cortona, &c. But this is an inconvenient, as well as expensive system, and the truest economy is, to cut down, or to go around such acclivities, whenever this is possible.*

The bad effects of this steepness are especially felt in winter, when ice covers the road, for the slippery surface causes danger in descending, as well as increased labor in ascending. The water of rains, also, runs down the road and gullies it out, destroying its surface, and causing a constant expense for repairs, oftentimes great enough to pay for a permanent improvement.

The loss of power on inclinations being so great as has been shown, it follows that it is very important never to allow a road to ascend or descend a single foot more than is absolutely unavoidable. If a hill is to be ascended, the road up it should nowhere have even the smallest fall or descent, for that would make two hills instead of one; but it should be so located and have such cuttings and fillings, as will secure a gradual and uninterrupted ascent the whole way.

In this point engineering skill can make wonderful improvements. Thus, an old road in Anglesea, laid out in violation of this rule, rose and fell between its extremities, 24 miles apart, a total perpendicular amount of 3,540 feet; while a new road laid out by Telford between the same points, rose and fell only 2,257 feet; so that 1,283 feet of perpendicular height is now done away with, which every horse passing over the road had previously been obliged to ascend and descend with its load. The new road is, besides, more than two miles shorter. Such is

* In Chapter IV., under the head of "Roads with Trackways," will be described a valuable palliation of the evils of steep ascents in cases where they cannot be avoided

one of the results of the labors of a skilful road-make., and many such improvements might be made in our American roads For a recent remarkable instance, see page 233.

UNDULATING ROADS.

There is a popular theory that a gently undulating road is less fatiguing to horses than one which is perfectly level. It is said that the alternations of ascent, descent, and levels call into play different muscles, allowing some to rest while the others are exerted, and thus relieving each in turn.

Plausible as this speculation appears at first glance, it will be found on examination to be untrue, both mechanically and physiologically; for, considering it in the former point of view, it is apparent that new ascents are formed which offer resistances not compensated by the descents; and in the latter, we find that it is contradicted by the structure of the horse. The question was submitted by Mr. Stevenson* to Dr. John Barclay of Edinburgh "no less eminent for his knowledge, than successful as a teacher of the science of comparative anatomy," and he made the following reply:—"My acquaintance with the muscles by no means enables me to explain how a horse should be more fatigued by travelling on a road uniformly level, than by travelling over a like space upon one that crosses heights and hollows; but it is demonstrably *a false idea*, that muscles can alternately rest and come into motion in cases of this kind. . . . . . Much is to be ascribed to prejudice originating with the man, continually in quest of variety, rather than with the horse, who, consulting only his own ease, seems quite unconscious of Hogarth's Line of Beauty."

Since this doctrine is thus seen to be a mere popular

* Report on the Edinburgh Railway

error, it should be utterly rejected, not only because false in itself, but still more because it encourages the making of undulating roads, and thus increases the labor and cost of carriage upon them.

GREATEST ALLOWABLE SLOPE.

A perfectly level road is thus seen to be a most desira ble object; but as it can seldom be completely attained, we must next investigate the limits to which the slopes of a road should be reduced if possible and determine what is the steepest allowable or *maximum* slope.

This depends on two different considerations, according as the slope is viewed as a descent or as an ascent, each of which it alternately becomes, according to the direction of the travel.

Viewed as a *descent*, it chiefly concerns the safety of rapid travelling, and applies especially to great public roads.

Viewed as an *ascent*, it chiefly concerns the draught of heavy loads, and relates particularly to routes for agricultural and other heavy transportation.

MAXIMUM SLOPE, CONSIDERED AS A DESCENT.

The slope should be so gentle, that when a heavy vehicle is descending, its gravity shall not overcome its friction so far as to permit it to press upon the horses. This limiting slope corresponds to the "angle of repose" of mechanical science; *i. e.*, the angle made with the horizon by the steepest plane down which a body will not slide of its own accord, its gravity just balancing its friction, so that the least increase of slope would overpower the resistance of the friction, and make the body descend. This "angle of repose" should therefore be the limit of the slope of a road, for on such an inclination a vehicle once set in motion would descend with uniform, unaccelerated velocity. This angle varies with the smoothness and hardness of the road, and also with the degree of friction of the axles of the carriage. On the very best class of broken-stone roads, kept in good order, and with a good carriage, it is considered by Sir Henry Parnell, from his experiments, to be 1 in 35, (or 151 feet to the mile) which should therefore be the maximum slope upon the best roads.* On such a slope a coach may be driven down, with perfect safety and complete control, at the speed of twelve miles per hour.

If the inclination be steeper than this, the danger of the descent is greatly increased, and the speed must be lessened. If it be so steep that a carriage cannot be safely driven down at a greater speed than four miles per hour, on every mile of such a slope there will be a loss of ten minutes of time, equivalent to two miles upon a level. To avoid such an inclination, a road-maker would therefore be justified, by considerations of time-saving, in adopting a level route three times as long as the steep one.

When inclinations are reduced to this limit of 1 in 35, there is little loss of power, compared with a perfect level, in either direction of the travel; for the increased labor of ascending is compensated in a great degree by the increased ease of descending, while on a steeper slope this advantage is nullified by the necessity of the horses holding back the carriage to resist the excess of the force of gravity.

---

* On such roads Dr. Lardner considers the angle of repose to be as small as 1 in 40; while on roads not well freed from mud and dust, the friction increases the angle to 1 in 30; and on an inferior class of roads it is 1 in 20, or even steeper

MAXIMUM SLOPE, CONSIDERED AS AN ASCENT.

Suppose that a road is to be carried over a hill, which rises 100 feet in a horizontal distance of 500 feet, (*i. e.*, 1 in 5) and which cannot be avoided by any horizontal circuit within the limits of distance indicated on page 28. The question which presents itself is, how steep can the slope of a road up the side of this hill be most advantageously laid out, since, by adopting a zigzag line, the road may be made as long and therefore as gentle in the ascent as may be desired? The shortest line would run straight up the face of the hill, and this line would give the least amount of labor; but then this labor for horses would be impossible: and even if possible, the horses could not draw up the whole load which they had been drawing on the other parts of the road, nor could they descend it with safety. But, on the other hand, the road should approach this shortest line as nearly as other considerations will permit, since, if it should zigzag excessively for the purpose of lessening the steepness, it would be so long as to increase unnecessarily its cost and the time and labor of travel upon it. A medium and compromise between these two evils must therefore be found. What shall it be?

Supposing the load of a horse on the level portions of the road to be as much as he can regularly and constantly draw, his power of drawing it up an ascent will depend upon how much *extra* exertion he is capable of putting forth. This is not very accurately ascertained or defined, and depends very much on the length of the ascent, but may be assumed at double his usual exertion.* Now a horse drawing a load on a level road of the best character,

---

* Gayffier, p. 9

such as has been previously considered, is obliged by the resistance of the friction to exercise against his collar a pressure of about one thirty-fifth of the load. If he can just double this exertion, he can lift one thirty-fifth more, and the slope which would force him to lift that proportion would be (as was shown on page 32) one of 1 in 35. On this slope he would therefore be compelled to double his ordinary exertion, and on this supposition it would be the maximum slope allowable, considered as an ascent.

These two methods of determining the maximum slope (by considering it as an ascent and as a descent) are entirely independent of each other.* If they give different results, the smallest one, or the least slope obtained, must be adopted; for, if it be disadvantageous to employ a slope steeper than 1 in 35, it must *à fortiori* be still more so, to employ one steeper than 1 in 30, or 1 in 20; though even greater slopes are too often met with.

Upon most of our American roads the resistance of friction would be found to be nearer $\frac{1}{20}$ than $\frac{1}{35}$, and 1 in 20 would therefore be their maximum slope with their present condition of surface. But as it is to be hoped that in this respect they will, before long, be greatly improved, in which case they would demand more and more gentle slopes, we should anticipate this desirable consummation, by giving in advance to all new lines of road at least, if not to the faulty old ones, *slopes not exceeding* 1 *in* 30, which seems to be a just medium.

---

* They give identical results in this case, only because the extra exertion happened to be taken as doubled. Suppose it to be tripled. The horse can lift $\frac{2}{35}$ more, which corresponds to a slope of 1 in 17½. Horses can indeed for a short time exercise a tension of *six* times the usual amount, but the above assumption of double is more dependable, though it cannot be fixed with the precision which is desirable.

The *maximum* established by *L'administration des Ponts et Chaussées*, the French government board of engineers of roads and bridges, is 1 in 20. This, however, was fixed at a time when the usual surface of roads was much inferior to its present condition.

The great Holyhead road, made by Telford through the very mountainous district of North Wales, has 1 in 30 for its maximum, except in two cases, (one of 1 in 22, and a very short one of 1 in 17) and in them the surface of the road was made peculiarly smooth and hard, so that no difficulty is felt by loaded vehicles in ascending. On the old line of road, the inclinations had been sometimes as great as 1 in 6, 1 in 7, &c.

On the great Alpine road over the Simplon pass, (which rises to a height of a mile and a quarter above the level of the sea) the slopes average 1 in 22 on the Italian side, and 1 in 17 on the Swiss side, and in one case only become as steep as 1 in 13.

In the state of New York several turnpike companies are limited by law to a maximum slope of "eighteen inches to a rod," *i. e.* 1 in 11. But this limit ought not to be even approached in practice.

On our "National" or "Cumberland" road the slopes in many places are much too great, and its superintendent, Capt. Wever, writes* that "if the road had been very considerably elongated in order to effect a graduation at angles not exceeding three degrees, or 1 in 19, (and for the maximum, two degrees, or 1 in 29, would be better) the road could be travelled in as short a space of time as it now is, and the power used could move double the burden it now can; thus rendering the road, for commercial purposes, doubly advantageous."

If the ascent be one of great length, it will be advantageous to make steepest the lowest portion of it, upon which the horses come with their full strength, and to

* Report to United States Chief Engineer, 1828.

make the slopes gentler towards the summit of the ascent, to correspond to the continually decreasing strength of the fatigued horses.

MINIMUM SLOPE.

A true level has been thus far considered to be a most desirable attribute, and one to be earnestly sought for, in establishing a perfect road. This principle must be qualified, however, by the announcement that there is a *minimum*, or least allowable slope, which the road must not fall short of, as well as a *maximum* one, which it must not exceed. If the road were perfectly level in its longitudinal direction, its surface could not be kept free from water without giving it so great a rise in its middle as would expose vehicles to the danger of overturning. But when a road has a proper slope in the direction of its length, not only do the side-ditches readily discharge the water which falls into them, but every wheel-track that is made, becomes also a channel to carry off the water.

The *minimum* slope (flatter than which the road should not be) is assumed by an experienced English engineer to be one in eighty, or 66 feet to the mile. The minimum established in France by the *Corps des Ponts et Chaussées* is .008, or one in a hundred and twenty-five, or 42 feet to the mile. An angle of one-half a degree is often named in this connection; it equals one in a hundred and fifteen. In a perfectly level country the road should be artificially formed into gentle undulations approximating to the minimum limit.

Finally, then, we arrive at this conclusion, that the longitudinal slopes of a road should be kept, if possible, between 1 *in* 30 *and* 1 *in* 125, *never steeper than the former, nor nearer to a level than the latter.*

TABLES OF INCLINATIONS.

There being three different methods of specifying degrees of inclination, (viz. by the angle made with the horizon, by the proportion between the ascent and the horizontal distance, and by the ascent per mile) it is frequently desirable to compare the different expressions. The following tables show the values which correspond to each other.

Angles.	Inclinations.	Feet per mile.
$\frac{1}{2}$°	1 in 115	46
$\frac{3}{4}$°	1 in 76	69
1°	1 in 57	92
1$\frac{1}{2}$°	1 in 38	138
2°	1 in 29	184
2$\frac{1}{2}$°	1 in 23	231
3°	1 in 19	277
4°	1 in 14	369
5°	1 in 11	462

Inclinations.	Angles.	Feet per mile.
1 in 10	5° 43′	528
1 in 13	4° 24′	406
1 in 15	3° 49′	352
1 in 20	2° 52′	264
1 in 25	2° 18′	211
1 in 30	1° 55′	176
1 in 35	1° 38′	151
1 in 40	1° 26′	132
1 in 45	1° 16	117
1 in 50	1° 9′	106
1 in 100	0° 35′	53
1 in 125	0° 28′	42

3. WHAT ROADS OUGHT TO BE AS TO THEIR CROSS-SECTION.

The *cross-section* of a road is the view which it would present if cut through at right angles to its length, one of the portions being removed. It comprises the following subjects of investigation :

1. *The width of the road.*
2. *The shape of the road-bed.*
3. *Foot-paths, &c.*
4. *Ditches.*
5. *The side-slopes of the cuttings and fillings.*

Fig. 3.

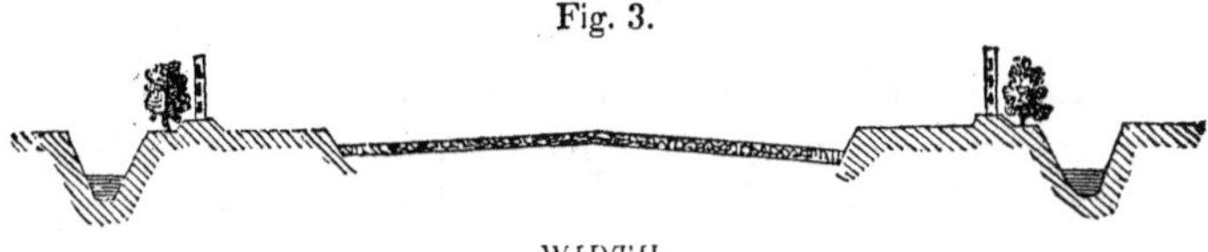

WIDTH.

The proper width for a road depends, of course, upon its importance, and the amount of travel upon it. Its *minimum* is about one rod, or 16½ feet, sufficient to enable two vehicles to pass each other with ease. For ordinary town roads a good width is from 20 to 25 feet. A width of 30 feet is fully sufficient for any road, except one which forms the approach to a very populous city.

Any unnecessary width (such as is often adopted in a spirit of public ostentation) is injurious, not only from its waste of land, but from its increase of the labor and cost of keeping the road in repair; each rod in width adding two acres per mile to the area covered by the road.

In the state of New York, by the revised statutes, "All public roads, to be laid out by the commissioners of highways of any town, shall not be less than *three rods* wide."

This is to be the width between fences; and no more

of it need be worked, or formed into a surface for travelling upon, than is deemed necessary.

The same laws declare, "It shall be the duty of the commissioners of highways to order the overseers of highways to open all roads to the width of *two rods at least*, which they shall judge to have been used as public high ways for twenty years."

It is also ordered that "all private roads shall not be more than three rods wide."

Turnpike-roads are obliged by the statute to be "laid out not less than *four rods* wide," and "*twenty-two feet* of such width to be bedded with stone," &c. When a precipitous locality renders the full width impracticable, "*twenty-two feet*" is the minimum width permitted.

Where a road ascends a steep hill-side by zigzags, it should be wider on the curves connecting the straight portions. The width of the roadway may be increased about one-fourth, when the angle between the straight portions of the zigzags is from 120° to 90°; and the increase should be nearly one-half, when the angle is from 90° to 60°.*

The *Roman* military roads had their width established, by the laws of the Twelve Tables, at twelve feet when straight, and sixteen when crooked; barely sufficient for the army, baggage, and military machines.

The *French* engineers make four different classes of roads.†

The first class comprises such as pass from the capital of one country to that of another. Their width is 66 feet, of which 22 in the middle are stoned or paved.

Those of the second class pass from the metropolis of a country to its other great cities. Their width is 52 feet, of which 20 in the middle are stoned.

Those of the third class connect large towns with each other

* Mahan, p. 282.

† Gayffier, p. 90

and with first-class roads. Their width is 33 feet, with 16 feet in the middle stoned.

The fourth class contains common town roads. Their width is 26 feet, with the same middle causeway as the last.

In *England*, the prescribed width for turnpike-roads at the approach to populous towns is 60 feet. The limits of by-roads are, for carriage-roads, 20 feet; for horse-roads, 8 feet; and for foot-paths, 6½ feet.*

Telford's Holyhead road, a model road for a hilly country, has the following width in the clear within the fences: 32 feet on flat ground; 28 feet when there are side-cuttings less than three feet deep; and 22 feet along steep ground and precipices.

The *United States National* or Cumberland road has 80 feet in width cleared, but the road itself is only 30 feet.

The broken-stone road between Albany and Troy is 32 feet wide, besides two sidewalks of 8 feet each.

The "Third Avenue" of the city of New York is 60 feet wide between the sidewalks, each of which occupies 20 feet: 26 feet of its middle are stoned.

Broadway, New York, is 80 feet wide between the houses, of which 19 feet on each side are occupied by the foot-pavements, leaving 42 feet for the carriage-way.

When broken-stone roads are adopted, it is usual, for the sake of a saving in the first cost, to make only a certain width or "causeway," in the middle of the road, of the harder material, and to form the sides, or "wings," of the natural earth, (or of broken stone, if the causeway be a pavement) which will be preferable in summer and for light vehicles and horsemen.† Sixteen feet for the middle and twelve for the sides is a common proportion.

If the stoned part be made narrower than just wide enough for two carriages to pass upon it, it should be made only wide

---

* Roads and Railroads, p. 73.

† A serious objection to this plan is, that the wheels which cross the road, and are alternately on the stone and on the earth, will deposite earth upon the stone surface, to the great deterioration of its advantages.

enough for one; for any intermediate width will be a waste of all the surplus beyond what one requires.

If the road is to be made wider than two vehicles require, (which strictly is only 12 feet) it should be enlarged at once to 23 feet; for any intermediate width will cause unequal and excessive wear, and therefore be false economy: an unexpected conclusion, which results from an investigation of Gayffier, pages 184–8.

It would be preferable to place the harder material on the sides of the road, instead of on the centre; for the drivers of heavily-laden vehicles will generally keep them on the sides of the road, so that they can walk on the foot-paths; and if this part be not of the hardest material, it will soon be cut up and rutted by the heavy wagons following each other in the same track.*

SHAPE OF THE ROAD-BED.

In forming the road-bed, or travelled part of the road, the first and most important point, in a flat country, is to raise it above the level of the land through which it passes, so that it may be always perfectly free from water; a precaution which is one of the most essential requisites for keeping a road in good condition. Roads are often placed in a hollow-way, (or even a trench is dug, when better materials are to be added) and their surface is allowed to remain so low, that they form excellent gutters to drain the adjacent fields, at the expense of the comfort, labor, and time of all who travel them. Even the best ditches cannot always secure them from the land-springs, (which will sometimes pass under the ditches by fissures which form inverted siphons) and the only effectual means will be the raising of the surface by an embankment of

* Parnell, p. 129

two or three feet. The excavations for the ditches should invariably be thus applied.

The necessary *elevation* having been established, the *shape* of the road-bed, at right-angles to its length, or its "transverse profile," must be decided upon.

The road must not be flat, but must "crown," or be higher in its middle than at its sides, so as to permit the water of rains to rapidly run off into the side ditches. If originally flat, it is soon worn concave, and its middle becomes a pool, if it be on level ground; or a water-course, if it be on an inclination. In the former case, the road becomes mud; in the latter, the smaller materials are washed away, and the larger stones left bare. Both these evils are of continual occurrence on our country roads, but may be easily prevented, by shaping the road according to the instructions to be presently given.

The usual, though improper, shape given to a road in order to make it crown, has been a convex curve, approaching a segment of a circle, or a flat semi-ellipse.

Fig. 4.

Though recommended by high authorities, it is very faulty, in consequence of its slope not being uniform, (the proportion between arcs and versed sines constantly changing) and giving too little inclination near the middle, and too much at the sides. From this peculiarity the following evils result:—

1. The water stands on the middle of the road, and washes away its sides.

2. It is worn down very unequally: for all carriages, to avoid the danger of overturning on the steep sides, will

take the middle of the road, which is the only part of it where they can stand at all upright; while the road ought, on the contrary, to be so formed as to induce vehicles to traverse it equally and indifferently in every part.

3. This excessive travel on the middle soon wears it into ruts and holes, so that more water will actually stand upon such an originally convex road than on one reasonably flat.

4. When carriages are forced to travel on the sides, they cause great additional wear to the road, from their constant tendency to slide down the sides, owing to the oblique angle at which the direction of gravity meets the surface.

5. As this sliding tendency is at right-angles to the line of draught, the labor of the horses and the wear of the wheels are both greatly increased.

6. Whenever vehicles are obliged to cross the road, and mount the central ridge, they must overcome the same resistance of gravity, as when they are drawn up a longitudinal hill.

The best transverse profile for a road on level ground, is that formed by two inclined planes, meeting in the

Fig. 5.

centre of the road, and having their angle slightly rounded by a connecting curve. The inclinations thus formed will be uniform, and the road will thus escape most of the evils incident to the curved profile.

The degree of inclination of these planes will depend on the surface of the road; being greatest where the road is rough, and lessening with its improvement in smoothness. It may also be somewhat less on a narrow road as the water will have a less distance to pass over. Its maximum is limited by the inconvenience which an excessive transverse slope would cause to carriages. A proper medium for a road with a broken-stone surface, is 1 in 24, or *half an inch to a foot.* Telford, in his Holyhead road, adopted 1 in 30, or 6 inches crown in a road of 30 feet; and McAdam 1 in 36, and even 1 in 60, or 3 inches in a 30 feet road. On a rough road the inclination may be increased to 1 in 20; and diminished on a road paved with square blocks to 1 in 40, or 1 in 50.

Up to these limits the transverse slope should increase with the longitudinal slope of the road, which it should always exceed, in order to prevent the water running too far down the length of the road, and gullying it out; for the water of rains runs off from the middle of a road in the diagonal of a rectangle, the sides of which are proportioned to the steepness of the two slopes, longitudinal and transverse.

If these slopes be equal, the rectangle becomes a square, and the direction of the escaping waters makes an angle of 45° with the direction of the road. If the transverse slope be double the longitudinal, the waters in their diagonal course make an angle of 63½° with the road, as in the figure. If the road be level longitudinally, they run off at right angles.

Fig. 6.

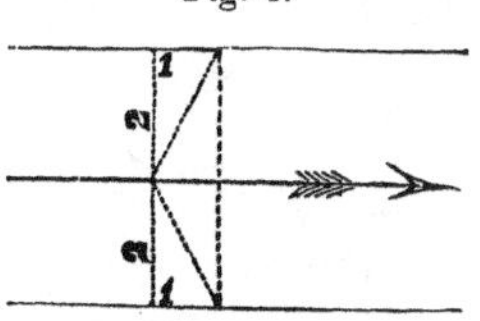

On a steep side-hill, the transverse profile should be a single slope, inclining inwards from the outer edges of the road to the face of the hill. The ditch should be on the side of the hill, and its waters be carried at proper intervals under the road to its outside This form is particu-

Fig. 7.

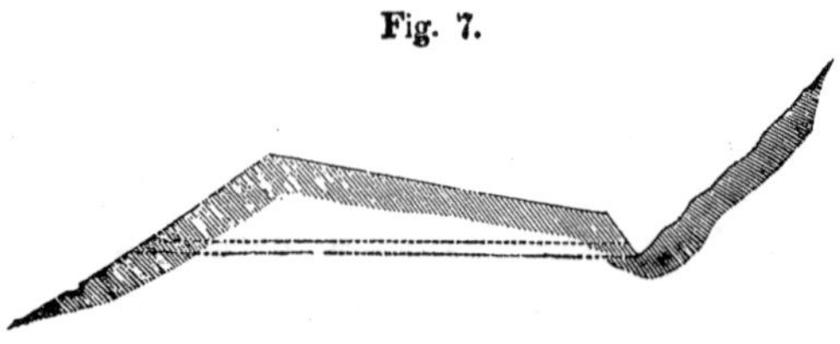

larly advantageous when the road curves rapidly around the hill, since it counteracts the dangerous centrifugal force of the vehicles. It may, therefore, be also adopted on the curves of a road in embankment.

Through *villages*, where space must be economized, and the side ditches dispensed with, the middle portion

Fig. 8.

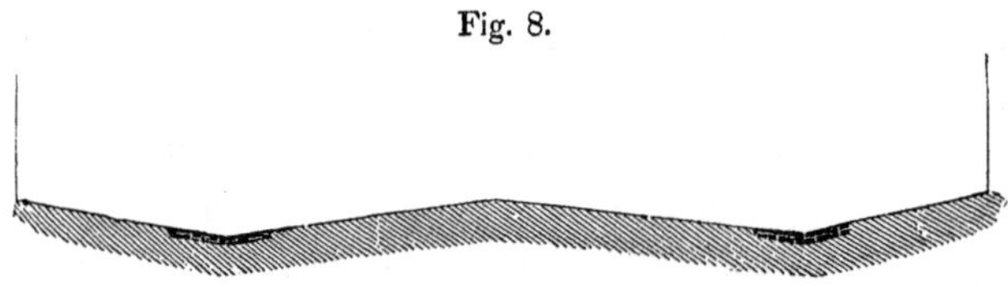

of the road is made to descend each way from the centre as usual, but the sides slope upwards towards the houses. Two furrows, or shallow water-channels, are thus formed, which should be paved to a width of two feet on each side of their middle. This form may also be used on a hill-side.

A frequent, but very bad shape, is hollow in its middle, in which the waters run. Its faults are, that carriages slide down towards each other, especially in frosty weather, and that the large stream in the middle washes away the road. It should never be used except when the width is greatly contracted, and when it is absolutely impossible to obtain room for ditches.

Fig. 9.

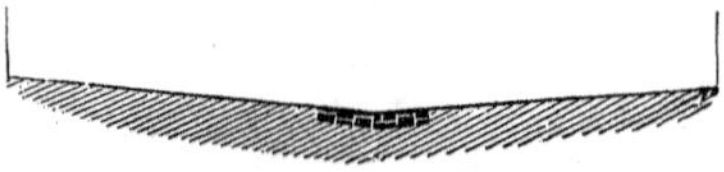

FOOTPATHS, &c.

On each side of the carriage-way should be flat mounds, raised six inches above the road. Sods, eight inches wide and six inches thick, should be laid against these mounds in such a manner as to form a sloping edge. The water which falls on the surface of the road runs along the bottoms of these sods, in the " side channels" formed by them, till it passes off under the mounds into the ditches. These mounds, in a great road of thirty feet width, should be six feet wide, and their surfaces should be inclined 1 inch in a yard. One of them should be covered with gravel for a footpath, and the other be sown with grass-seed. Their general adoption would greatly increase the safety of night-travelling, the accidents in which often occur from running on high banks or into ditches. They are not high enough to overturn a coach when one wheel runs upon them, but they indicate at once that the carriage is leaving the road.

Outside of the footpaths should be *fences*, (or hedges, where the climate will permit) and outside of the fences should be the *ditches*. These mounds, ditches, &c., are shown in Fig. 3.

DITCHES.

The drainage of a road by suitable ditches is one of the most important elements in its condition. All attempts at improvement are useless till the water is thoroughly got rid of, and a bad road may often be transformed into a good one, by merely forming beside it deep ditches, sufficiently inclined to carry off immediately all the water which falls upon it. Even if the water does not stand on the surface so as to form mud, if it filtrates from the higher land beside it, and from springs under it, and is not well drained off, it will weaken the substratum of the road so as to render it incapable of bearing heavy loads, and will be absorbed into the upper stratum by capillary attraction. If the road have a covering of broken stones, the water penetrating into it makes them wear away very rapidly by assisting the vibrating motion of their fragments, as lapidaries grind down the hardest stones by their own dust, with the aid of water.

The ditches should lead to the natural water-courses of the country; and should, if possible, have a minimum slope of one in a hundred and twenty-five, corresponding with the "minimum slope" of the road, though less will suffice if the bottom be truly cut and kept free from grass. They should generally be sunk to a depth of three feet below the surface of the road. Their size will be regulated by their situation, being greater where they intercept the water from side-hills rising above the road, and also where the country is humid. A width of one foot at bottom, with side-slopes depending on the nature of the soil, will generally suffice. In wet soils the ditches should be so wide and deep, that the earth taken from them may be sufficient to raise the bed of the road between them three feet higher than the natural surface.

There should be a ditch on each side of the road on level ground, or in cuttings, and on the upper side of the road, where it is on a hill-side. The water from the side channels must be carried into these, and the contents of the ditches must pass under the road to the natural water-courses by means of *drains*, *culverts*, &c., as will be explained in Chapter III. under the head of "Mechanical Structures."

SIDE-SLOPES OF THE CUTTINGS AND FILLINGS.

These are designated by the ratio of the base to the perpendicular of the right-angled triangle, of which the

Fig. 10.

slope is the hypothenuse, the base being always named first, and the perpendicular being the unit of measure. Thus, if a cutting of ten feet in depth goes out twenty feet, as in the figure, its slope is said to be 2 to 1; if it goes out but five feet, it is said to be ½ to 1.

The Slopes of Cuttings or Excavations vary with the nature of the soil, being made for economy as steep as its tenacity will permit. Solid rock may be cut vertically, or at a slope of ¼ to 1. Common earth will stand at 1 to 1, or at 1½ to 1; the latter is safer. Gravel requires 1½ to 1. Some clays will stand at 1 to 1; while some, originally sloped 2 to 1, have slipped till they have assumed a slope of 6 to 1. The proper degree of slope is best determined by observing that at which the earth in question naturally stands. Heavy clayey earth will assume a slope of ¾ to 1, and very fine dry sand of nearly 3 to 1; these are the extremes in ordinary cases.

Deep cuttings should not, however, be made with less slopes than 2 to 1, (even though they would stand steeper) so that the sun and wind may freely reach the road to keep it dry. The south side of excavations may be made even 3 to 1, when the extra earth can be profitably used in a neighboring embankment.

When the lower part of a cutting is in rock, and has a steep slope, and the upper portion in earth has a much flatter one, a wide "bench," or offset, should be made, where the change of slope takes place.

Fig. 11.

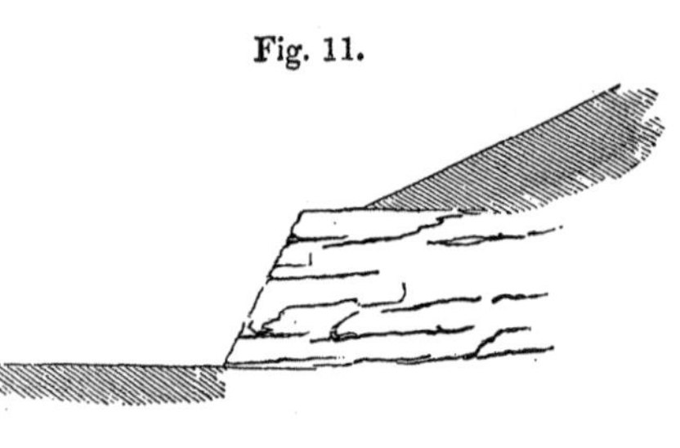

The following Table shows the angle with the horizon made by slopes of various proportions of base to height.

Slopes.	Angles.
$\frac{1}{4}$ to 1	75° 58′
$\frac{1}{2}$ to 1	63° 28′
$\frac{3}{4}$ to 1	53° 8′
1 to 1	45°
$1\frac{1}{4}$ to 1	38° 40′
$1\frac{1}{2}$ to 1	33° 42′
$1\frac{3}{4}$ to 1	29° 44′
2 to 1	26° 34′
3 to 1	18° 26′
4 to 1	14° 2′
5 to 1	11° 19′
6 to 1	9° 27′

Fillings or Embankments have less variety than cuttings in the nature and condition of their materials, and therefore have less variety of slope, which is usually $1\frac{1}{2}$ to 1, or 2 to 1; though some clays (which should, however, never be employed, if their use can be avoided) require 3 or 4 to 1, when more than four feet high.

CURVED SIDE-SLOPES.

The customary form of the side-slopes of cuttings and fillings—that of an inclined plane—is not the form of most perfect equilibrium and stability. To secure this, the slope may be steep near its top, with its upper angle rounded off, but must widen out at its bottom, where the pressure is the greatest. This is the natural face which an excavation assumes when left to itself, as shown in

Fig. 12.

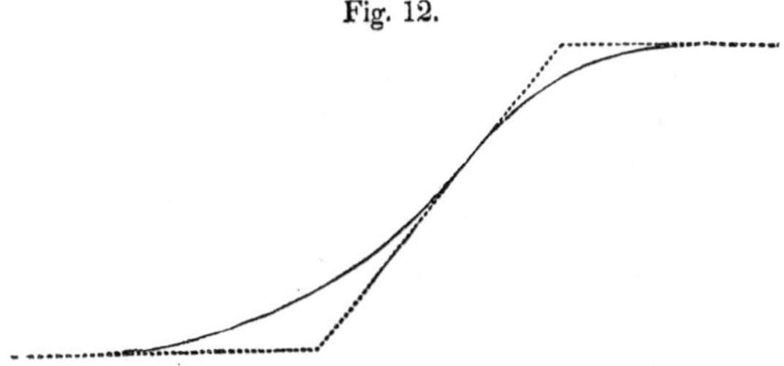

the figure. Its top, or salient angle, becomes convex; and its bottom, or re-entering angle, is filled up into a concavity, thus forming a curve of contrary flexure. If side-slopes were originally formed into this shape, they would be much more permanent, and the elements, rain, gravity, &c., would then work with man, and assist the labors of art, instead of destroying them, as when the usual form is employed. This curve of stability is moreover that of beauty, coinciding with Hogarth's "line of grace."

This plan is not known to have been ever put into practice, though the walls supporting a bank, particularly for a quay, are sometimes made concave outwardly; and the dam of the Croton Aqueduct has, for its outer profile, somewhat such a curve as has been above recommended

4. WHAT ROADS OUGHT TO BE AS TO THEIR SURFACE.

QUALITIES DESIRABLE.

The surface of a road ought to be as SMOOTH and as HARD as possible, so as to reduce to their smallest possible degree the resistances of *elasticity*, *collision*, and *friction*.

Smoothness is not only essential to comfort, but even more so to economy of labor, of carriage-wear, and of road wear. Carriages passing over a smooth road are not only drawn more pleasantly, and with less exertion of animal strength, but also do much less damage to the road, than when it has hollows into which the wheels fall with the momentum of sledge-hammers, each blow deepening the hole and thus increasing the force of the next blow.

Hardness is that property of a surface by which it resists the impression of other bodies which impinge upon it. It is essential to the preservation of smoothness, except in the case of elastic surfaces.

RESISTANCES TO BE LESSENED.

Elasticity.—A road may be perfectly smooth, both before and after a vehicle has passed over it, but if it sink in the least under the passage of a wheel, this yielding presents before the wheel a miniature hill, up which the vehicle must be raised with all the loss of power demonstrated on page 32. If the depression were one inch, and the wheel four feet in diameter, an inclined plane of 1 in 7 would be formed, and one-seventh of the entire weight would need to be lifted up this inch. A road surface of caoutchouc, or India-rubber, of the most perfect smoothness, would therefore be the worst possible for traction, though very pleasant for passengers. The wheels would always be in depressions, and the horses would be always pulling up hill. An elastic bottom for a road, such as a boggy substratum, would for this reason cause great waste of draught. A solid, unyielding foundation is therefore one of the first requisites for a perfect road.

Collision.--The resistance of collision is occasioned by the hard protuberances, inequalities, stones, and other loose materials of a road against which the wheels strike, with great loss of momentum and waste of the power of draught; for the carriage must be lifted over them by the leverage of the wheels. It is, therefore, most important that such obstacles should be as few and as small as possible, the resistance being proportional to their size, as appears in the investigation which follows.

The power required to draw a wheel over a stone or any obstacle, such as S in the figure, may be thus calculated. Let P represent the power sought, or that which would just balance the weight on the point of the stone, and the slightest increase of which would draw it over. This power acts in the direction CP with the leverage of BC or DE. Gravity, represented by W, resists in the direction CB with the leverage of BD. The equation of equilibrium will be $P \times CB = W \times BD$, whence

Fig. 13.

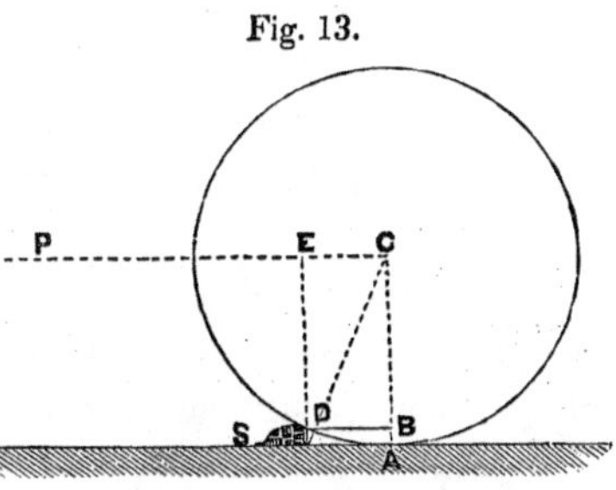

$$P = W\frac{BD}{CB} = W\frac{\sqrt{CD^2 - BC^2}}{CD - AB}.$$

Let the radius of the wheel $= CD = 26$ inches, and the height of the obstacle $= AB = 4$ inches. Let the weight $W = 500$ lbs., of which 200 lbs. may be the weight of the wheel, and 300 lbs. the load on the axle. The formula then becomes

$P = 500 \frac{\sqrt{676 - 484}}{26 - 4} = 500 \frac{13.85}{22} = 314.3$ lbs. The pressure at the point D is compounded of the weight and the power, and equals $W \frac{CD}{CB} = 500 \times \frac{26}{22} = 591$ lbs., and therefore acts with this great effect to destroy the road in its collision with the stone, in addition to its force in descending from it. For minute accuracy, the non-horizontal direction of the draught, and the thickness of the axle, should be taken into the account.

The power required is lessened by proper springs to vehicles, by enlarged wheels, and by making the line of draught ascending.

The resistance produced by the hollows between the stones of a pavement is of a different nature. According to the investigations of M. Gerstner, the resistance arising from such a surface is directly proportional to the load, to the square of the velocity, and to the ratio of the width of the cavity to the radius of the wheel; and inversely proportional to the width of the paving stones.

Friction.—The resistance of friction arises from the rubbing of the wheels against the surfaces with which they come in contact, and will always exist, however the surface may be improved. Its two extremes may be seen on a road of loose gravel, and on a railroad. It is greatly increased when the surface is covered with mud, or other loose material, into which the wheel may sink, and thus give a wider contact. The degree in which it is influenced by the surface, may be shown by rolling an ivory ball successively over a carpet, a fine cloth, a smooth floor, and a sheet of ice; the distances to which the same force will impel it over these surfaces increasing in the order in which they have been named.

The surface of a road may be improved by the various methods of diminishing the friction to be examined in

Chapter IV., such as "Macadamizing" the road, or covering it with a layer of finely broken stones; paving with smooth stone blocks; covering with planks; or laying wheel-tracks of stone, wood, or iron.

The friction on all these surfaces is different, and can be determined only by experiment. The instrument used for measuring it is called a *Dynamometer.* It resembles in principle and general construction the "spring-balances" in common use, in which the application of a weight compresses a spiral spring, the shortening of which, as shown by a properly graduated scale, indicates the amount of weight applied. In the *dynamometer* the power takes the place of the weight of the spring-balances, one end of the instrument being connected with the carriage, and the other with the horses, and the force which they exert to overcome the friction being shown by the index.

Sir John Macneill has greatly improved the instrument, by adapting to it a piston working in a cylinder full of oil, which lessens the vibrations of the index, and enables its indications to be read with more ease and precision. He has also added to it a contrivance for making the instrument itself record the degree of force exerted at each moment of motion. It likewise registers the distance passed over, and the rises and falls of the road.*

This valuable instrument affords a means of ascertaining the exact power required to draw a carriage over any line of road; it will thus enable one line of road to be compared with another, and their precise amount of difference in case of draught, to be determined; it will show the comparative value of the different methods of improving the surface; and it will enable a registry to be kept from year to year of the state of a road, showing where and how much it has improved or de-

* For a full description of th's instrument, see *Parnell*, pp. **327–347.**

teriorated, and therefore how judiciously, or the contrary, the funds expended on it have been applied.

The following are the results of experiments made with this instrument on various kinds of road. The wagon employed weighed 21 cwt., and the resistance to draught was as follows:—

* On a gravel road, laid on earth—per 21 cwt.,	147lbs.	=	$\frac{1}{16}$
* On a broken-stone road, " "	65	=	$\frac{1}{36}$
* " on a paved foundation, "	46	=	$\frac{1}{51}$
* On a well-made pavement, "	33	=	$\frac{1}{71}$
† On the best stone track-ways, per gross ton,	$12\frac{1}{2}$	=	$\frac{1}{179}$
‡ On the best form of railroad, "	8	=	$\frac{1}{280}$

From the above experiments we infer, in round numbers, taking the maximum load on a gravel road for the standard, that a horse can draw—

On the best broken-stone road,	3	times as much,
On a well-made pavement,	$4\frac{1}{2}$	times as much,
On the best stone track-ways,	11	times as much,
On the best railways,	18	times as much.

Poncelet§ gives the following relations of the friction to the pressure, for wheels with iron tires rolling on different surfaces:—

On a road of sand and gravel,		$\frac{1}{16}$
On a broken-stone road,	in ordinary condition,	$\frac{1}{25}$
	in perfect condition,	$\frac{1}{67}$
On a pavement in good order,	at a walk,	$\frac{1}{54}$
	at a trot,	$\frac{1}{42}$
On oak planks not dressed,		$\frac{1}{98}$

The most complete series of experiments upon the friction of vehicles have been recently made by *M. Morin.*‖ Some of the most important results are given below, in a tabular form. The fractions express the relation of the force of draught to the total load, vehicles included.

---

* Parnell, pp. 43, 73. † Ibid. p. 107.
‡ Lecount, p. 219. § Mécanique Industrielle, p. 507.
‖ Aide-Mémoire de Mécanique, p. 337.

CHARACTER OF THE ROAD.	CHARACTER OF THE VEHICLE.					
	Carts.	Trucks (of $2\frac{1}{2}$ tons.)	Diligences (of five tons.)		Carriages with seats hung on springs.	
New road, covered with gravel five inches thick,	$\frac{1}{12}$	$\frac{1}{9}$	$\frac{1}{8}$		$\frac{1}{8}$	
Solid causeway of earth, covered with gravel $1\frac{1}{2}$ in. thick,	$\frac{1}{16}$	$\frac{1}{11}$	$\frac{1}{10}$		$\frac{1}{10}$	
Causeway of earth in very good condition,	$\frac{1}{41}$	$\frac{1}{20}$	$\frac{1}{26}$		$\frac{1}{26}$	
Oaken platform,	$\frac{1}{70}$	$\frac{1}{46}$	$\frac{1}{41}$		$\frac{1}{42}$	
Broken-stone road.			Walk.	Trot.	Walk.	Trot.
Very dry and smooth,	$\frac{1}{75}$	$\frac{1}{54}$	$\frac{1}{48}$	$\frac{1}{41}$	$\frac{1}{49}$	$\frac{1}{42}$
Moist or dusty,	$\frac{1}{53}$	$\frac{1}{38}$	$\frac{1}{34}$	$\frac{1}{27}$	$\frac{1}{34}$	$\frac{1}{27}$
With ruts and mud,	$\frac{1}{33}$	$\frac{1}{24}$	$\frac{1}{21}$	$\frac{1}{18}$	$\frac{1}{22}$	$\frac{1}{19}$
Deep ruts and thick mud,	$\frac{1}{19}$	$\frac{1}{14}$	$\frac{1}{12}$	$\frac{1}{10}$	$\frac{1}{12}$	$\frac{1}{10}$
Pavement, dry,	$\frac{1}{90}$	$\frac{1}{65}$	$\frac{1}{57}$	$\frac{1}{38}$	$\frac{1}{59}$	$\frac{1}{39}$
Pavement, muddy,	$\frac{1}{69}$	$\frac{1}{50}$	$\frac{1}{44}$	$\frac{1}{33}$	$\frac{1}{45}$	$\frac{1}{34}$

From the above table it is apparent how important is the condition in which the best-made road is kept, and how greatly the labor of draught is increased by mud or dust on its surface. The character of the vehicle is also seen to have great influence on the degree of friction.

The principal general results, deduced by M. Morin from the elaborate experiments above referred to, are given on the following page.

DEDUCTIONS FROM MORIN'S EXPERIMENTS.

1. The resistance, or "Traction," is directly proportional to the *load*, and inversely proportional to the diameter of the *wheel*.

2. Upon a paved, or a hard Macadamized road, the resistance is independent of the width of the *tire* when it exceeds from 3 to 4 inches. On compressible roads, the resistance diminishes when the breadth of the tire increases.

3. At a walking pace, the traction is the same, under the same circumstances, for carriages with *springs*, or without them.

4. Upon *hard* Macadamized and upon paved roads, the traction increases with the *velocity;* the increments of traction being directly proportional to the increments of the velocity, above a speed of about 2¼ miles per hour; but it is less as the road is more smooth, and the carriage less rigid, or better hung.

5. Upon *soft* roads of earth, or sand, or turf, or roads freshly and thickly gravelled, the traction is independent of the *velocity*.

6. Upon a well-made and compact *pavement* of hewn stones, the traction at a walking pace is not more than three-fourths of that upon the best Macadamized road under similar circumstances: at a trotting pace it is equal to it.

7. The *destruction of the road* is in all cases greater as the diameters of the wheels are less and it is greater in carriages without than w'th springs.

5. WHAT ROADS OUGHT TO BE AS TO THEIR COST.

A *minimum* of expense is, of course, highly desirable; but the road which is truly cheapest is not the one which has cost the least money, but the one which makes the most profitable returns in proportion to the amount which has been expended upon it.

To lessen the cost of the construction of a road, while striving to attain the attributes which we have found to be desirable, we should endeavor to avoid the necessity of making high embankments, or deep excavations, or any rock-cuttings; the cuttings through the hills should just suffice to fill up the valleys crossed; the line of the road should be carried over firm ground and such as will form a good surface if no artificial covering be used; or if it is to be Macadamized, it should pass near some locality of good stone; and it should be so located as to require but few and small mechanical structures, such as bridges, culverts, retaining walls, &c.

COMPARISON OF COST AND REVENUE.

The more nearly, however, the road is made to approximate towards "what it ought to be," the more difficult will it be to satisfy the demands of economy. Some medium between these extremes must therefore be adopted, and the choice of it must be determined by the amount and character of the traffic on the road which it is proposed to make or to improve. For this purpose an accurate estimate is to be made of the cost of the proposed improvement, and also of the annual saving of labor in the carriage of goods and passengers which its adoption will produce. If the latter exceed the interest of the former, (at whatever per centage money for the investment can be obtained) then the proposed road will be "*what it ought to be as to its cost.*" From these considerations it will appear that it may be truly *cheaper* to expend ten thousand dollars per mile upon a road which is an important thoroughfare, than one thousand upon another road in a different locality.

"How to estimate the cost of a road" will be considered at the end of Chapter II., which treats of its "Location." Under the present head, we will examine how we may estimate the probable profits of a road, and from the comparison of the two estimates determine how much the projectors of an improved road would be justified in expending upon it.

AMOUNT OF TRAFFIC.

Let us suppose that it is proposed to improve a road in any way, whether by Macadamizing its surface, by shortening it, or by carrying it around a hill which it now goes over. The first point to be ascertained is the quantity and nature of the traffic which already passes over the line. This may be most accurately found by stationing men to count and note down all that passes in a given time of average activity; and from a sufficient number of such returns, well classified, deducing the annual amount.

COST OF ITS TRANSPORTATION.

The cost of conveying this amount of traffic is next to be calculated. To simplify the question, we will neglect the gain in speed, and consider only the saving in heavy transportation. Assume that over the road, thirty miles in length, 50,000 tons of freight are annually carried, and that the average friction of its surface (as determined by a dynamometer) is $\frac{1}{20}$ of the weight. The annual force of draught required is therefore 2500 tons, or 5,000,000 lbs. If the average power of draught of a horse at 3 miles an hour for 10 hours a day be taken at 100 lbs.,* there would be required $\frac{5,000,000}{100}=50,000$ horses working at 3 miles per hour. At this rate they would traverse the road in 10 hours, or a working day, and the total amount of labor would equal 50,000 days' work of a horse, or $37,500, taking 75 cents for the value of one day's work.

PROFIT OF IMPROVING THE SURFACE.

Suppose now that the road is to be *macadamized*, or *planked*, or in any way to have the friction of its surface reduced to $\frac{1}{50}$. The total force of draught will then be $\frac{50,000 \times 2000}{50} = 2,000,000$ lbs. $=$ 20,000 horse power, at 3 miles per hour, for 30 miles, or 10 hours = 20,000 days' work of a horse. This is a saving from the former amount of 30,000. Taking the value of the day's work of a horse at 75 cents, $22,500 would be the actual saving of labor in each year, by the improvement proposed, which amount the carriers could afford to pay, (either in tolls, or in ma-

* The power of a horse at different velocities is very variable, and, in spite of many experiments, is not yet ascertained with the precision desirable. The usual conventional assumption is 150 lbs. moved 20 miles a day at the rate of 2½ miles per hour. This is equivalent to Watts' horse-power of 33,000 lbs. raised 1 foot in 1 minute. Tredgold's experiments give 125 lbs. moved 20 miles a day at 2½ miles per hour. Smeaton gives 100 lbs. moved at same rate; and Hachette 128 lbs. Numerous careful experiments on an English railway (detailed in "Laws of Excavation and Embankment on Railways," page 105) give 110 lbs. moved 19.2 miles per day at the rate of 2.4 miles per hour. Gayffier (page 178) fixes the power for a strong draught-horse at 143 lbs. for 22 miles per day at 2¾ miles per hour; and for an ordinary horse, at 121 lbs. for 25 miles per day at 2½ miles per hour. As the speed of a horse increases, his power of draught diminishes very rapidly, till at last he can only move his own weight.

king the improvement themselves) for their diminished expenditure on horses. If money were borrowed at 6 per cent., $375,000 would be the amount which could be expended in making the improvement, supposing the data to have been correctly assumed. If the improvement can be made for any amount less than this, the difference will be so much clear gain.

PROFIT OF LESSENING THE LENGTH.

Next, suppose that the improvement is only *shortening the road a mile*, by a new location of part of it. One-thirtieth of the original distance, and therefore labor, is saved, or $\frac{50,000}{30} = 1667$ days' work of a horse $=$ $1,250 $=$ interest of $20,833. Add to this the amount which the construction of this extra mile would have cost, and if the proposed improvement can be made for the sum of the two, or even a little more, it should be at once carried into effect; for, besides the saving in the original cost and in the annual labor, there is also that of time, and of the former cost of repairs of the extra mile, which is now dispensed with.

PROFIT OF AVOIDING A HILL.

If the improvement be *avoiding a hill*, the resistance of gravity is to be compared with that of friction. Suppose that a certain road ascends a hill which is a mile long, and has an inclination of 1 in 10, and descends the other side which has the same slope, and that a level route can be obtained by making the road a mile longer. It is demanded how much may be expended for this purpose. Suppose that the friction on this road is $\frac{1}{40}$, and that 50,000 tons, as before, pass over it annually. On the original road of two miles, the force of draught required to overcome *friction* is $\frac{50,000 \times 2000}{40 \times 100} = 25,000$ horse power, at 3 miles per hour, or $\frac{25,000 \times 2}{3} = 16,667$ hours for the 2 miles = 1667 days' work of a horse. To overcome the *gravity* of the loads on the inclination of 1 in 10 requires $\frac{50,000 \times 2000}{10} = 10,000,000$ lbs. for 1 mile = 333,333 lbs. for 30 miles = 3333 days' work of a horse. The descent of a mile on the other side of the hill is not a compensation, for a horse will have no more to take down the descent than he had dragged up the ascent. The total annual labor to overcome both friction and gravity on these two miles is therefore 1667 + 3333 = 5000 days' work of a horse.

Upon the new road proposed, there is no inclination to overcome, but an extra mile of length. The force of draught upon it due to friction is $\frac{50,000 \times 2000}{40} = 2,500,000$ lbs. for 3 miles = 250,000 lbs. for 30 miles = 2500 days' work of a horse. The saving of labor is therefore 5000 — 2500 = 2500 days' work of a horse = \$1875 = interest of \$31,250, which amount (deducting cost of repairs of the extra mile) may be expended in making the new road.

These calculations have been made for extreme cases, in order to make the principle more striking, but the advantages deduced from them have fallen short of the truth, since only the original amount of traffic has been considered, while all experience shows that this is very greatly increased by any improvement in the means of transport particularly by the increased speed, which is an incidental advantage which we have not taken into account. This increase of traffic cannot, however, be determined in advance, by mathematical calculation, though we can readily see from how wide a belt of country the inhabit ants might profitably avail themselves of the improved road, and will do so eventually; but how many of them will at once profit by it depends on considerations of taste, feeling, and prejudices, which are beyond the power of numbers.

CONSEQUENT INCREASE OF TRAVEL.

To ascertain from what distances to the right or left on either margin, the improved road might expect to attract travel to itself from other thoroughfares by the cross roads, the following course of reasoning may be employed.

Let AB be a portion of the improved road, connecting the points A and B. Let C be a town connected with the other two points by the old unimproved roads CA and CB. It is required to determine whether the travel from C to A can with the least cost (the cost being compounded of time and labor) go to A by the old road CA, or take the old cross-road CB to the nearest point B of the improved road, and then follow the latter to A.

Fig. 14.

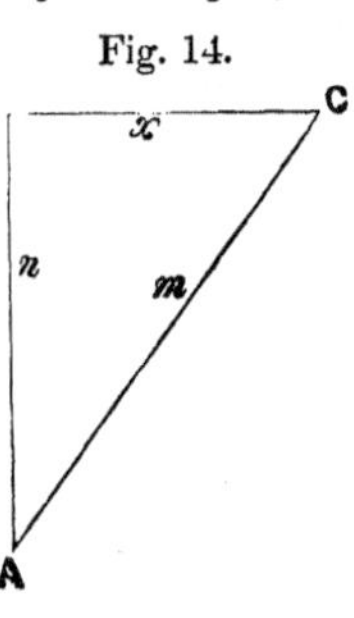

The first point is to ascertain the *ratio* of improvement of the new road compared with the old, or its ratio of diminution of cost of travel. For simplicity of calculation let us call this ratio *two*. Denote the miles in AC by m, in AB by n, and in BC by x. The relative cost of travel over the line AC will also be m, over BC it will be x, but over AB it will be only $\frac{n}{2}$. If, then, $x + \frac{n}{2} < m$, it will cost less to make the circuit from C to A through B; and both routes will be equal in cost when $x + \frac{n}{2} = m$. In this calculation, therefore, the hypothenuse equals the perpendicular and half the base!

The preceding method will decide the question for any one place, but the following plan may be resorted to for the purpose of marking out on the map the entire area, from within which travel may be expected to be attracted to make use of the improved road.

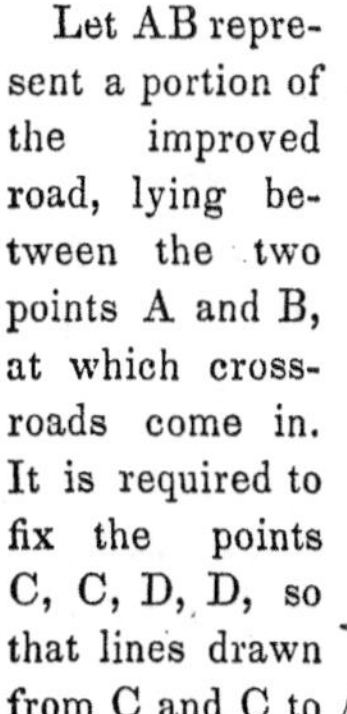

Let AB represent a portion of the improved road, lying between the two points A and B, at which cross-roads come in. It is required to fix the points C, C, D, D, so that lines drawn

Fig. 15.

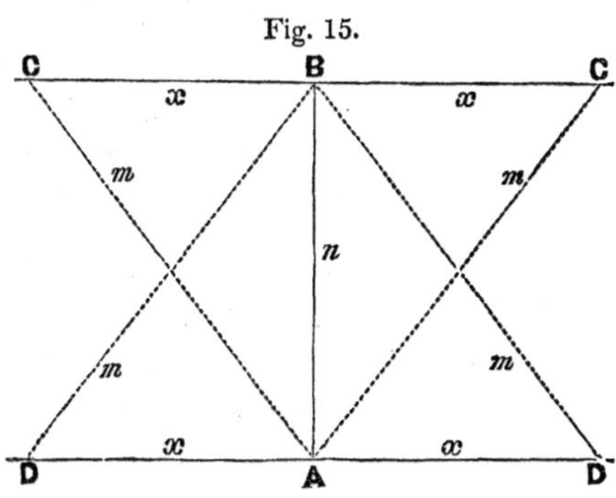

from C and C to A, and from D and D to B, shall define this tributary area. BC or AD is to be found in terms of AB; i. e. x in terms of n.

By the preceding investigation,

$$x + \frac{n}{2} = m.$$

But in the right-angled triangle ABC,

$$m = \sqrt{(x^2 + n^2.)}$$

Substituting in first equation, we get

$$x + \frac{n}{2} = \sqrt{(x^2 + n^2;)}$$

whence is obtained the value,

$$x = \tfrac{3}{4} n.$$

Therefore from A and B set off, at right angles to AB, BC, and AD, each equal to $\frac{3}{4}$ AB; join AC and BD; and the area included will be that within which it would cost less for the inhabitants to use the improved road, though with increased distance, than to pursue the direct but unimproved road.*

* *Lecount,* Treatise on Railways, p. 12.

CHAPTER II.

THE LOCATION OF ROADS.

"I do not know that I could suggest any one problem to be proposed to an engineer, which would require a greater exertion of scientific skill and practical knowledge, than laying out a road."—DR. LARDNER, in 1836.

THE *location*, or *laying out*, of a road, consists in determining and marking out on the ground those points through which the road should pass, in order to satisfy, as nearly as possible, the requirements of "what a road ought to be."

These requirements, so far as they affect the location of a road, are, in recapitulation, as follows:

As to direction—that the road should be as straight as possible, but that straightness should be considered subordinate to easiness of grade.

As to slopes—that the road should be as level as possible; that it should avoid unnecessary undulations; and that its slopes should not exceed 1 in 30, nor fall below 1 in 125.

As to cost—that the amount of excavation, embankment, mechanical structures, &c., should be the least which will make the road "what it ought to be," in reference to the quantity of traffic upon it.

If the country through which the road is to pass should be a plain of uniform surface, a straight line joining the two *termini*, and running along the surface of the ground would satisfy all these conditions at once. In most cases wever, the ground is so uneven, hil'y, and undulating, to present very great difficulties in the way of a proper cation. The *shortest* line would pass over the tops of hills and the bottoms of valleys, and would thus be often so steep as to be impassable. The *most level* line would often increase the distance too much by its necessary windings; as would also the *cheapest* line, which seeks to avoid all cuttings and fillings. It is generally impossible to unite all these requirements, and to secure all the good qualities and valuable attributes of the ideally perfect road; and the best line will therefore be a compromise between them all. Great skill is consequently required to select the best possible line among these conflicting claims, and this skill is more often needed in our new and rapidly expanding country than in England and other long-settled regions, where the lines of all important roads have been long since established; though even there many miles of old roads are yearly abandoned, and new lines substituted for them, in order to make a slight saving of distance, or to diminish the height to be overcome.

Two distant points of departure and arrival being given, it is required to determine the best line for a road connecting them.

In many cases the best general route for the desired road can be determined with perfect certainty without going upon the ground, by simply examining a map of the district upon which merely the courses of the streams are laid down. From them an instructed and skilful eye can deduce all the elevations and depressions of the country with great precision and accuracy. To do this, however, requires a knowledge of so much of Physical Geography as explains the manner in which nature has disposed the inequalities of the surface of the earth.

1. ARRANGEMENT OF HILLS, VALLEYS, AND WATER-COURSES

Hills and valleys at first glance appear to the ignorant, and even to the better informed, to be utterly without system, order, or arrangement; but they have in reality been disposed by nature with a great degree of symmetry, and their forms and positions are found to be the result of the uniform action of natural laws, and to be capable of being traced out and understood with comparative ease.

Hills being the great antagonists and natural enemies of the road-maker, he must endeavor to find out their weak points, and to learn where he can best attack and penetrate them, and most easily overcome their opposition to his improvements. *Water-courses* being his guides and chief assistants, he must study their habits and principles of action, and learn what are the causes which produce their seeming vagaries of direction.

Hills are most usually found constituting *chains*, or *ridges*, though sometimes collected in *groups*, and at others *detached*, or *isolated*. The chains are usually made up of several parallel *ranges*, and often send forth branches or *spurs* in transverse directions. Sometimes they are merely the slopes of a *table-land* in which their summits merge. To form a proper conception of a *range* of hills, imagine in the midst of a plain an elongated mass of the form of the roof of a house. The two faces of this represent the *slopes* of the range; their intersection is the *ridge*, their bases are the *feet*, the distance from one foot to the other is the *breadth*, and from one extremity to the other the *length;* the vertical elevation of the ridge above either foot is its *relative height*, and above the sea its *absolute height.* All water which falls upon the slopes descends thence in a well-defined track which corresponds with *the line of greatest slope*, the direction of which it is therefore important to determine.

LINE OF GREATEST SLOPE.

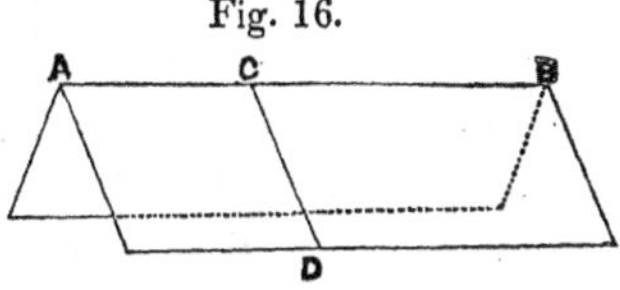

Fig. 16.

If the ridge AB of a range of hills be horizontal, and its opposite slopes inclined planes cutting each other in that horizontal line, a spherical body, allowed to roll down freely from any point C of the ridge, will descend in the line CD at right angles to the horizontal line AB; this line CD being its nearest possible approach to the vertical line in which it tends to move in obedience to the law of gravity. CD is therefore the *line of greatest slope*, and consequently of quickest descent. It is this line which water tends to follow in its search for the shortest path of descent.

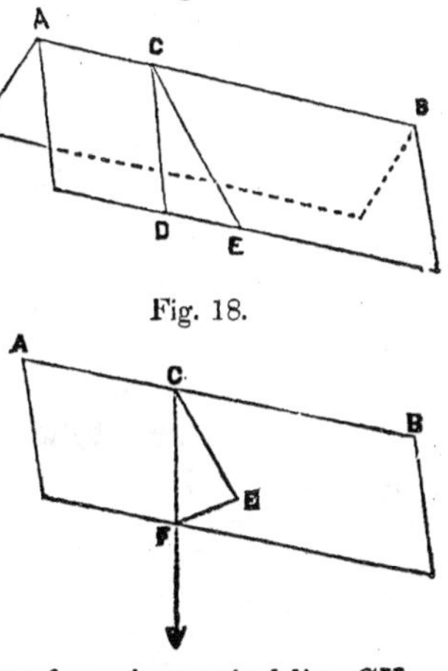

Fig. 17.

Fig. 18.

If the ridge AB be inclined, the path down which the sphere will roll is no longer CD at right angles to AB, but another line CE, diverging in the direction of the slope of the ridge. To determine its precise position, from any point C, Fig. 18, let fall a vertical line CV, and, from any point F of this vertical, raise a perpendicular to the plane of the slope, meeting it in E. Draw CE, and it will be the line of greatest slope required; for it is at the least possible distance from the vertical line CV.

The same result might be otherwise obtained by raising at C a perpendicular to the plane of the slope, and from any point therein letting fall a vertical line, which will intersect the slope at some point E, which is to be joined to C as before.

When the slopes are not planes, the constructions are more complicated, as the "lines of greatest slope" then become curves.*

The waters which have fallen upon the mountain-tops from time immemorial, have hollowed out for themselves, or have adopted for their passage, channels which follow the *lines of greatest slope*, whose directions we have just investigated. In descending the slopes of a range of hills, they thus form "principal" valleys, the directions of which, as we have seen, are perpendicular to the ridge when it is horizontal, and, when it is inclined, share its general inclination. These streams thus divide the range or chain into *ramifications* or *branches*, having approximately the same direction as themselves. The line in which the opposite slopes of two of these adjoining "branches" intersect each other, and which thus marks out the lowest line of a valley, is called a *thalweg*.† The foot of one of the opposite slopes which enclose a valley is generally parallel to the foot of the other in all its sinuosities, a projecting point of the one corresponding to a receding cavity in the other. This symmetry is, however, sometimes replaced by alternate widenings and contractions.

The main ridge is cut down at the heads of the streams into depressions called *gaps*, or *passes*; the more elevated points are called *peaks*. They are respectively the origins of the valleys and of the branches on both sides of the principal slope. In the gaps are often found swamps,

* Gayffier, p. 3.

† A German word, (signifying "the road of the valley") which has been naturalized in the French language, and might be conveniently added to our engineering vocabulary in English

fed by the rain which falls on the peaks between which they lie. In these the streams take their rise, and thence run in contrary directions down the opposite slopes of the ridge. The intermediate point, from which they start and diverge, is called the *culminating point* of the pass.

Thus the "Notch" of the White Mountains is the "cul minating point" from which diverge the Saco and the Ammonoosuc, the one emptying into Long-Island Sound and the other into the Atlantic. So, too, from the various culminating points in the Alleghany chain, streams run, on the one side towards the Atlantic, and on the other to the great lakes and to the Mississippi. From the culminating points of the Rocky Mountains, the slightest impulse would turn the nascent stream either into the head-waters of the Missouri and thence into the Gulf of Mexico, or into the head-waters of the Columbia and thence into the Pacific Ocean. The same phenomena, on a miniature scale, are repeated on every ridge after every shower.

A river of the largest class marks the lowest points (or the *thalweg*) of a "principal" valley. On each side of it is a bounding ridge, which is itself pierced by "secondary" valleys, through each of which runs a stream of less magnitude, its waters emptying into the first-named river, of which it is a tributary. The ridges which form the valleys of each of these lateral streams are in their turn furrowed by valleys of the third class; their banks by the valleys of streams of still less importance; and so on.

The "principal" valley is a trunk, from which, and from one another, the lesser valleys and streams ramify, like the branches of a tree, or like the veins of the body · meeting it at angles approaching more nearly to a right angle in proportion as the ridge of the slope which they furrow approaches to a horizontal line.

Fig. 19.

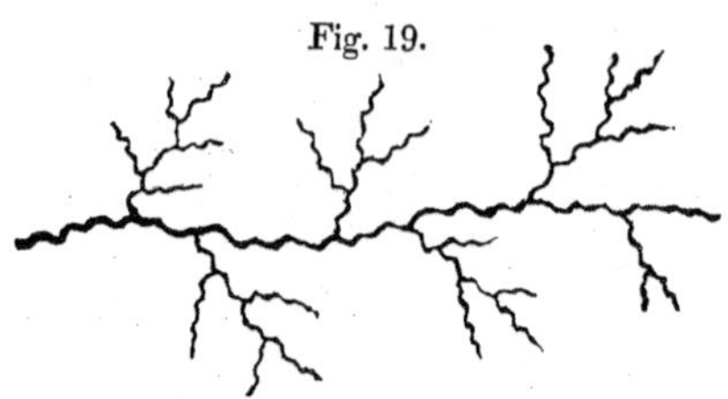

INFERENCES FROM THE WATER-COURSES.

We thus see how an accurate map of the streams of any district may enable us to deduce from them the position of the valleys, their lowest points, and the lines of greatest slope; for with these the water-courses coincide. The position of the ridges which form the valleys is a necessary corollary, as well as their lines of greatest slope.

Having determined these, we can profit by the following fundamental principles:

1. If a principal ridge is met by two secondary ridges at the same point, the point of meeting is a maximum of height.

2. If a principal ridge is met by two *thalwegs* at the same point, the point of meeting is a minimum of height.

3. If a principal ridge is met at the same point by a secondary ridge and a *thalweg*, nothing can be inferred.*

The following examples† show more in detail some of the inferences which may be drawn from the map:

Fig. 20.

If, on any portion of a map, the streams appear to diverge from any point, as A, that point must be the common source of the streams, and therefore the highest part of that region.

The converse is likewise true: if the streams all converge towards some

* Jullien, ii. 293.

† Mahan, p. 278.

Fig. 21.

point, as B, that will be the lowest spot of the district embraced within the map.

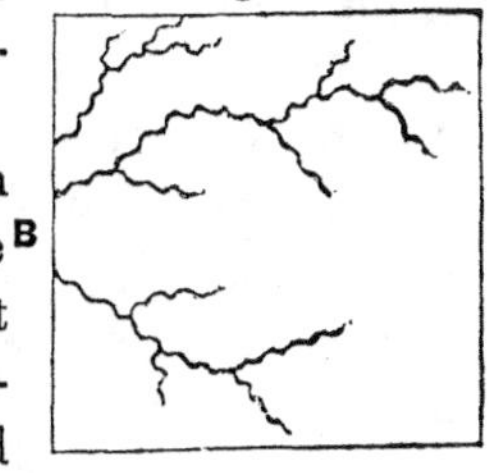

If two streams are seen to flow in opposite directions from the same point, as C, it may be inferred that this spot is at the head of the respective valleys of these streams, and supplies them with water, and that it must be fed by higher ground beside it; or, in other words, that there is a ridge of hills separating the heads of the two streams, and that there is a depression or indentation in this ridge at the point C, which is therefore the natural and proper location for a road connecting the two valleys.

B

Fig. 22.

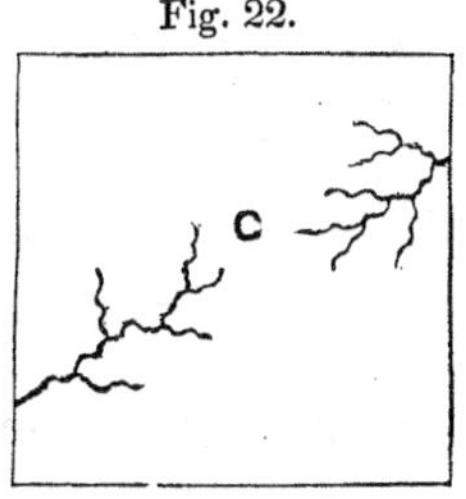

If two streams are parallel to each other, and flow in the same general direction, this circumstance simply indicates that the ridge which divides them has the same general inclination and direction as the streams. But if any of their smaller tributaries approach each other at their sources, as at D, this indicates a depression of the main ridge at that point, and marks it out as the lowest and easiest spot for the crossing of a road, as in the preceding case.

Fig. 23.

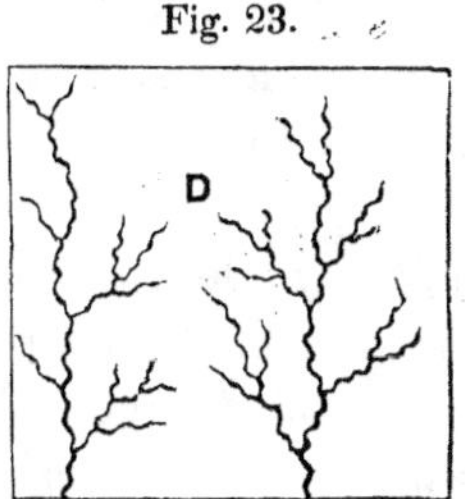

If two streams have been flowing in parallel courses, but at a certain point E diverge from each other,

Fig. 24.

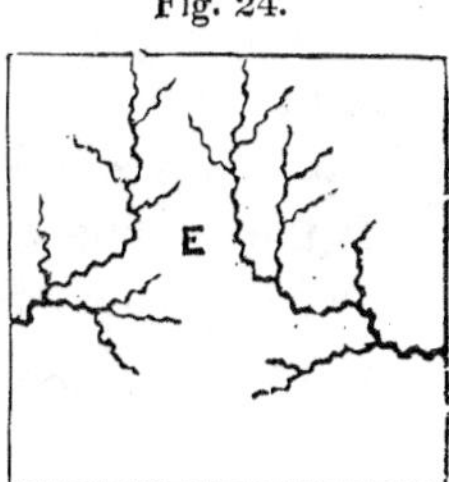

that spot is the lowest point of the ridge between them.

Fig. 25

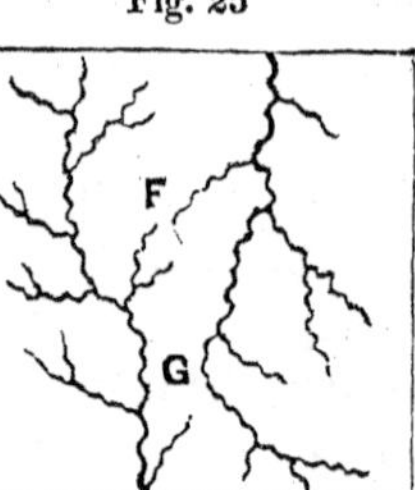

If two streams are generally parallel in their courses, but flow in opposite directions, the low points in the ridge between them will still be shown by the approach to each other, as at F, of the branches or secondary streams; or by the principal streams approaching each other at any point, as at G.

Having thus become acquainted, by the aid of the map, with the principal features of the ground, we are prepared to plan, if not the precise location of the road, at least the proper course for the preliminary explorations upon the ground. Long lines of road usually follow the valleys of streams, and thus secure moderate grades and find the lowest passes of the ridges to be crossed. In this way the Simplon road crosses the Alps, by ascending the valley of the Saltine to its head, and then descending that of the Doveria. So, too, the Boston and Albany railroad finds an easy grade from Worcester to Springfield in the valley of the Chickapee river, and then winds through the mountains, up the valley of the Westfield, till it reaches the head-waters of the Housatonic upon the other side of the ridge. The Utica and Schenectady railroad never quits the valley of the Mohawk. In short, all roads strive to avail themselves of such facilities. If they cannot, and if the map shows that their general course is transverse to the directions of the streams, instead of with them, it may be at once predicted that they will be steep in their ascents and descents, or exceedingly expensive in their construction.

These principles having been established, and all possible information obtained from the map, the *Reconnais sance* may be commenced.

2. RECONNAISSANCE.

This is a rapid preliminary survey of the region through which the road is to pass, and is generally made by the eye alone without instruments. It is intended to be only an approximation to accuracy, and to serve to determine through what points routes should be instrumentally surveyed. The road-maker must examine the country, map in hand, visit and identify the points selected on the map, and see whether his closet decisions have been correct. He must go over the ground backward and forward in opposite directions, for it will often appear quite different, and convey very dissimilar impressions, according to the point from which it is viewed. Thus, a hill which one is descending may seem to have a very easy slope, while it may appear very steep to one ascending it. No time or labor should be spared in these first explorations, as they will save much expense in the subsequent surveys, which in their turn should be thoroughly executed, to secure the route most economical in construction. Indeed, the surveyor should become as perfectly acquainted with the face of the country as if he had passed his hand over every foot of it.

Certain points, called "ruling" or "guiding" points, will be found, through which the road must pass; such as a low gap in a range of hills, a narrow part of a river suitable for a bridge, a village, &c. But a road which is to be a thoroughfare between two places of great trade, should not be turned from its direct course to accommodate a small town, taxing for its benefit all who travel upon the road. "The greatest good of the greatest number" is here the rule. Still less should individual interest be allowed to operate, and the general interest of the community be sacrificed to the convenience or caprice of a single person. The permanent benefits to future generations should never be made subordinate and subservient to temporary and personal advantages.

Between these "ruling" points, the straight line joining them is to be marked out. The route adopted must vibrate on each side of this line, like an elastic cord, con tinually tending to coincide with it, except when deflected to the right or to the left by weighty reasons, such as the accidents of the ground supply. Thus, a swamp, which would render necessary an expensive causeway, is a sufficient cause for a wide deviation of the road to avoid it. The disadvantages of straight lines, which encounter and run over hills, have been explained in the preceding chapter. In the accompanying figure the upper sketch shows a *plan*, or map-view, of two roads, the one ACB *over* a hill, and the other ADB *around* it; and the lower sketch shows a *profile*, or side-view, of the respective heights of the same lines.

Fig. 26.

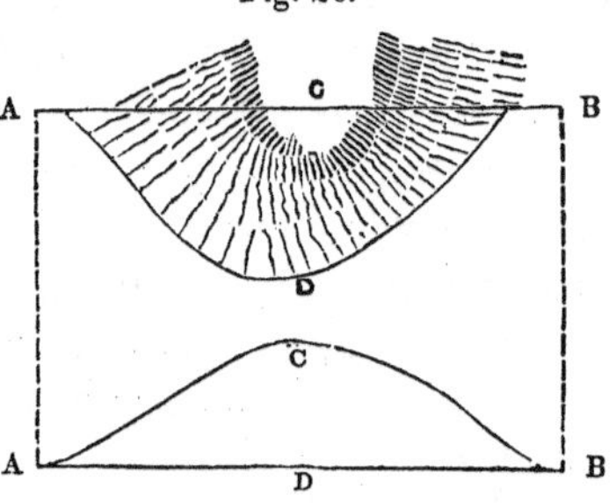

When there are many small valleys or ravines, with projecting spurs and ridges intervening, instead of making the road wind on the level ground, and follow all its sinuosities, as, ACCCCB, in the next figure, it will be better to make a nearly straight line, as ADDDB, cut through the projecting points in such a way that the earth dug out shall just suffice to fill the hollows. The gain by saving of distance may balance the cost of cutting and filling.

Fig. 27

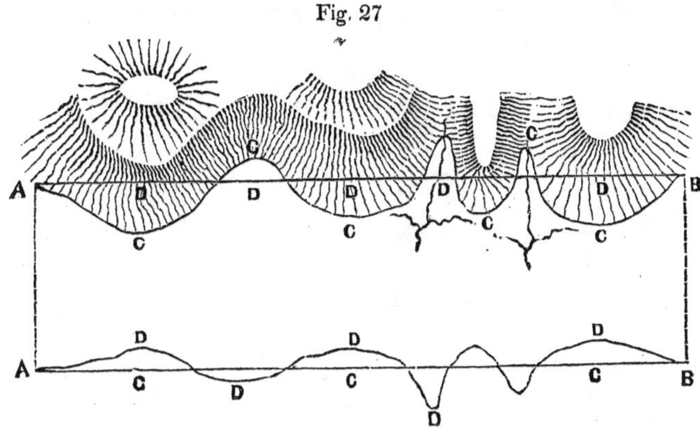

When the route follows the valley of a stream, it may conform to its sinuosities, if the turns are not too abrupt, and if the cuttings and fillings on a straighter line would be too expensive, but should approximate to the latter plan, if the importance of the road and the funds at command will justify the increased cost. The former plan, however, generally gives the cheapest and most level route ; and guided by this principle a blind man was for a long time the very best layer out of roads in the hilly regions of Yorkshire and Derbyshire. He followed the streams closely, and when they made too sharp bends, he sought in these arcs the straightest chords which passed over practicable ground.

When a valley is to be crossed, the route should generally deviate from the straight line ACB, (Fig. 28) and curve towards the head of the valley ADB, which there is usually shallower and narrower. If it deviated in the other direction, as AEB, it would increase the depth and width to be filled up, as is shown by the corresponding profiles

Fig. 28.

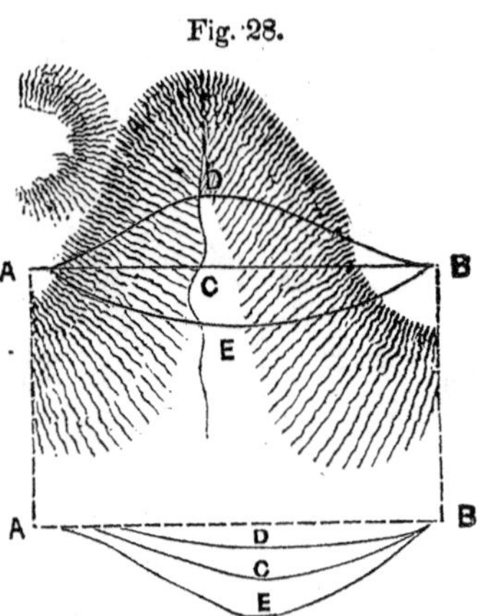

But sometimes the two sides of the valley approach each other at some point lower down, so as to render the space between their banks narrower though deeper; and if on measurement this area is found on the whole to be lessened, so as to require less embankment, the road should cross at that point instead of higher up.

Fig. 29.

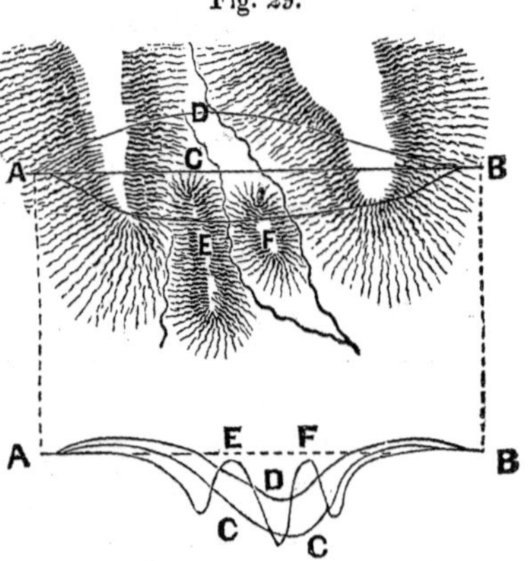

Another case in which a valley may, with advantage, be crossed down stream, is when in that part of the valley are found detached or isolated hills and ridges, as E and F, which may cause a great saving of embankment, on the line AEFB, compared with either the straight route ACB, or the up-stream one ADB, as is shown in the accompanying plan and profile, in which the same letters refer to corresponding lines.

When a road is to join two places on the opposite sides

of a ridge, we can profit by the observation that the streams, by the approach of their sources, show the lowest points of the ridge ; and of the various *passes* thus indicated, we should choose that one, the valleys of the streams from which run as nearly as possible in the direction of the required line ; and that one, also, which has the most uniform inclination—not easy at the foot d steep towards its summit, as is often the case.

When a road is to join two places situated on the same le of a mountain ridge, but half way down its side, a aight line between them would cross, in their deepest and dest parts, all the "principal" valleys which run down m every gap. One of two other plans must then be pted ; either to ascend, and carry the road, with neces- y windings, at the level of the culminating points of gap, where the valleys have only begun to be hollowed ut ; or to carry it at the foot of the ridge, where the val- s have run out to nothing, and merged themselves un- stinguishably in the plain. Either plan, in spite of the itial and final ascent and descent, is preferable to the ect line.

Fig. 30.

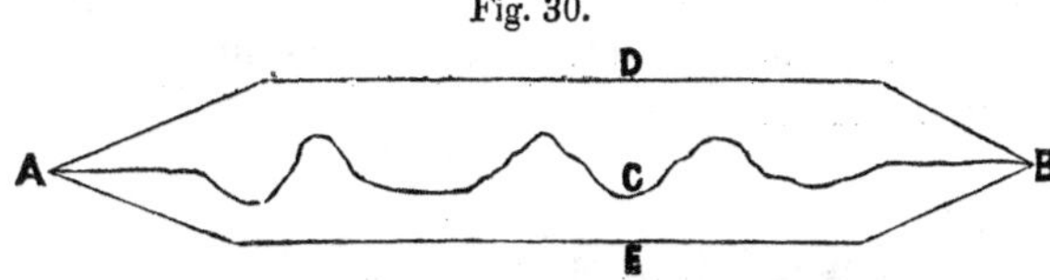

The respective profiles of the three plans would be somewhat as represented in the figure, in which ACB is the first plan, ADB the second, and AEB the last. The last line is generally taken, because there are more inhabitants at the foot of the ridge. It would properly run near the line of separation between the uncultivated slopes and the ploughed fields.

The location of a road is also influenced by the *geology* of a district, which must therefore be carefully studied This science will make known the probability of finding rock on cutting deep into a hill proposed to be crossed; in which case the cutting should be avoided, if possible, by a different location of the line. It will also determine the dips of the strata to be cut into, the angle at which they will stand, and their liability to slip; and therefor through which the line may best pass. If the road is t be covered with broken stone, or to be paved, a know ledge of the locality of the best materials might cause line approaching it to be preferred to one which left it a distance.

The *Reconnaissance* is to be made in accordance w the principles which have been enunciated, obtaining needful information from the residents of the region to b examined, and the details of its general course are to be marked out on the ground, thus establishing "Appro mate" or "Trial" lines. In a wooded country this is do by "blazing" the trees in the line selected, (removing chip from their sides with an axe;) and in a cleared co try by driving stout stakes at the most important points the line, such as all changes in its direction, and in t slope of the ground.

3. SURVEY OF A LINE.

When the different portions of a proposed line have been thus marked out, in order to form an accurate opinion of its merits, it is necessary to measure—

1. *Its distances.*
2. *Its directions.*
3. *Its heights.*

MEASUREMENT OF DISTANCES.

The length of a straight line, that is, the distance between its extremities, may be approximately estimated in a variety of ways, without the delay of actual measurement in detail.

Sound is a well-known means. Its velocity is 1100 an|t per second at the temperature of freezing.* If a gun fired by an assistant at one end of a line, an observer sic the other end, by counting the seconds which intervene str ween seeing the flash and hearing the report, and mulwi(ying their number by 1100, can estimate the distance fro|h considerable accuracy. If he have not a watch with ado:cond-hand, he can at once make a portable pendulum, sar| istening a pebble to a string, and making it swing in th(ular vibrations, each of which will be performed in an exact second, if the string be 39⅛ inches long; in half a ley.ond, if it be 9¾ inches long; and in a quarter second, d' :.s length be 2½ inches.

in This method is best adapted for considerable distances, dir hich there are good points for observation, such as nills on the two opposite sides of a wide valley.

or shorter distances, the *distinctness* with which difnt objects can be seen, is an approximate guide. Thus windows of a large house can generally be counted at e distance of 3 miles; men and horses can just be per:eived as points at about 1¼ miles; a horse is clearly distinguishable at ¾ mile; the movements of a man at ½ mile; and a man's head is plainly visible at ¼ mile.†

* For each degree of Fahrenheit above 32°, add one-half foot, and for each degree below, subtract one-half foot. A temperature of 60° would therefore give $1100 + \frac{28}{2} = 1114$ feet per second.

† Frome, p. 60.

The Arabs of Algeria define a mile as "the distance at which you can no longer distinguish a man from a woman." These distances of visibility will of course vary somewhat with the state of the atmosphere, and still more with individual acuteness of sight, but each person can modify them for himself.

For still less distances, an easy method is to prepare *a scale*, by marking off on a pencil what length of it, when it is held off at arm's length, a man's height appears to cover at different distances (previously measured with accuracy) of 100, 500, 1000 feet, &c. To apply this, when

Fig. 31.

a man is seen at any unknown distance, hold up the pencil at arm's length, making the top of it come in the line from the eye to his head, and placing the thumb nail in the line from the eye to his feet. The pencil having been previously graduated by the method above explained, the portion of it now intercepted between these two lines will indicate the corresponding distance.

If no previous scale have been prepared, and the distance of a man be required, take a foot-rule, or any measure minutely divided, hold it off at arm's length as before, and see how much a man's height covers. Then knowing the distance from the eye to the rule, a statement by the Rule of Three (on the principle of similar triangles will give the distance required. Suppose a man's height, of 70 inches, to cover 1 inch of the rule. He is then 70 times as far from the eye as the rule; and if its distance be 2 feet, that of the man is 140 feet. Instead of a man's height, that of an ordinary house, of an apple-tree, the length of a fence-rail, &c., may be taken as the standard of comparison.

Quite an accurate measurement of a line of ground may be made by *walking over it* at a uniform pace, and counting the steps. It is better not to attempt to make each of the paces three feet, but to take steps of the natural length, and to ascertain the value of each by walking over a known distance, and dividing it by the number of paces required to traverse it. An average length is 32 inches. An instrument, called a pedometer, has been contrived, which counts the steps taken by one wearing it, without any attention on his part. It is attached to the body, and a cord, passing from it to the foot, at each step moves a toothed wheel one division, and some intermediate wheelwork records the whole number upon a dial.

These methods are all approximations. For more accurate measurements a *chain* is employed. The usual surveyor's or Gunter's chain, is 66 feet or four rods long, and is divided into 100 links; but for the measurement of distances only, without reference to areas in acres, a chain of 50 or 100 feet is much preferable.

When *obstacles* are encountered on the line, rendering a direct measurement impossible, such as a house, a pond, a river, &c., resort must be had to some of the many ingenious contrivances to be found in the special treatises on surveying and engineering field-work. Two only of the best, which have the advantage of requiring no calculations, will be here given.

When the obstacle is one around which we can pass, such as a *house* or a *pond*, the following plan may be adopted. Let AB be the distance required. Measure from A obliquely to some point C, past the obstacle. Measure onward in the same line, till CD is as long as AC. Place stakes at C and D. From B measure to C, and from C measure onward in the same line, till CE is equal to CB. Measure ED, and it will be equal to AB, the distance required.

Fig. 32.

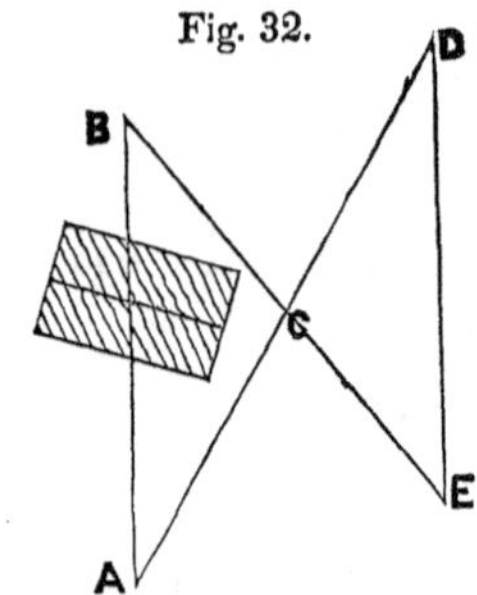

When the obstacle is a *river*, the following is the method to be employed. Let AB be the required distance. From A measure any line diverging from the river, as AD, and set a stake in its middle point C. Take any point E, in the line of A and B. Measure from E to C, and onward in the same line, till CF equals CE. Then find by trial the point G, which shall be at the same time in the line of C and B, and in the line of D and F. Measure the distance from G to D, which will equal the required distance from A to B. The lines which it is not necessary to measure are dotted in the figure.

Fig. 33.

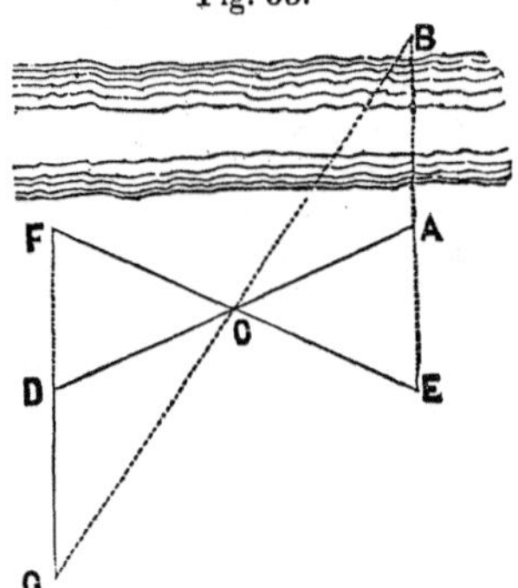

MEASUREMENT OF DIRECTIONS.

Having measured the lengths of the various portions of the line, by whatever method will give the degree of accuracy required, their directions are also to be examined, determined, and recorded.

These directions may be accurately determined, with reference to the adjoining portions of the lines, and therefore to any given line, by simple measurements with the chain, without the use of any of the usual complicated angular instruments.

Fig. 34.

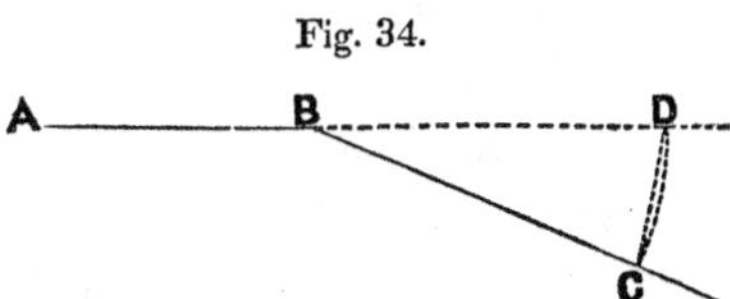

Let AB and BC represent two lines on the ground, meeting at any angle. It is required to determine the change in the direction of the line BC from that of AB, *i. e.*, the angle CBD, or the "angle of deflection." Set off from B equal distances, BC on the new line, and BD on AB produced, and measure DC, which is the chord of the angle required. To find the angle numerically, take half this measured chord, (which equals the sine of half the angle to radius BC) and divide it by BC. Find in a table of natural sines the angle corresponding to the quotient. Twice this is the angle CBD required. But even this brief calculation is needless for putting down the line upon paper, as it is only necessary to describe an arc from B as centre with BC as radius, and to set off CD of the proper length, the distances being taken from any one scale.

If the direction of a line be required independently of any other line upon the ground, it is usually referred to the direction of the meridian, *i. e.*, the line which passes through the north and south poles of the earth. The compass is the readiest means of obtaining this, although, in addition to its other inherent defects, it gives the angle made by the given line with only the *magnetic* meridian which is constantly changing, and from which the *tru* meridian in most places varies considerably.

To determine the *true meridian* (and therefore the angle which any line makes with it) without the use of the compass, the following is a simple and sufficiently accurate method. On the south side of any level surface erect an upright staff, shown, in horizontal projection, at A. Two or three hours before noon mark the extremity, B, of its shadow. Describe an arc of a circle with A, the foot of the staff, for centre, and AB, the distance to the extremity of the shadow, for radius. At about as many hours after noon as it had been before noon when the first mark was made, watch for the moment when the end of the shadow touches the arc at another point C. Bisect the arc BC at D. Draw AD, and it will be the true meridian, or north and south line, required.

Fig. 35.

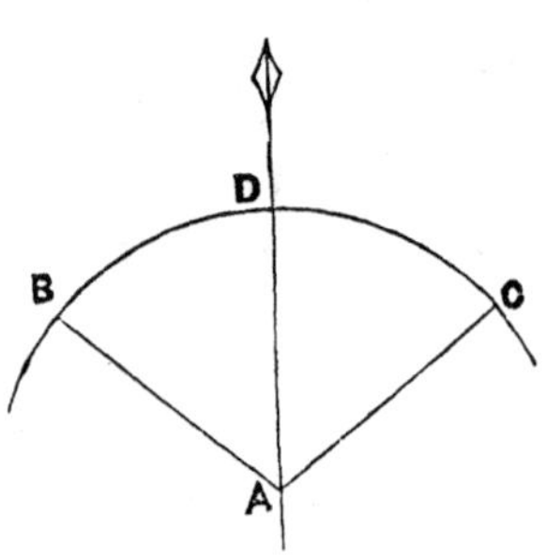

For greater accuracy, describe several arcs, mark the points in which each of them is touched by the shadow, bisect each, and adopt the average of all. The shadow will be better defined, if a piece of tin with a hole through it be placed at the top of the staff, as a bright spot will thus be substituted for the less definite shadow. Nor need the staff be vertical, if from its summit a plumb-line be dropped to the ground, and the point which this strikes be adopted as the centre of the arcs, through which the meridian line AD is finally to be drawn.*

* For the method of determining the true meridian by the north star and other methods, see Gillespie's Land Surveying, pp. 190–198.

MEASUREMENT OF HEIGHTS.

The relative heights of the different points, at which the line changes its slope, are next to be determined. The operation required for this purpose is called LEVELLING. It consists in finding how much each of these points is below any level line. The difference of their distances below it (measured perpendicularly to it) is the difference of their heights. The first step, then, is to discover means of getting a level line at any point desired

The principle, that a level line is everywhere perpendicular to the direction of gravity, furnishes the first method. Upon it depends the well-known "*Mason's level*," in which a straight edge AB is "level," or horizontal, when a line CD, exactly at right angles to it, is covered by a plumb-line attached to its upper extremity C.

Fig. 36.

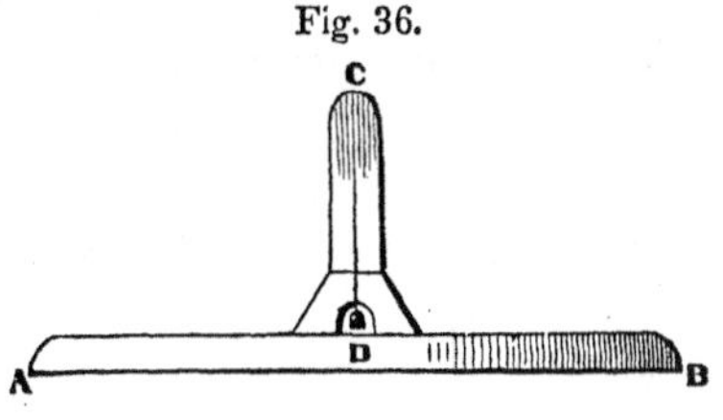

As this position of the level line is inconvenient, in practice, for long sights, by inverting the instrument we get the "*Plumb-line level.*" To construct it, at the middle of a straight edge, attach a bar, so that a line drawn through its middle is exactly at right angles to the straight edge. From the point of meeting suspend a plumb-line. Turn the instrument till the plumb-line covers the line drawn on the bar. Then will

Fig. 37.

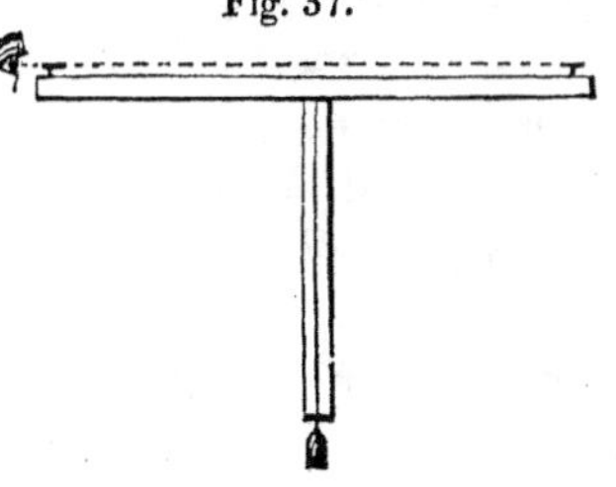

the straight edge be a *level line,* and by looking over its surface, or across sights, placed at equal heights above its ends, this level line may be produced by the eye, so as to pass over any point to which the straight edge is directed.

Fig. 38.

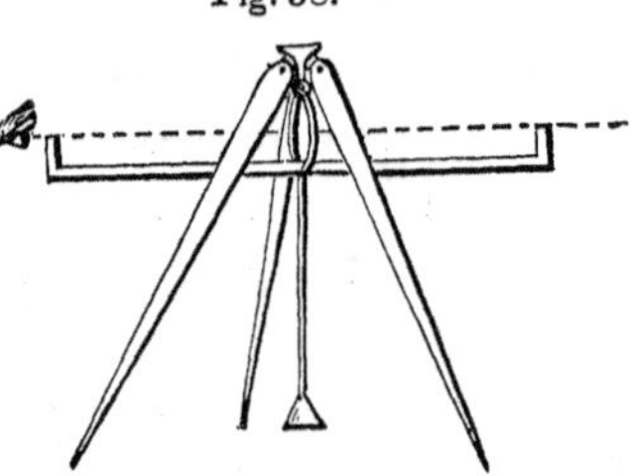

A modification of the plumb-line level, which has the advantage of being self-adjusting, is called the "*Pendulum level.*" As before, a straight edge and a bar are fixed at right angles to each other, but a heavy weight at the lower extremity of the bar keeps it always vertical, and, consequently, the straight edge always horizontal. The whole apparatus is suspended by a ring from the junction of three legs which move on pivots, so as to form a steady support on the most uneven ground. A "*tripod*" of this sort is generally employed for the support of all the instruments of surveying.

Fig. 39.

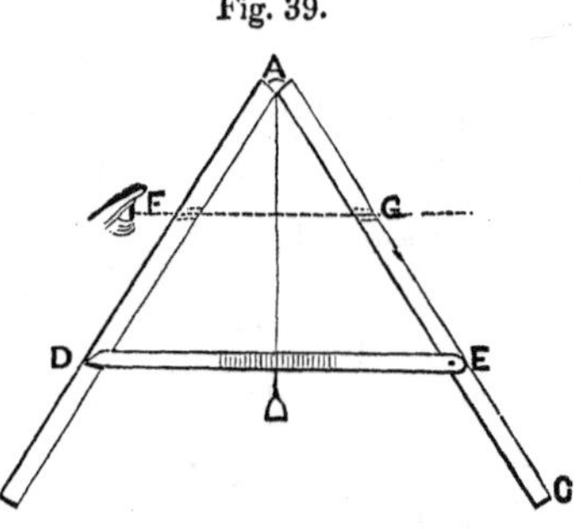

The "A *level*" is a portable and convenient modification of the mason's level. The legs AB, AC turn on a hinge at A, as does the bar DE at E, so that all three may be folded up into a stout rod. When the plumb-line corresponds with the middle of the bar DE, the feet of the instrument are on the same level At F and G are fixed two sights, at equal distances from the feet, so that when the latter are level, the line, obtained by looking through these sights, is level also. The use of the other divisions on the bar DE will be explained under the head of "Grades."*

Another simple instrument depends upon the principle that "water always finds its level." If a tube be bent up at each end, and nearly filled with water, the surface of the water in one end will always be at the same height as that in the other, however the position of the tube may vary. On this truth depends the "*Water level.*" It may be easily constructed with a tube of tin, lead, copper, &c., by bending up, at right angles, an inch or two of each end. In these ends cement thin vials, with their bottoms broken off, so as to leave a free communication between them. Fill the tube and the vials, nearly to their top, with colored water. Cork their mouths, and fit the instrument, by a steady but flexible joint, to a tripod.

Fig. 40.

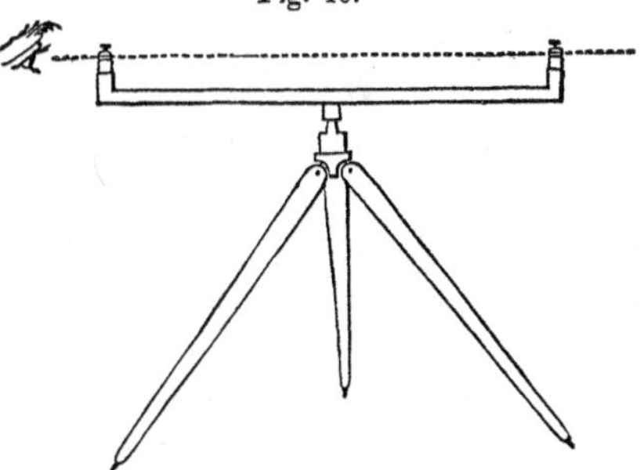

To use it, set it in the desired spot, place the tube by eye nearly level, remove the corks, and the surfaces of the water in the two vials will come to the same level. Looking across them, we get the level line desired Sights of equal height, floating on the water, and rising above the tops of the vials, would give a better-defined line.

The "*Spirit level*" consists essentially of a curved glass

* See Simms on Drawing Instruments, 2d edition, p. 146.

Fig. 41.

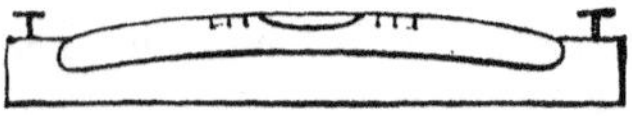

tube filled with alcohol, but with a bubble of air left within, which always seeks the highest spot in the tube, and will therefore by its movements indicate any change in the position of the tube. To prepare the tube for use, it is placed with its convexity uppermost, and supported either in a block, or by suspension; and when the bottom of the block, or the sights at each end of it, coincide with some level line previously established, marks are made on the tube at the extremities of the air bubble. The instrument is then ready for use; for whenever the bubble, by raising or lowering one end, has been brought to stand between the original marks, (or, in case of expansion or contraction, at equal distances on either side of them) the sights will be on the same level line.

When, instead of the sights, a telescope is made parallel to the level, and various contrivances to increase its delicacy and accuracy are added, the instrument becomes the engineer's spirit-level, and is out of the reach of the unprofessional readers for whom this volume is chiefly designed.* The same is the case with the "French reflecting level."

By whichever of these various means a level line has been obtained, the subsequent operations in making use of it are identical. Since the "water level" is easily made and tolerably accurate, we will suppose it to be employed. Let A and B represent the two points, the difference of the heights of which is required. Set the instrument on any spot from which both the points can be seen, and at such a height that the level line will pass above the highest one. At A let an assistant hold a staff graduated into feet, tenths, &c. Turn the instrument towards the staff.

* For its description and adjustments, see Gillespie's Levelling, Part I., chap. IV.

look along the level line, and note what division on the staff it strikes. Then send the staff to B, direct the instrument to it, and note the height observed at that point. If the level line pro-

Fig. 42.

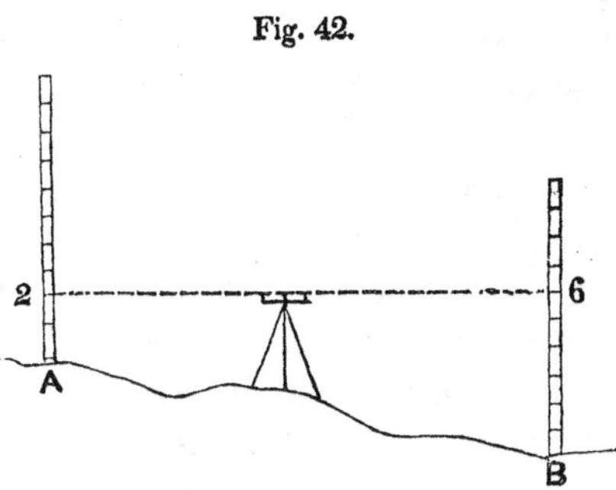

duced by the eye passes 2 feet above A and 6 feet above B, the difference of their heights is 4 feet. The *absolute* height of the level line itself is a matter of indifference. If the height of another point, C, not visible from the first station, be required, set the instrument so as to see B and C, and proceed exactly as with A and B. If C be found to be 3 feet above B, it will be $4-3=1$ foot below A. If C be 1 foot below B, as in Fig. 43, it will be $4+1=5$ feet below A. The comparative heights

Fig. 43

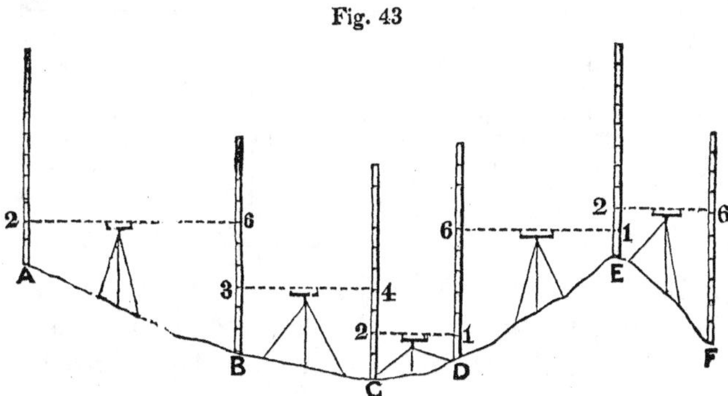

of a series of any number of points, can thus be found in reference to any one of them.

The beginner in the practice of levelling may advan

.ageously make in his 'field-book" a sketch of the heights noted, and of the distances, putting down each as it is observed, and imitating, as nearly as his accuracy of eye will permit, their proportional dimensions.* But when the observations are numerous, they should be kept in a tabular form, such as that which is given below. The names of the points, or " stations," whose heights are de manded, are placed in the first column ; and their heights, as finally ascertained, in reference to the first point, in the last column. The heights above the starting point are marked +, and those below it are marked —. The back-sight to any station is placed on the line below the point to which it refers. When a back-sight exceeds a fore-sight, their difference is placed in the column of " ascents ;" when it is less, their difference is a " descent." The following table represents the same observations as the preceding sketch, and their careful comparison will explain any obscurities in either.

Stations.	Distances.	Back-sights.	Fore-sights.	Ascents.	Descents.	Total Heights.
A						0.00
B	100	2.00	6.00		— 4.00	— 4.00
C	60	3.00	4.00		— 1.00	— 5.00
D	40	2.00	1.00	+1.00		— 4.00
E	70	6.00	1.00	+5.00		+ 1.00
F	50	2.00	6.00		— 4.00	— 3.00
		15.00	18.00		— 3.00	

The above table shows that B is 4 feet below A ; that C is 5 feet below A ; that E is 1 foot above A ; and so on. To test the calculations, add up the back-sights

* In the figure, the limits of the page have made it necessary to contract the horizontal distances to one-tenth of their proper proportional size.

and fore-sights. The difference of the sums should equal the last "total height."*

The level line obtained by any of these instruments is a tangent to the surface of the earth, and therefore diverges from the surface of standing water, which presents a curve corresponding to that of the earth. The difference between the lines of true and apparent level, is 8 inches at the distance of a mile; but since it varies as the *square* of the distance, it is very insignificant in sights of ordinary length, (one-eighth of an inch for a sight of one-eighth of a mile) and may be completely compensated by setting the instrument midway between the points whose difference of level is desired; a precaution which should always be taken, when possible. If the ground renders sights of unequal length unavoidable, a balance should be struck as soon as possible, by adopting corresponding inequalities in the contrary direction.

The heights observed along the length of the road, which give its "longitudinal section," should be taken at every change of slope; and at every hundred feet, when the line is finally located.

It is also necessary to take them at right angles to its length, in order to obtain the "transverse" or "*cross sections.*" These are required for the subsequent calculations of the "cutting and filling," and to enable the engineer to see what would be the effect upon these, of moving the line to the right or to the left. The right page of the note-book is usually devoted,

Fig. 44.

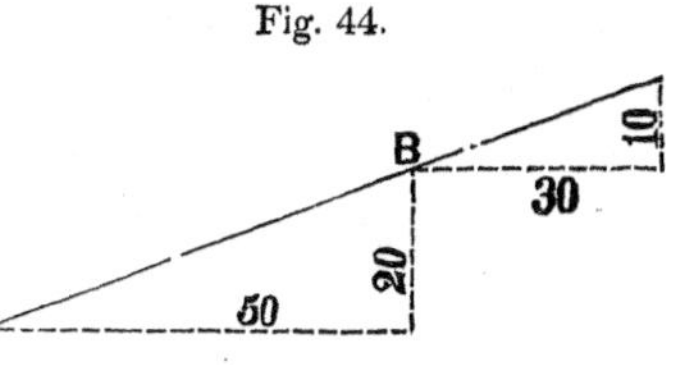

* For another form of Levelling Field-book, see page 145*.

in part, to the cross-sections, taken in reference to any station, as B. In this example, on the right, the ground rises 10 feet in going out 30 ; on the left, it falls 20 in a "distance out" of 50.

These cross-sections should be taken at every change of longitudinal slope. At every change of slope transversely, single heights and "distances out" should be taken. The future calculations of cubical contents will be facilitated by observing the following rules :—

1. Take a cross-section whenever either edge of the road passes from excavation to embankment, or *vice versâ*.

2. When the road is partly in excavation and partly in embankment, ascertain the "distance out" at which the grade, or level of the base, cuts the surface of the ground.

3. Take heights at each edge of the base, *i. e.* at distances on each side of the centre line, equal to the half width of the base of the road.

4. Take the intermediate cross-sections at some decimal division of 100 feet.

The *Mountain Barometer* is an instrument of great value for the rapid determination, with approximate accuracy, of the heights of the leading points in an extensive district of country.*

The temperature of *boiling water* supplies another easy means of approximation. The degree of heat at which water boils diminishing as the height increases, tables have been constructed from observation, with the aid of which the height of a place may be calculated from the temperature at which water there boils.†

---

* Gillespie's Levelling, p. 79.

† See Silliman's Journal, 1846, pp. 134-5.

4. MAPPING THE SURVEY.

The *lengths*, *directions*, and *heights* of the different portions of the line having been ascertained, they are next to be represented on paper, in such a way as to convey to an instructed eye a complete idea of the ground over which the route passes. This idea will be as accurate as could be obtained from actual examination, and much more easily embraced by the mind ; the details being made subordinate to the leading features.

The mapping of a line comprehends two distinct branches :

1. The plot.
2. The profile.

THE PLOT OF A LINE.

This represents the lengths and directions of the different portions of a line, projected on a horizontal plane, as they would be seen by an eye looking down upon them from a great height directly above them. If the lengths have been measured horizontally, as is usual, they will require no farther reduction. Before commencing the plot, its "*scale*" must be determined, *i. e.*, what proportion the representation is to bear to the reality, or how many feet of the line each foot of the plot is to represent. If one foot of the plot represent 1000 feet of the line, 100 feet of the latter will occupy one tenth of a foot on the plot, and so on. Any convenient scale may be assumed, but must be carefully preserved unchanged in the same plot. The changes in the direction of the line, or the angles of deflection of its adjacent parts, may be most easily laid down, as explained on page 91, by describing an arc from the angular point with the same radius used on the ground, (taken to the proper scale) and setting off on the arc, as a chord the proper distance measured in like manner.

If the deflection had been measured by an angular instrument, (which, however, the preceding method dispenses with) it would have been laid down on the paper by a "protractor," the most usual form of which is a small brass semicircle, divided into degrees similar to those on the instrument.

Upon the *plot*, it is usual to represent the hills and valleys in the vicinity of the line ; but since they are supposed to be seen horizontally projected in a "map-view," as they would appear to an eye looking down upon them from an infinite height, they cannot be drawn with the rises and falls of the front view in which we usually see them, but must be represented by some artificial and conventional method. They are accordingly supposed to be cut by a number of equidistant horizontal planes, and the horizontal "contour curves" of intersection to be drawn upon the map, the intervals between them being filled up by short hatchings perpendicular to the curves.* Hills, represented on these principles, are indicated by numerous diverging lines, shorter, nearer, and heavier, in proportion as the hill is steeper, and *vice versâ*. See the examples on pages 83–4.

All water-courses must also be carefully represented on the plot; and the nature of the surface, whether pasture, ploughed land, swamp, woods, &c., together with the detached objects upon it, such as houses, mills, churches, &c., be indicated by certain arbitrary signs. For our purposes they are not necessary, but may be found, if desired, in any topographical manual.

* For fuller details, see Gillespie's "Levelling, Topography, and Higher Surveying," Part IV.

THE PROFILE OF A LINE.

This represents, to any desired scale, the heights and distances of the various points of a line, projected on a vertical plane. It thus gives a "side-view" of its ascents and descents. Any point on the paper being assumed for the first station, a horizontal line is drawn through it; the distance to the next station is measured along it to the required scale; at the termination of this distance a vertical line is drawn; and the given height of the second station above or below the first is set off on this vertical line. The point thus fixed determines the second station, and a line joining it to the first station represents the slope of the ground between the two. The process is repeated for the next station, &c.

But the rises and falls of a line are always very small in proportion to the distances passed over; even mountains being merely as the roughnesses of the rind of an orange. If the distances and the heights were represented on a profile to the same scale, the latter would be hardly visible. To make them more apparent it is usual to "exaggerate the vertical scale" tenfold, or more, *i. e.*, to make the representation of a foot of height ten times as great as that of a foot of length. Take, for instance, the example on page 98. Let one inch represent one hundred feet for the distances, and ten feet for the heights.

From A draw a horizontal line. Measure on it one inch, representing one hundred feet of length. Thence draw downwards a vertical line. Measure on it four-tenths of an inch, representing four feet of height. This fixes the point B. Join A to B. This line AB represents the slope of the ground. Next, along the horizontal line, measure six-tenths of an inch farther, representing sixty feet

Fig. 45.

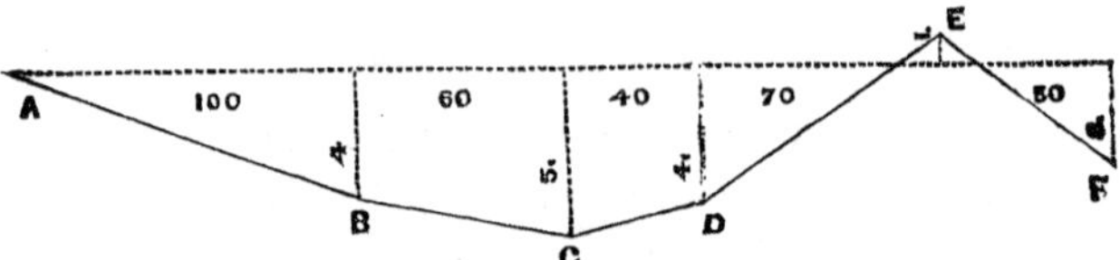

of length. Measure on a vertical line thence drawn, five-tenths of an inch, representing five feet of height. This fixes C. Join C to B'. Proceed in like manner for all the levels.

The distances may be written horizontally in their appropriate places, and the heights or depths of the ground (above or below the *datum* line) vertically, along the lines which represent them, as in the figure.

5. ESTABLISHING THE GRADES.

The *grade* of a line is its longitudinal slope, and is designated by the ratio between its length and the difference of height of its two extremities. The ratio of these two quantities gives it its name, as we have seen; the road being said to have a grade of one in thirty when it rises or falls one foot in each thirty feet of length. When the "profile" of a proposed route has been made, a "grade-line" is drawn upon it (usually in red) in such a manner as to follow its general slope, but to average its

Fig. 46.

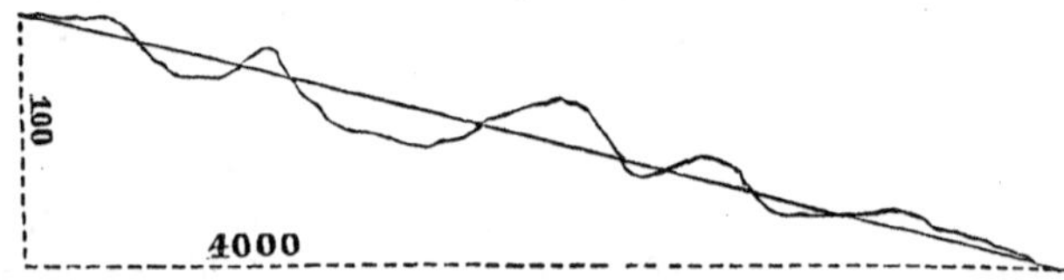

irregular elevations and depressions, as in the figure. The ratio between the whole distance and the height is then to be calculated. If, as in the figure, it rise 100 in 4000, the grade is one in forty, flatter than our assumed limit of one in thirty, and the line will be a satisfactory one, if on calculation it be found that the cuttings about equal the fillings. If either be much in excess, the grade is altered to equalize them, as will be explained under the next head. But if the grade be found steeper than the limit, as when it ascends the face of a hill with a rise of 100 feet in 1500, or a slope of one in fifteen, either the hill must be cut down, or, which is generally preferable, the length of the line must be increased so as to equal 100 × 30 = 3000. The best method of obtaining this increased length, or "development," (whether by a zigzag or by a single oblique line) will depend upon the manner in which the line meets the face of the hill, whether at right angles or obliquely, and should be determined by geometric constructions upon the plot, such as those which follow, modified if necessary by the features of the ground.

Problem. To fix the position of a line joining two given points, so that it shall ascend with a given grade a slope steeper than this grade, and shall also be the shortest possible line which will fulfil this condition.*

Case 1. *When the straight line joining the two points meets the slope at right angles.*

Fig. 47.

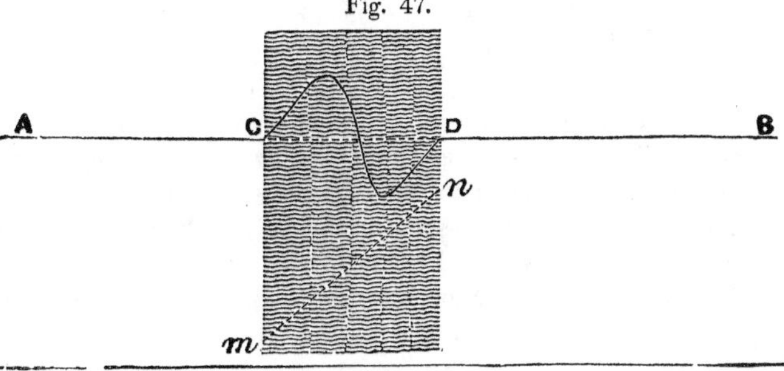

* Gayffier, p. 13

Let A and B be the given points, and let the top and bottom lines of the slope to be ascended be considered parallel. Let *mn* represent the length which the road up the hill must have to ascend with the proper grade. Join the given points by a straight line, and between the points C and D, at which this line meets the top and bottom line, establish a zigzag, of a sufficient number of turns to make its entire length equal to *mn*, the "development" required; which in the instance last supposed would be 3000 feet, the straight line CD being only 1500.

The road which ascends the Catskill mountain makes seven such zigzags or tacks. Their angles should be rounded off by curves, as explained in a following article on "Final Location." At these curves the width of the road should be increased, as directed on page 46.

Case 2. When the straight line meets the slope obliquely, and the two given points are very distant from each other.

Fig. 48.

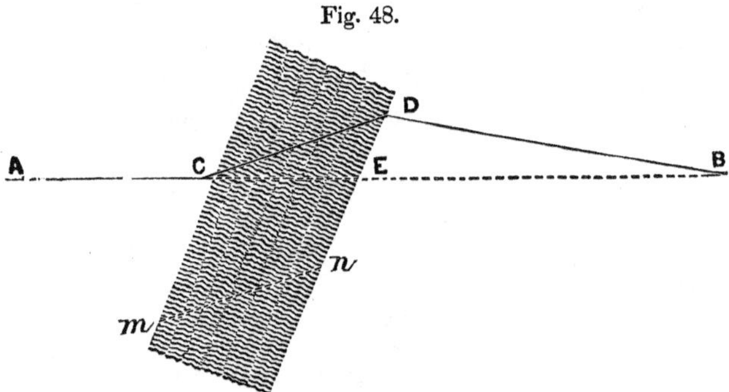

Let A and B be the given points. Between the top and bottom lines of the slope draw a line *mn* at such a degree of obliquity as will make its length equal to the development required, which, in the instance supposed, is 3000 feet. The straight line AB would be too steep between C and E. Therefore from the point C draw a line CD, parallel and therefore equal to *mn* Join DB, and the line ACDB will be the one desired.

A zigzag between C and E would give a longer line; for, comparing the parts of the line thus obtained with those of the other, we find AC common to both; the zigzag CE equal to CD by construction; and EB *longer* than DB, because farther from the perpendicular.

The construction above directed is merely approximately true, becoming perfectly so only when the points A and B are infinitely distant from each other. The strict construction is that which follows.

Case 3. *When the straight line meets the slope obliquely, and the two given points are near each other.*

From the given points A and B draw perpendiculars to the nearest edges of the slope. The line joining the feet of these perpendiculars will be less than, equal to, or greater than, the developed line *mn*, according to the steepness of the slope, and the degree of obliquity with which it is met by the straight line which joins the given points. Three sub-cases, requiring different constructions, are thus formed.

Sub-case 1. *When the line joining the perpendiculars is shorter than the developed line* mn.

Fig. 49

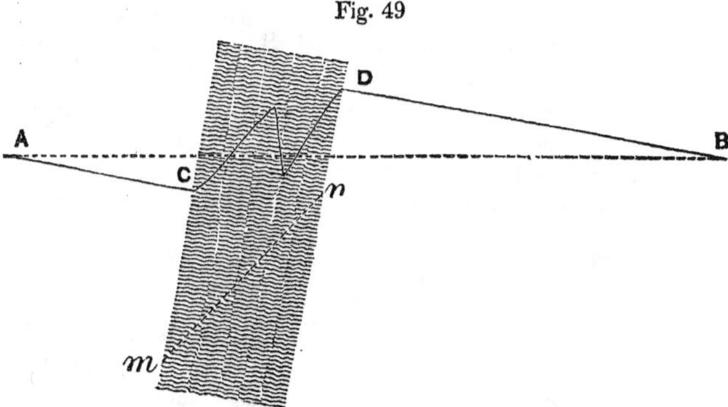

From A and B draw AC and BD perpendicular to the edges of the slope. Join C and D by a zigzag line, equal in length to the developed line *mn*. Then will the line thus obtained fulfil the conditions required; the length of the zigzag being equal to the necessary length *mn*, and the lines AC and BD being perpendicular to the top and bottom of the hill, and therefore the shortest possible distances to it.

Sub-case 2. *When the line joining the perpendiculars is equal to the developed line* mn.

Fig. 50.

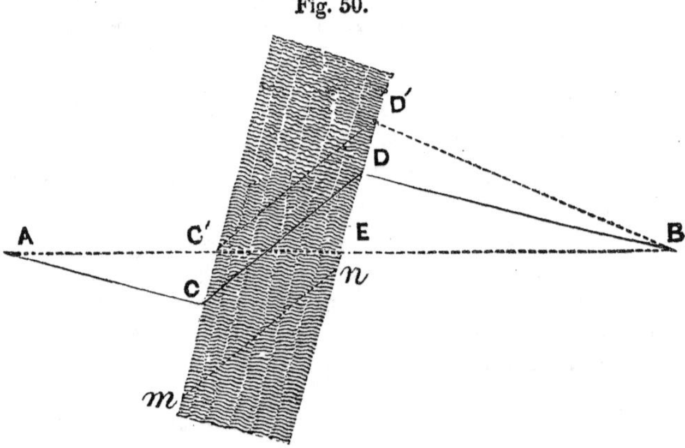

Draw the perpendiculars AC and BD, as before, and join their feet by the line CD. Then will the line ACDB be shorter than any other line, (as AC′D′B, obtained by the construction of Case 2) for AC and BD, being perpendiculars, are the shortest possible, and CD has a constant length, wherever it may be placed.

A zigzag line from C′ to E would not produce the shortest line, for the same reasons as in Case 2.

Sub-case 3. *When the line joining the perpendiculars is longer than the developed line* mn.

From A, Fig. 51, draw AE parallel and equal to *mn*. Join EB. From the point D (where this line intersects the edge of the slope) draw DC parallel and equal to AE. Join CA. Then will ACDB be the shortest line required.

For, AE, being equal to *mn*, cannot be shortened, and EB is a straight line, and therefore the shortest possible line, as is

Fig. 51.

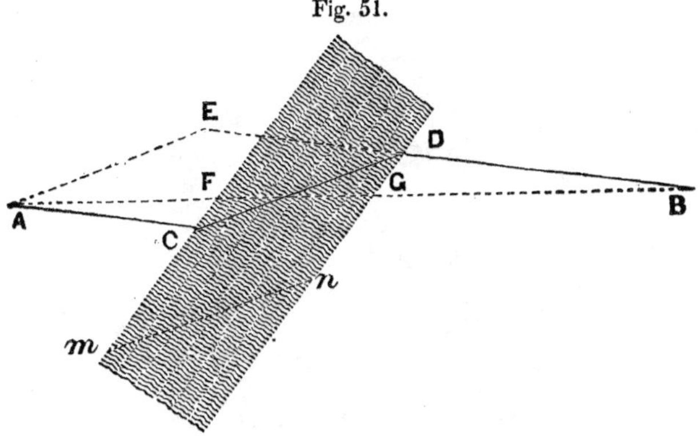

consequently the whole line AEDB. But this line is in the wrong place, and its parts require to be transported parallel to themselves. By this operation is formed the line ACDB, which has all its parts equal to those of the former line, and which is therefore the shortest possible.

It might seem preferable to adopt the direct line AB, and to ascend the hill by a zigzag from F to G; but this would not give the shortest line; for AF and GB are longer respectively than AC and DB, because farther from the perpendiculars.

When the lines AC and DB, obtained by the construction above directed, fall beyond the perpendiculars let fall from A and B upon the top and bottom of the slope, this result shows that this construction is inapplicable, and that the case is one in which it is proper to adopt the perpendiculars, and to join them by a zigzag of the proper length.

Case 4. When two neighboring slopes are separated by a level space; whether a valley, or a table-land on the ridge of a hill.

Between the top and bottom lines of one slope draw the line *mn*, equal in length to the developed line with which that slope must be crossed; and in like manner on the second slope, draw the line *pq*.

Then from A draw AE, parallel and equal to *mn*. From E draw EF parallel and equal to *pq*. Join FB. The line AEFB

Fig. 52.

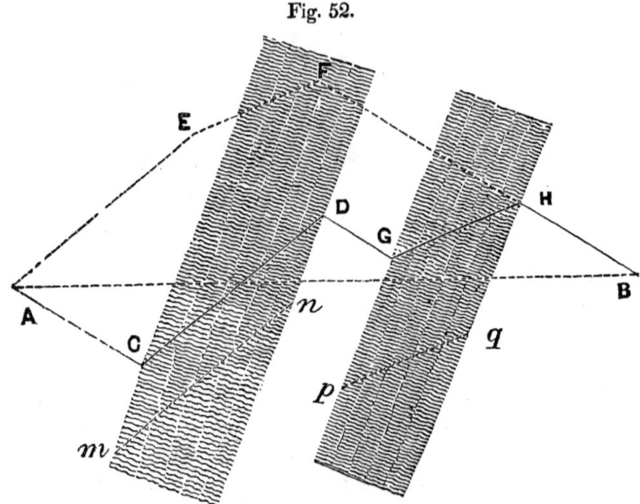

is the shortest line possible, for the same reasons as was AEB in the preceding sub-case 3. But its parts require to be differently arranged without changing their length, which is effected thus. From H draw HG parallel and equal to FE. From G draw GD parallel to HF, and terminating at the edge of the next slope. From D draw DC parallel and equal to EA. Join CA by a line which will be parallel to FH. This new line ACDGHB is equal to the former line AEFB, and therefore is the shortest line required.

If the space between the two slopes was a valley, in which there was a given point to be passed through, as a bridge, the problem would divide itself into two others, such as have been solved in the preceding cases.

Grades may be approximately estimated upon the ground, (without measuring distances and heights) by a slight addition to the "plumb-line level," described on page 93. Connect the horizontal and vertical bars by oblique braces. To prepare it for use, depress or elevate the sights, so that their line coincides with an ascent or de scent of one in thirty, or any other grade previously estab lished by levelling. Mark the point at which the plumb-line now cuts the oblique braces. Do the same for other grades, the more varied the better, and the instrument will thus become a *clinometer*, or grade-measurer. When it is placed upon any slope, and its sights directed to any object (such as a target on a rod, or a paper in a cleft stick) at the same height above the surface as its upper edge, that division on the brace which is cut by the plumb-line will indicate the inclination of the slope. The A level described on page 94, may be used in a similar manner, a scale having been in the same way formed on the bar DE.

An extempore *clinometer* may be made with a sheet of paper, a thread, and a pebble. Take a sheet of paper of any shape, double it, and a straight line is formed; double it again along the straight line, and four right angles are obtained. Cut out one of the right angles, and double it so as to bring the sides of the right angle together, and it will be bisected, forming two angles of 45°. Fold this in three equal parts, and angles of 15° will be formed; repeat the last operations, and angles of 5° will be obtained. The subdivision may be carried as far as desired. To use the instrument, form a plumb-line by tying a pebble to the end of a thread, and attach it at the centre of the angles. Holding the right angle to the eye, if the grade be descending, or the opposite corner if it be ascending, turn the paper till its edge is in the line which passes from the eye to some object at the same

Fig. 53.

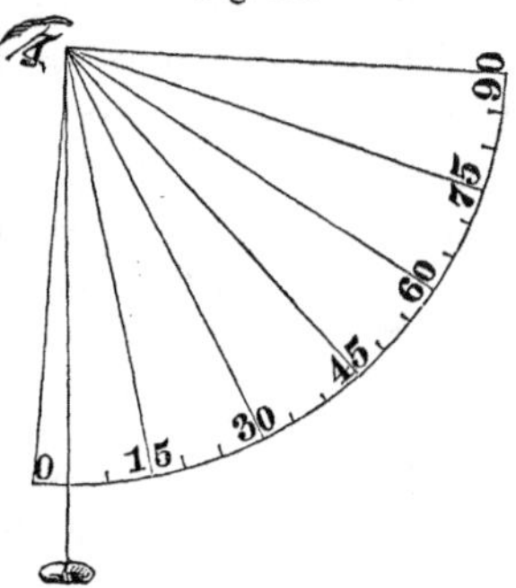

height above the surface. The plumb-line will then indicate the angle of the slope. In the figure it strikes 5°, equivalent, by the table on page 44, to 1 in 11.

6. CALCULATING EXCAVATION AND EMBANKMENT

The proper grade-line having been thus determined, and drawn on the profile, (which shows the heights of the ground over which the line passes) the difference between the height of the ground and that of the grade-line at any point, will of course represent the depth of cutting, (or excavation) or the height of filling (or embankment) as the case may be, at that point. This depth, or height, in feet and decimal parts of a foot, should be written in red figures (*cotes rouges*) at the proper points of the profile. With these, knowing also the intended width of the road and the inclination of the side-slopes, the cubical contents of the excavations and embankments, or the amount of "earth-work," may be accurately calculated.

The cost of the road will depend in a great degree upon the quantity of the "earth-work" to be done, and may be greatly lessened by making the amount of excavation precisely equal to that of embankment, so that what is dug out of the hills may just suffice to fill up the hollows. It is therefore very important for economy to calculate these amounts with accuracy before the final location of the line, so that if they are found to be unequal, the position and grades of the line may be changed to produce the equality desired.

This accurate calculation is necessary, after the final location, for another reason; inasmuch as the contractors, who usually perform the work, are paid, not by the day, nor in the lump, but by a certain price per cubic yard, the exact determination of the number of which is therefore required to ascertain their just dues.

PRELIMINARY ARRANGEMENTS.

For calculating the cubical contents of the solid mass of earth cut out or filled in, four different methods are in common use. All four, however, require the same preliminary arrangements and preparations, which will therefore be now given.

Figure 54 is a plan (on a scale of 800 feet to the inch) and figure 55 a profile (on a vertical scale of 80 feet to the inch) of an old line of road, which it is desired to improve by cutting down the hill and filling up the hollow, so as to form a single slope, with a uniform grade, from A to B.* The distances between the stations are written horizontally; the heights of the ground above the datum line are written along the vertical lines which represent them; and at the extremities of these vertical lines are placed the numbers which represent the depths of cutting or filling at those points, and which are equal to the differences between the heights of the ground and of the "grade-line," or new road

SECTIO-PLANOGRAPHY.

A method of representing the cuttings and fillings upon the plan, devised by Sir John Macneill, has been named '*Sectio-planography*." Usually the plan and the profile are drawn separately, and when the former varies much from a straight line, it is difficult for an unpractised

* S'mms on Levelling, Am. edition, p. 81

Fig. 55. Fig. 54.

PROFILE.
46.
561
59.2
+18.
858
53.9
+20.
825
26.9
0
820
0.9
-19.
825
4.9
-8.
330
10.
PLAN.

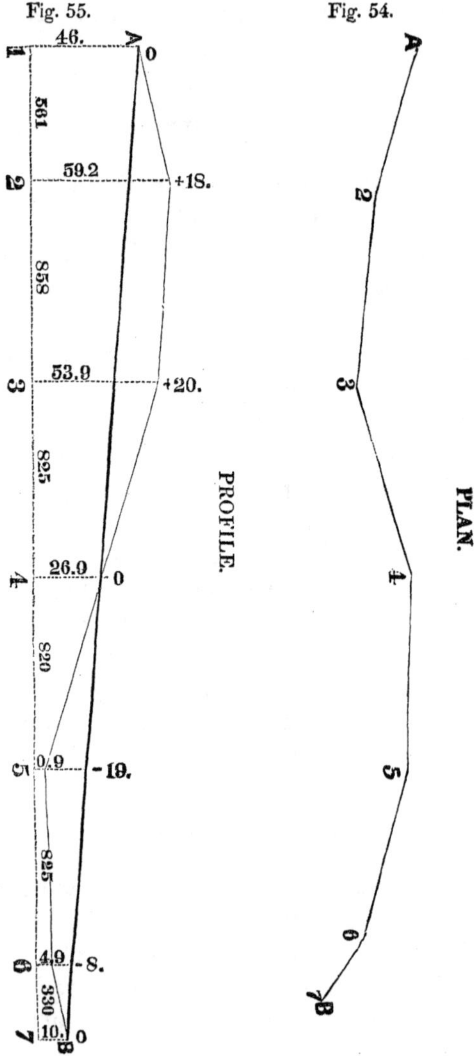

eye to discover the corresponding points of the plan and profile; as the latter is formed by placing all the distances along a straight base line, and therefore fills a longer space than the winding plan; and as the two are also frequently drawn far apart, or even on different sheets. In the improved method, the depths of cutting and of filling at each station are set off on one side or the other of the *plan* of the line, as laid down upon the map, so that all the information desired, with regard to any portion of the line, may be found on that very spot. The accompanying figure shows its application to the preceding example, the "plan" of which has been intentionally made very winding.

Fig. 56

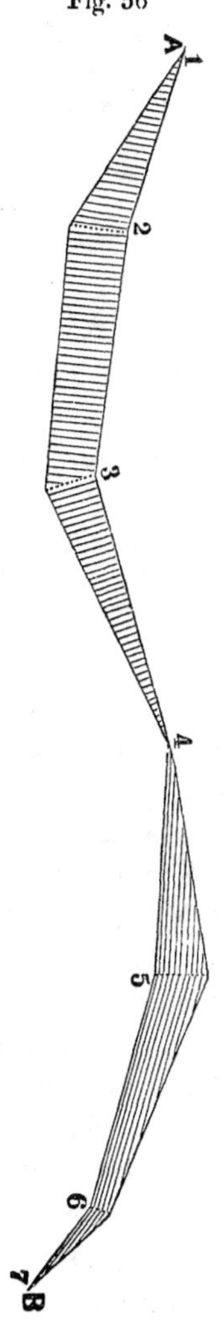

To make the distinction more striking, the cuttings may be shaded with lines perpendicular to the line of the road, and the fillings with lines parallel to it; or the former may be colored red, and the latter blue.*

TABULAR ENTRIES.

The *data* of the profile, with those deduced therefrom, should be presented in a tabular form, such as that which follows, and which refers to figure 55.

* See Simms on Sectio-Planography.

1	2	3	4	5	6	7
Station.	Distances.	Height of ground above datum line.	Rise or fall of the grade line for each distance.	Height of grade above datum.	Cut. +	Fill. −
1		46.0		46.0	0	
2	561	59.2	—4.8	41.2	18.	
3	858	53.9	—7.3	33.9	20.	
4	825	26.9	—7.0	26.9	0	0
5	820	0.9	—7.0	19.9		19.
6	825	4.9	—7.0	12.9		8.
7	330	10.0	—2.9	10.9		0
	4219		36.0			

The *first* column contains the number of the station, or point, the height of which above the datum line is contained in the *third* column. The *second* column records the distances between the stations.

The *fourth* column shows the rise or fall of grade for each distance, obtained by a simple proportion, the whole distance and difference of height being known. Thus, 4219 : 561 : : 36 : 4.8.

The *fifth* column shows the height of the grade line, *i. e.* of the road as improved, above the datum line at each station. The numbers in it are obtained by subtracting successively the fall between two stations, as recorded in the fourth column, from the height of the grade line at the preceding station. Thus, $46 - 4.8 = 41.2$.

The *sixth* and *seventh* columns show the depths of cutting or filling at any station. They are the differences between the height (from column 3) of the ground at any station, and the height of the grade-line (from column 5) for the same station.

The station (No. 4) at which the cutting ends, and the filling begins, is called a "Point of Passage."

CUBICAL CONTENTS.

With these data the calculation may be commenced, and the end areas—or middle areas, or both, according to the method adopted—be sought, and the cubical contents thence deduced. The details of these calculations are of great importance to the practical engineer, but occupy so much space, that they have been transferred to the *Appendix*. The following are their results.

"*Averaging end areas*" is the most usual method of calculation in this country, but gives a result which always *exceeds* the correct amount, in a greater or less degree, according to the inequality of the end areas. In the present example, its error in excess is 10,000 cubic yards, which amount, beyond what was justly due for the work done, would be paid by a company or town by which this improvement should be made, if their engineers should adopt this method of calculation.

The calculation by "*Middle areas*" gives an amount which *falls short* of the true one, by a deficiency equal to exactly half of the excess of the preceding method.

"*The Prismoidal formula*" alone gives the *correct* amount; which, in this example, is 2,200,968 cubic feet of excavation, and 1,541,152 of embankment.

The fourth method, by "*Mean Proportionals*," gives a result always less than the true one, and exceedingly erroneous when one of the end areas is nothing.

The substitution of the correct Prismoidal method for the erroneous ones which are so frequently employed, is demanded by every consideration of accuracy, economy, honesty, and justice; and the full calculations in the *Appendix* show that the additional labor required is too trifling to be a reasonable obstacle.

BALANCING THE EXCAVATION AND EMBANKMENT.

When the quantity of excavation on any given portion of the road exceeds that of the embankment, the excess is called "*Surplus*," and must be deposited, upon the adjoining land, in masses called "*Spoil-banks*."

When the excavation is insufficient to make the embankment, the deficiency is called "*Wantage*," and must be supplied from extra "*Side-cuttings*" in the neighboring fields.

Both these cases are expensive and otherwise objectionable; it is therefore very desirable to make the excavation and embankment "balance" each other, so that the earth dug out may just suffice to fill up the hollows. If the calculations show much disparity in the two amounts, the location of the line must be changed in some way, so as to effect the desired equality.

This equalization must, however, be restrained within certain limits; for it should evidently be abandoned, when, in order to find sufficient excavation to make the embankment, it would be necessary to go to such a distance that the cost of transport would exceed the cost of making side-cuttings for the embankment, and of depositing the distant excavation in spoil-banks. The comparison of the price of transport with that of excavation and of land, will therefore determine the distance within which the balancing must be established.

SHRINKAGE.

The equality recommended must be taken with an important qualification, dependent on the fact that earth transferred from excavation to embankment shrinks, or is compressed, so as to occupy, on the average, *one-tenth* less space in bank than it did in its natural state, 100 cubic yards " shrinking" into 90.

Rock, on the contrary, occupies more space when broken, its bulk increasing by about one-half.

In experiments made on a large scale, by Ellwood Morris, C. E.,* the *shrinkage* of light sandy earth was $\frac{1}{8}$ of its volume in excavation; of yellow clayey earth $\frac{1}{10}$; and of gravelly earth $\frac{1}{12}$. The *increase* of hard sandstone rock, quarried in large fragments, was $\frac{5}{12}$ of its volume in excavation; and of blue slate-rock, broken into small fragments, $\frac{6}{10}$.

Upon some of the public works of the state of New York, the usual allowance has been for the shrinkage of gravel and sand 8 per cent.; for clay 10 per cent.; for loam 12 per cent.; for mucking, or surface soil, 15 per cent.; and for clay "puddled" 25 per cent. The increase of bulk of rock was taken at one-third, or sometimes at one-half; though some experiments showed that one yard of slate-rock made from 1.75 to 1.8 cubic yards of embankment.

These considerations lead us to modify the requirements of equality in the excavations and embankments, and to adjust them so that the former shall exceed the latter by about 10 per cent.

CHANGE OF GRADE.

We will now take up the example on page 114, in which we find the excess of the excavation over the embankment, or its " surplus," (according to the correct calculation, of which the result is given, on page 117) to be 659,816 cubic feet. We must therefore change the grade,

* Journal of the Franklin Institute, October, 1841.

so as to lessen the excavation, and increase the embankment, till the former exceeds the latter by only one-tenth of itself. The grade line AB (figure 55) might be raised either at A or at B. The latter is preferable, since it will increase the gentleness of the slope. The height which it should be raised at B might be calculated in advance, but the complication of the resulting formula is so great, that it will be better to *assume* some height, (which an experienced eye can do with considerable accuracy) and having found, by a simple proportion, the changes in the cuttings and fillings at each station, to recalculate the whole cubic contents. If the desired difference has not been attained in the result, it will at least be a guide by which a second assumption can be made with a very close approximation to precision.

Consider the grade to be raised three feet at station 7. A proportion between the sum of the distances from station 1 to 7, and that to any other station, will give the change of cutting or filling at that station.

For station	2,	4219 : 561	: : 3 :	0.4
" "	3,	4219 : 1419	: : 3 :	1.0
" "	4,	4219 : 2244	: : 3 :	1.6
" "	5,	4219 : 3064	: : 3 :	2.2
" "	6,	4219 : 3889	: : 3 :	2.8

The place of station 4, *i. e.* the "Point of passage," is changed by the elevation of the grade line AB, and removed towards station 3, to some new station 4′; see Fig. 57. A problem here presents itself, to find the distance between the old and new points of passage, knowing the slope of the grade and that of the ground. Call the former m to 1, and the latter n to 1. Let the elevation of the new grade line above the old point of passage $= h$ feet; and the distance required $= d$.

An inspection of the figure shows that the heights of the two right-angled triangles, whose bases are d, are respectively

Fig. 57.

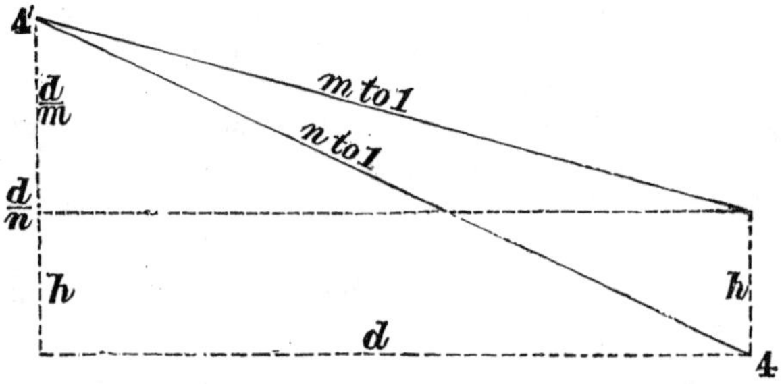

$\frac{d}{n}$ and $\frac{d}{m}$. It is also evident that $\frac{d}{n} = \frac{d}{m} + h$; whence is obtained the general formula,

$$d = h\,\frac{mn}{m - n}.$$

In the present case (see table on page 116) $m = \frac{4219}{36 - 3} = 128$, $n = \frac{825}{53.9 - 26.9} = 31$; and $d = 1.6 \times \frac{128 \times 31}{128 - 31} = 65$.

The new station 4′ is therefore distant from station 3, $825 - 65 = 760$ feet, and from station 5, $820 + 65 = 885$ feet.

Adopting these new distances, and changing the cuttings and fillings in accordance with the elevations of grade obtained by the proportions on the preceding page, they will stand thus:

Station.	Distance.	Elevation of grade.	New Cut.	New Fill.
1		0.0	0	
2	561	0.4	17.6	
3	858	1.0	19.0	
4′	760	1.6	0	0
5	885	2.2		21.2
6	825	2.8		10.8
7	330	3.0		3.0

The calculations being repeated with these data, it will be found that the excavation will amount to 2,048,000 cubic feet, and the embankment to 1,980,000; an apparent surplus of 68,000 cubic feet; but since, in order to allow for the shrinkage, there should be an excess of 205,000, it appears that there is really a *Wantage*, and that the grade has been raised too much; so that an elevation of only $2\frac{1}{4}$ feet at B would probably produce the desired balance.

TRANSVERSE BALANCING.

When the road lies along the side of a slope, so that it is partly in excavation and partly in embankment, it is ne-

Fig. 58.

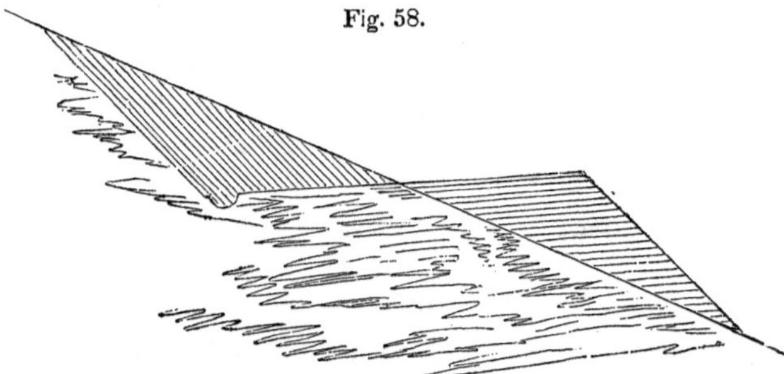

cessary so to place its centre line, that these two parts of its cross-section may balance. When the ground has a uniform slope, the desired end would be obtained (if the side slopes were the same for excavation and embankment, and if no "shrinkage" existed) by locating the centre line of the road on the surface of the ground. In other cases, as when the side of the excavation slopes 1 to 1, and that of the embankment 2 to 1, a formula to determine the position of the centre line of the road may be readily established.

If earth be wanted for a neighboring embankment, the amount of excavation may be easily increased by moving the road farther into the hill, with the additional advantage of lessening its liability to slip. The line may be thus changed on the map, according to the notes of cross-sections in the level book, and be subsequently moved, by a corresponding quantity, on the ground.

When the slope of the ground is very steep, the transverse balancing must be disregarded, and the road made chiefly in excavation, to avoid the insecurity of a high embankment, as will be explained under "Construction."

DISTANCES OF TRANSPORT.

The *equality* of the masses of excavation and embankment is not the only consideration. The *distances* to which it is necessary to transport the earth which is moved, must also be taken into the account. Suppose that a mass of earth, whose surface is ABCD, is to be

Fig. 59

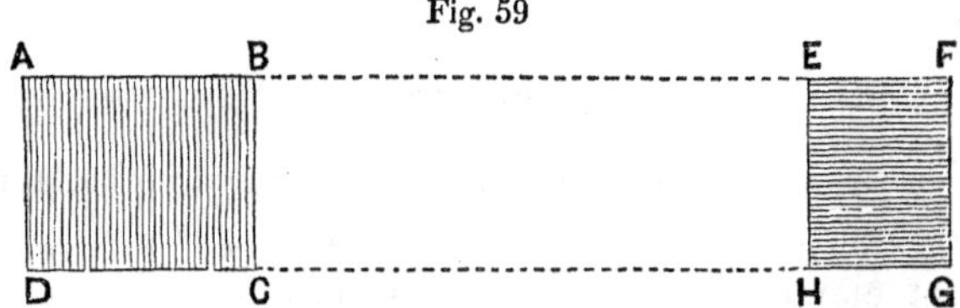

removed to the embankment whose surface is EFGH, and which has a thickness sufficient to make the two masses equal. The *mean distance* of the transport is required. Conceive the mass ABCD divided into a very great number of smaller masses. The sum of the products of these portions, by the distance which each of them is actually moved, will equal the product of the sum of the portions (*i. e.* of the whole mass) by the mean distance. The mean distance therefore equals the above sum of products divided by the whole mass.*

In such cases as usually occur on a road, in which the

* Gayffier, p. 122.

cubes of excavation and embankment are comprised between two parallel planes, whose horizontal traces are ABEF and DCHG, and in which sections made by other planes, parallel to the first, cut off equal partial volumes, we know, from the principles of mechanics, that the mean distance desired is equal to the distance of the centres of gravity of the two volumes. In the simple example above, the mean distance of transport would be the distance between the centres of the two rectangles.

The methods of apportioning the excavations among different embankments, which ought to be adopted in more complicated cases of various distances of transport, in order to attain the minimum of labor and expense, will be considered in the next chapter, which treats of actual "Construction."

7. ESTIMATE OF THE COST OF A ROAD.

A minute and careful estimate of every possible source of expense in the construction of a road, is a very important element in determining its location. The principal items are the Earthwork, Land, Mechanical structures, Engineering, and Contingencies.

EARTHWORK.

The amount of "Earthwork," or excavation and embankment, is supposed to have been determined by the preceding calculations. Its cost per cubic yard depends on the *wages* of labor, the *quality* of the earth, and the *distance* over which it is moved.

Wages. The daily wages of an ordinary laboring man vary of course with the locality and the season, and range from 50 cents to $1 25. In making the estimate, it must not be overlooked, that if wages are at that time unusually low, they will be likely to rise, if the work be so large in amount as to make the demand for labor exceed the supply.

Quality. The amount of labor required, for breaking up and removing any given volume of earth, will of course depend upon its degree of compactness and cohesiveness, which is termed its "*quality.*" This is estimated by the proportion between the number of picks in use, and the number of shovels which these picks will keep constantly employed. Thus, if the earth be so loose that it can be shovelled up without being loosened by the pick, it is called "earth of one man." If it be so hard as to require one picker, or "getter," to be constantly employed, to keep one shoveller, or "filler," at work, it is called "earth of two men." If one "getter" can keep two "fillers" busy, it is "earth of 1½ men;" its nomenclature being formed by dividing the total number of men employed by the number of volumes of earth removed. The "quality" of earth can be accurately determined only by actual experiment, though it may be estimated with tolerable precision by an experienced eye. In deep cuts, borings should be made, or shafts sunk, to ascertain the nature of the lower strata. In this examination a knowledge of the geological arrangement of the district will be of great assistance.

An average laborer can shovel into a cart, in a day of ten hours, from 10 to 14 cubic yards, *measured in the embankment*, of earth previously loosened with a pick or plough. Of hard and firm gravelly earth, or gravel and clay mixed, he can load 10 cubic yards; of loam, (sand and clay) 12 cubic yards; and of sandy earth 14 cubic yards.* To loosen the earth will cost from 1 to 8 cents

* Journal of Franklin Institute, September, 1841.

per cubic yard; the hardest earth requires to be picked; the others may be ploughed; and some sandy earth does not require any loosening, but may be shovelled up at once. At wages of one dollar per day, the cost of shovelling into a cart would therefore be from 7 to 10 cents per cubic yard; to which the cost of loosening must be added. If it were "earth of two men," it would cost double. The excavation of rock will cost from 50 cents to $1.00, according to its hardness, and the disposition of its seams.

The following table shows the number of cubic yards which can be loosened, loaded, &c., by an average laborer, in a day of 10 hours.*

NATURE OF THE WORK.	CHARACTER OF THE MATERIAL.					
	Common Earth.	Loose and light earth.	Mud.	Clay and stony earth.	Compact Gravel.	Rock, (blasted)
Digging up, or Loosening.	18 to 23	16		9	7 to 11	2.4
Excavation; including throwing 6 to 12 feet.	8 to 12	8	7 to 16	4		2.2
Loading in barrows.	22		8		19	
Transport by barrows; per hundred feet.	20 to 33				24 to 28	
Loading in carts.	16 to 48				17 to 27	10
Spreading and Levelling.	44 to 88		25		30 to 80	

* Deduced from the experiments of M. Ancelin.

The cost per cubic yard of each kind of labor will be readily obtained by dividing the day's wages by the number of cubic yards in the table.

The cost of throwing with the shovel is usually one-third of that of digging up.

From 90 to 120 square yards of surface of embankment can be "trimmed" in a day.

When the net cost of performing any work has been ascertained, one-twentieth of it should be added for the cost of tools, superintendence, &c.; and one-tenth of the whole for the profits of the contractor.

Distance. The third element in the price of earthwork is the distance to which the excavations must be removed. If the road be on a side-hill, and be so located that the excavation from its upper side can be at once thrown over to make the embankment on its lower side, the cost will be little more than that of the simple excavation. But usually large amounts of earth require to be removed considerable distances, with great increase of expense. The methods to be employed will vary with the circumstances of the case, as will be explained under the head of "Construction."

The comparative cost, per cubic yard, (according to experiments made at Fort Adams, Newport, R. I.,) of excavating earth, and removing it to various distances, with wheelbarrows, one-horse carts, and ox-carts, is given by the following table, which includes the cost of loosening, filling, and dropping; and estimates a laborer's wages at $1.00 per day of 10 hours; a horse, cart, and driver, at $1.34 per day of 9 hours; and an ox-team and driver $1.60 per day of 9 hours. The earth was ploughed up at a cost of $\frac{2}{3}$ cent per cubic yard.

DISTANCES IN FEET.	COST IN CENTS PER CUBIC YARD.		
	Wheelbarrow.	One-horse cart.	Ox-cart.
30	5.5	8.2	8.6
60	6.9	8.4	8.8
90	8.2	8.6	8.9
120	9.5	8.8	9.1
150	10.9	9.0	9.3
180	12.2	9.2	9.4
210	13.5	9.4	9.6
240	14.8	9.6	9.8
300	17.5	10.0	10.1
450	24.2	11.0	10.9
600	30.8	12.0	11.8
900	44.1	14.0	13.4
1200	57.4	16.0	15.1
1500	70.7	18.0	16.8

From the preceding table it appears, that, with its data, the cost, after loading, of removing the earth 100 feet, was, in barrows, 4.43 cents per cubic yard; in one-horse carts, .66 cent; and in ox-carts, .56 cent.

Some accurate experiments on the Birmingham and Gloucester Railway* make the cost in barrows, per 100 feet, at $1.00 per day, $\frac{100}{29}$ = 3.4 cents; the experiments of M. Ancelin† give $\frac{100}{33}$ to $\frac{100}{20}$ = 3 to 5 cents; the American translator of Sganzin‡ $\frac{100}{18}$ = $5\frac{1}{2}$ cents.

It is usual in barrow-work, to consider any *vertical* transport of the earth as costing eighteen times as much as the same number of feet of horizontal distance; though from accurate experiments it seems that the ratio should be as 24 to 1 for barrows, and as 14 to 1 for horse carts.§

* Laws of Excavation and Embankment on Railways, p. 136.

† See page 126 ‡ Page 110. § Gayffier, p. 146.

The cost of transport by any method will be expressed by the formula—

$$\frac{P\ (2D + d)}{L \times C},$$

in which P = price of day's work of the vehicle and its driver.

D = mean distance of transport.

d = distance which could have been gone over in the time consumed in each filling and emptying.

L = the distance which would be gone over in a day by the vehicle, proceeding without interruption at its average pace; usually between twenty and twenty-five miles, or between 100,000 and 130,000 feet.

C = the cubic contents of the load, expressed in fractional parts of a cubic yard.

If $P = 134$, $D = 1500$, $d = 1000$, $L = 100{,}000$, and $C = \frac{1}{2}$, the formula becomes $\frac{134\,(3000 + 1000)}{100{,}000 \times \frac{1}{2}} = 10.7$ cents.*

The complete cost, with one-horse carts, of excavating earth, transporting it, and forming an embankment, is very completely expressed in a *formula* enunciated in the Journal of the Franklin Institute for September, 1841, by Ellwood Morris, C. E.

The average pace of a horse carting embankment is taken at 100 feet of trip, and back, per minute; and the time lost in loading, dumping, &c., at four minutes per load.

For the variable quantities the following symbols are employed:—

a = number of feet in the average haul, or "lead," of the embankment.

* For a table thus calculated, see Marlette, p. 91.

b = number of hours worked per day.

c = daily wages of laborer, in cents.

d = " " cart, including driver and all expenses of carting.

e = cost of loosening materials, in cents, per cubic yard; ranging from one to eight, as stated on page 125.

f = number of cart-loads required to form a cubic yard of bank. Usually 3 on a descending road, $3\frac{1}{2}$ on a level, and 4 on an ascending road.

g = number of cubic yards which a medium laborer will load into a cart per day, ranging from ten to fourteen, as stated on page 125.

Then the minutes in the day's work $= 60\,b$;

The minutes consumed in each trip $= \frac{a}{100}$;

The number of trips, or loads hauled per day, is

$$= \frac{60b}{\frac{a}{100} + 4};$$

The number of cubic yards hauled per day, is

$$= \frac{60b}{f\left(\frac{a}{100} + 4\right)};$$

The cost of hauling, per cubic yard, is

$$= d \div \frac{60b}{f\left(\frac{a}{100} + 4\right)} = \frac{df\left(\frac{a}{100} + 4\right)}{60b} \qquad [A].$$

Adding to this the cost of excavation $= \frac{c}{g}$, that of loosening $= e$, and that of trimming $= 1$ cent, we obtain for the total cost of a cubic yard of embankment,

$$e + \frac{c}{g} + \frac{df\left(\frac{a}{100} + 4\right)}{60b} + 1 \qquad [B].$$

Applying it to an actual case, in which $a = 1000$, $b = 10$, $c = 125$, $d = 175$, $e = 2\frac{1}{2}$, $f = 3\frac{1}{2}$, $g = 12$, the formula [A] for the cost of hauling, becomes—

$$\frac{175 \times 3\frac{1}{2}\left(\frac{1000}{100} + 4\right)}{60 \times 10} = 14.3 \text{ cents};$$

and the formula [B], for the total cost of a cubic yard of embankment, becomes—

$$2.5 + \frac{125}{12} + 14.3 + 1 = 28.2 \text{ cents}.$$

The actual cost, with these data, on an amount of 22,000 yards, was 27.9 cents, differing from the calculation only three-tenths of a cent; and on a total amount of 150,000 yards, the actual and calculated costs in no case differed more than one cent.

An easy approximate rule for the average cost of hauling one cubic yard any distance on a level, with such carts and rates of travel as those above referred to, may be deduced from formula A :—

For 300 feet divide the wages of cart and driver by 24			
500	"	"	" 19
1000	"	"	" 12
1500	"	"	" 9
2000	"	"	" 7
2500	"	"	" 6
3000	"	"	" 5

The greater the distance of the haul, the less is the proportional cost, in consequence of less time being lost in filling and dropping.

In excavation and embankment with the *scraper* or *scoop*, (the use of which will be explained under the head of Construction) the number of cubic yards moved per day of ten hours, a distance expressed by a feet (adding vertical height to horizontal distance) $= \frac{4200}{a + 93\frac{1}{2}}$.* If the wages of scraper and driver be denoted by c cents, and cost of loosening by d, the cost per cubic yard $= d + \frac{c\,(a + 93\frac{1}{2})}{4200}$. When $a = 55$, $c = 275$, and $d = 1$, the cost becomes—

$$1 + \frac{275\,(55 + 93\frac{1}{2})}{4200} = 1 + 9.7 = 10.7 \text{ cents.}$$

When an *embankment* is made of earth carried beyond a certain distance, (usually 100 feet in the direction of the length of the road) it is paid for *twice;* once as excavation and once as embankment, according to prices previously stipulated; but when carried less than this distance, (as when thrown from the upper to the lower side of a road which is half in cutting and half in filling) only one price, that of the excavation, is estimated for; and the amount of embankment in this situation must be subtracted from the total amount, before multiplying this by the embankment price. If a portion of an embankment is required to be made of some peculiar material, which can be obtained only from a greater distance than the other materials of the bank, a separate and higher price should be estimated for it.

In our estimate, thus far, we have determined only the cost of the excavation and embankment.

The *land* to be occupied by the road is another important item. The quantity to be taken having been calcu

* Journal of Franklin Institute, October, 1841.

lated, with due allowance for the extra width of the cuttings and fillings, is to be reduced to acres in agricul tural districts, and to square feet in towns and villages. Its value, if not settled by agreement, must be determined by appraisers, who are, however, naturally too much inclined to favor the interests of private individuals to the prejudice of the company, or public body, which constitutes the opposite party, subjecting them to the payment of extravagant compensations.

The cost of *fencing* will vary with the locality.

The *mechanical structures,* as bridges, culverts, &c, if numerous and large, add greatly to the cost of the road; but, if important, must be confided to a professional engineer.

The stonework is usually paid for by the cubic yard, but in some parts of the country by the "perch," of 25 cubic feet.* Wood is paid for by the cubic foot, or "solid measure," when no one of its dimensions is as small as some conventional limit, which is usually 4 inches; but "board measure" (one-twelfth of the former) is employed when the wood is 4 inches, or less, in any of its dimensions. "Running measure," referring to length only, is used for simple constructions, which have small and regular cross-sections, as ditches, piles, &c.

The *Engineering* expenses, including laying out, superintendence, office-work, &c., are usually estimated at 10 per cent. upon the amount of the other items.

Every possible source of expense should be taken into the account, and an ample price for each allowed; but, finally, at least 10 per cent, upon the total amount, must be added for *Contingencies.*

* More precisely 24¾ feet, its standard being a rubble wall, 16½ feet long, and 18 inches thick.

Even then the actual expense will generally exceed the estimate.* For this opprobrium of the engineering profession there are many causes.

The *price of labor*, as the work proceeds, particularly if it be one of magnitude, may rise far above what it was at the time of the estimate.

In a deep cutting, *rock* may be found, where earth was expected, and the cost of that part of the excavation will therefore be increased tenfold.

Many *improvements* in the plan of the work are suggested and adopted as it proceeds ; almost always with an increase of cost.

Finally, it must be confessed that many incidental expenses, trifling in themselves, but considerable in their aggregate, rarely fail to be overlooked in the original estimate.

8. FINAL LOCATION OF THE LINE.

When the preceding operations of measuring, mapping, and calculating, have been performed upon each of the various lines of communication between the two extremities of the route, which have been considered worth surveying, their relative merits are to be examined. One may be shorter ; another more level ; a third may require less earthwork, and so on. The good and bad points of each route are to be compared by the principles laid down on pages 68 and 69, and that one adopted which will enable the most labor to be performed on it with the least

* On the twenty principal railroads in England, the average proportion of the actual cost to the original estimate was as 2¾ to 1. The least variation was 62 per cent excess; in the greatest, the cost was nearly six times the estimate.

number of horses, provided the expense of its construction fall within the limits established by calculation, or by necessity.* The persons who are to make the selection and decision should have before them,

1. A general map of the localities.
2. A profile of each line.
3. Cross-sections at short intervals.
4. The calculations of excavation and embankment
5. Drawings of the bridges, culverts, &c.
6. Specifications of all the works.
7. Amounts of stone-work, timber, &c
8. Analysis of the prices of each.
9. Estimate of cost.
10. Estimate of revenue.
11. Descriptive memoir.

The *final location* of the line adopted is then to be made. It consists chiefly in—

1. Rectification of the straight portions of the line
2. Laying out its curves.
3. Staking out its side-slopes.

RECTIFICATION.

The minor irregularities, bends, and zigzags of the line (caused in part by the transverse balancing) may often be removed by substituting for them one straight line, which will be the average of their deviations on either side. A flagstaff being placed at one end of the line, an observer, at the other end, by signals directs assistants to place "in line" the rods which they bear; and the points thus found are marked by stakes, which are usually driven at every hundred feet. In the case of long lines, through a coun-

* Parnell, pp. 322, 433.

try of forests, the use of the compass, or some other angular instrument, is almost indispensable, for it is still an unsolved problem in engineering, how, without the aid of these, the Romans attained the wonderful straightness with which they carried their roads over thickly-wooded hills and valleys, with such lofty disdain of the effects of gravity.

Fig. 60.

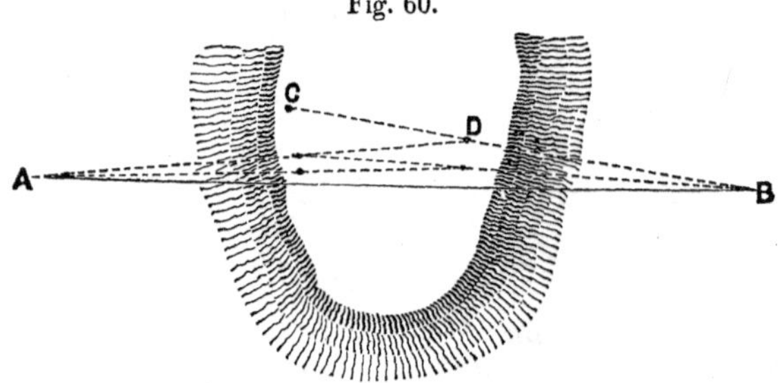

When a hill rising between two points, as A and B, prevents one being seen from the other, two observers C and D may place themselves on the ridge, as nearly as possible in the line between the two points, and so that each can at once see the other and the point beyond. C looks to B, and by signals puts D "in line." D then looks to A, and puts C in line. C repeats his operation, and so they alternately "line" each other, continually approximating to the straight line between A and B, till they at last find themselves both exactly in it.

When a wood, or some such obstacle, intervenes between the two points, as in Fig. 61, a different method must be adopted. The direction from A to B not being exactly known, leaving a rod at A, set up another at C, as nearly in the desired line as possible. Go on as far as the two rods at A and C can be seen, and set up another at D, "in line" with A and C Go on beyond D, and place another rod,

Fig. 61.

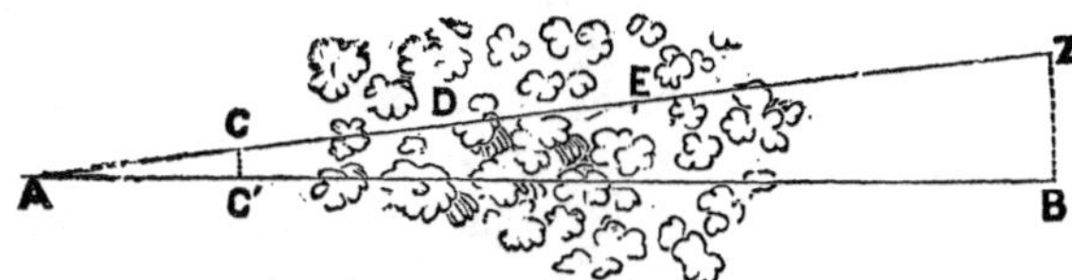

E, in line with D and C; and so proceed, producing the straight line till it arrives at Z, opposite B. Measure the distance ZB, and move the stakes C, D, E, &c., towards the true line by a quantity proportional to their distances from A. Thus if AZ be 1000 feet, and ZB, the final divergence, be ten feet, a stake C, 200 feet from A, should be moved two feet to C′, in order to bring it into the true line AB; and so, proportionally, with the rest.

CURVES.

The angles, which are formed by the meeting of the straight lines established in the *approximate* location of the road, must be rounded by curves, to which the straight lines must be tangents at their points of junction.

On every curve there is an unavoidable *loss of power* in the deflection of vehicles from the straight line which all bodies in motion tend to follow; and there is *danger* from the effects of the centrifugal force. The resistance is inversely as the radius of the curve, *i. e.*, greater as the radius of the curve is smaller; for the force required to draw a carriage around a curve may be considered as composed of two portions; one equal to the force which would be required to draw it over a straight line of the same length as the curve, and the other dependent on the additional power necessary, at each instant, to draw it into the curved line from the tangent in which it tends to move. A certain amount of force being required to produce the entire change in direction, the smaller the radius of a curve, the less space and time is given for the exercise of this force, and a larger share of it must therefore be exerted at each moment, with a great increase of labor and danger.

It is therefore very important that every road-curve should have as great a radius as possible. It should never be less than one hundred feet.

When a curve is necessary upon a steep grade, the inclination should be flattened at that place in order to compensate for the additional resistance of the curve. On this account a zigzag line up a hill is more objectionable than an oblique ascent by a straight line.

The curves which are employed to unite straight lines are usually either circular or parabolic arcs.

CIRCULAR ARCS.

Having given two straight lines meeting at C, (or which would so meet, if produced) it is required to mark out on the ground an arc of a circle to which these lines shall be tangents.

Fig. 62.

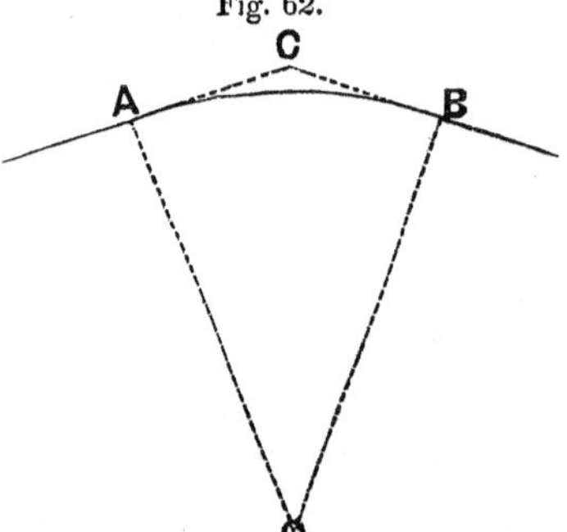

The simplest mode for an arc of small radius would be to find the centre, by erecting perpendiculars to the tangent lines at equal distances, A and B, from their point of meeting C. The intersection, O, of the perpendiculars would be the centre, from which the arc might be swept with a cord of proper length. But curves are often employed with a radius of one or more miles, so that this method would seldom be practicable The curve must therefore be traced independently of its centre.

In practice, instead of a circle, a polygon is marked out, with sides or chords each one hundred feet long. Stakes are set at the ends of each of these chords, and are therefore in the circumference of the desired circle. The chords themselves, in circles of large radius, will nearly coincide with the arcs.

The question is now, in what manner to fix the position of these chords. Two methods are in common use ; one by "angles of deflection," and the other by "versed sines." The former is generally employed upon railroads, but requires the use of an angular instrument.* The latter dispenses with this, and is therefore the one which will be here explained.

Fig. 63.

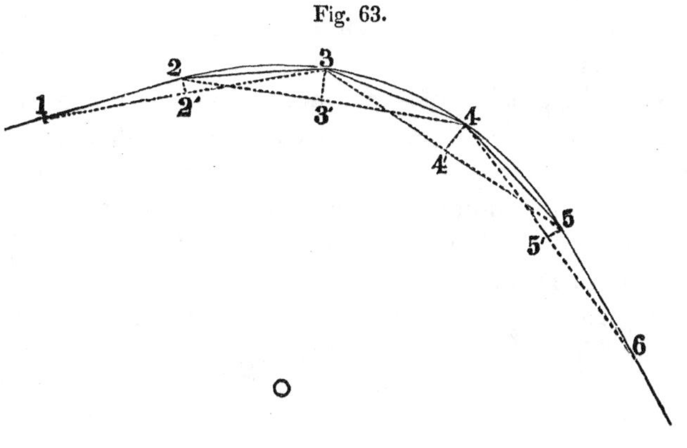

* For its details and other methods of running curves, see Appendix C.

The stations are supposed to be at equal distances (each of which is usually a chain of 100 feet) and the versed sine to be given, or to have been found by trial. Assume it at two feet, and let station 2 be the point at which the c irve is to begin. From station 2 measure inward, towards the centre', half the versed sine (or one foot) to 2′, and place there a rod. Stretch out the chain from 2, and bring its farther extremity into the line of 2 and the back station 1, and it will fix station 3, at which a stake is to be driven. From 3 measure inward the full versed sine to 3′ ; draw on the chain till its extremity is in line with 3′ and 2, and it fixes station 4. So proceed, measuring inward the full versed sine, at each station, till you arrive at the station (5, in the figure) where it is desired to end the curve, and to pass off on a tangent. There only *half* the versed sine is to be used. Station 6 is thus found, and the line 5.....6 gives the direction of the final tangent, as 2.....1 gave that of the initial one. The stations 2, 3, 4, 5, thus found, will be points in the cir cumference of a circle to which the lines 1.....2 and 5.....6 are tangents.

To find approximately intermediate points, measure outwards from the middle of each chord, a secondary versed sine = one fourth of the original versed sine. If more points are required, measure from the middle of the new chord, a tertiary versed sine = one fourth of the secondary one ; and so on.

The versed sine has been thus far supposed to be known. To calculate it from the angle of two meeting lines, the following problems are required.

Problem 1. To find the radius of the circular arc which unites two straight lines meeting at a given angle, the distance from their intersection to the initial and final points of the curve being also given.

Fig. 64.

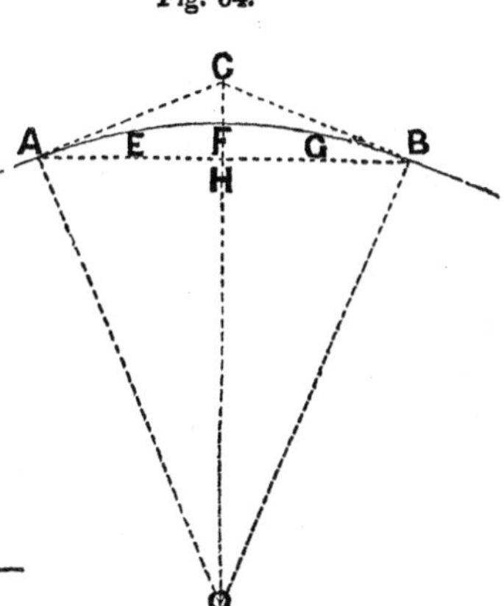

In figure 64, ACB is the given angle, and A and B the initial and final points, at equal distances from the point of intersection. The triangle CBO, right-angled at B, gives

$$BO = \frac{\text{tan. } BCO \times BC}{\text{rad.}}, \textit{ i. e.}—$$

The required radius is equal to the natural tangent (to radius unity) of half the given angle, multiplied by the distance from the intersection to the beginning or ending of the curve.*

Fig. 65.

Problem 2. To find the versed sine, having given the radius.

Given the radius OA or OF, and any two equal chords, AE, and EF, required the versed sine ED.

$$ED^2 = AE^2 - AD^2$$

$$AD^2 = AO^2 - DO^2 = AO^2 - (EO - ED)^2 =$$

$$= AO^2 - (AO - ED)^2 =$$

$$= AO^2 - AO^2 + 2AO \, . \, ED - ED^2 = 2AO \, . \, ED - ED^2.$$

$$\therefore ED^2 = AE^2 - 2AO \, . \, ED + ED^2$$

$$2AO \, . \, ED = AE^2$$

$$ED = \frac{AE^2}{2AO};$$

i. e., the versed sine is equal to the square of the chord, divided by twice the radius. When AE = 100 feet, the versed sine is equal to 5000 feet divided by the number of feet in the

* AC and AB being known, Radius $OA = \frac{AC \times AH}{CH}$

radius. When AE = 66 feet, the versed sine equals 2178 feet divided by the radius.

When the lines, which are to be united by a curve, do not actually meet, the angle which their directions form may be readily calculated; but after a little practice it will be easier to assume some versed sine; to run a trial curve with it; and after ascertaining whether it be too large or too small, to assume another nearer the proper one, and so proceed.

Compound Curves. The above method supposes that the curve has the same radius, or degree of curvature, throughout, and that it unites the two tangents at equal distances from their intersection. But it is often required to increase or to lessen the degree of curvature, and thus to form a "compound curve," as in the figure. To effect this, at the station where the change is to be made, use, for measuring inward, half the sum of the old and new versed sine, and thence proceed with the new one only. Thus, if 2 feet has been the original versed sine, and the features of the ground which is next to be passed over require a curve of 6 feet versed sine to be employed, at the desired point use a versed sine of 4 feet, and thenceforward one of 6 feet. If the curvature is to be lessened, the same rule applies.

Fig. 66.

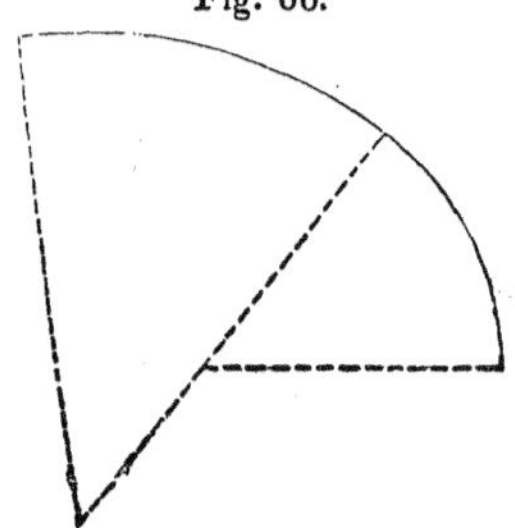

Reversed Curves. It is sometimes necessary to reverse the direction of a curve, and to commence curving in a contrary manner, without allowing a straight line to in-

* It is often desirable to know how far the curve will depart from the intersection of the tangent lines. In figure 64, the distance required $= FC = OC - OF = \sqrt{(OA^2 + AC^2)} - OA$

Fig. 67

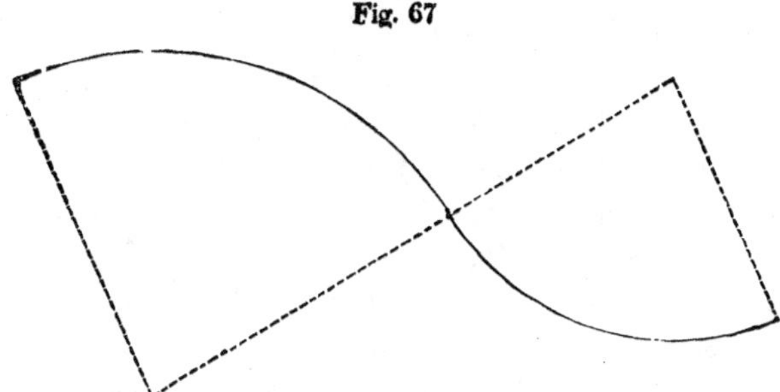

tervene. At one chain beyond the point at which it is desired to make the change, place a stake in the line of the two last, and at it begin to use the proper versed sine in the contrary direction.

PARABOLIC ARCS.

The following method furnishes an easy means of obtaining a Parabolic curve.

Fig. 68.

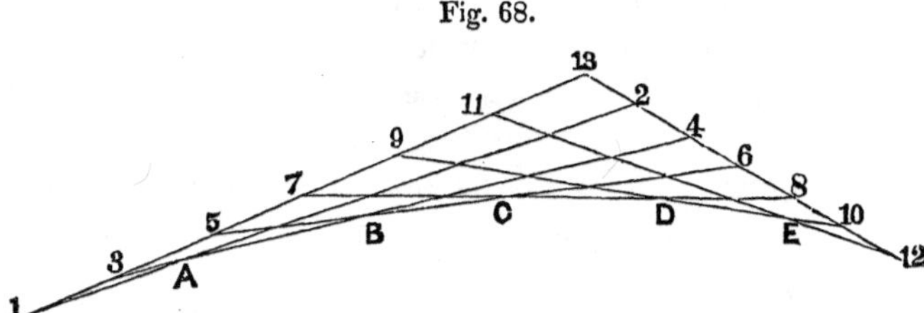

Divide the two tangent lines 1....13, and 13....12, (whether of equal or different lengths) into the same number of equal parts, as many as may be thought necessary. Number the points of division on one tangent with the odd numbers 1, 3, 5, &c., up to the vertex; and on the other tangent number them, from the vertex, with the even numbers 2, 4, 6, &c. Join the points 1 and 2, 3 and 4, 5 and 6 and so on; and the inner intersections A, B, C, D, E, will be points in the curve desired.

To fix the points of this curve upon the ground, tall stakes must be placed at each of the points of division of one of the tangent lines, and two men be stationed on the other. One places himself at station 1, and directs his eyes to station 2. The other places himself at 3, and looks to 4. A third man, holding a rod, is moved, by alternate signals from each of the others, till he comes to a point which is in both their lines of sight at once. This will be the point A. The man at 1 now passes to 5, and looks to 6, the other remaining at 3. The rodman, being again placed in both their lines of sight, thus fixes the point B. The remaining points are similarly determined.

The Parabolic curve, though little used in this country, is generally preferred in France, and has the following advantages over a circular arc.

It approaches nearer to the intersection of the tangent lines; and as they are supposed to have been originally placed on the most favorable ground, the less the curve deviates from them, the less increase of cutting and filling will it cause. The more numerous the divisions, the nearer does it approach the tangents.

Its curvature is least at its beginning and its ending, so that its deviation from the straight line is less strongly marked.

It can join two straight lines of unequal length, as in the figure; while a circular arc, of constant radius, requires both the tangents to meet it at equal distances from their intersection.

SETTING GRADE PEGS.

The line of the road having been marked out by the methods which have now been given, and stakes set at the end of every chain, small "level pegs" are then to be driven beside them, with their tops at the surface of the ground, and their heights above or below the intended height of the road (i. e. its "grade line") are to be ascertained by a levelling instrument, and the corresponding "Cut" or "Fill" marked upon the large stakes.

Another form of the levelling field-book, better adapted for this work than that given on page 98, though less safe for beginners, is presented below. It refers to the same stations and levels, noted in the previous form of page 98, and shown in fig. 43.

Sta.	Dist.	B. S.	Ht. Inst. above Datum.	F. S.	Total Heights.
A					0.00
B	100	2.00	+2.00	6.00	—4.00
C	60	3.00	—1.00	4.00	—5.00
D	40	2.00	—3.00	1.00	—4.00
E	70	6.00	+2.00	1.00	+1.00
F	50	2.00	+3.00	6.00	—3.00
		15.00		18.00	—3.00

In the above form it will be seen that a new column is introduced, containing the Height of the Instrument, (i. e. of its line of sight,) not above the ground where it stands, but above the *Datum*, or starting-point, of the levels. The former columns of "Ascent" and "Descent" are omitted. The above notes are taken thus. The height of the starting-point or "Datum," at A, is 0.00. The Instrument being set up and levelled, the rod is held at A. The Backsight upon it is 2.00; therefore the height of the Instrument is also 2.00. The rod is next held at B.

The Foresight to it is 6.00. That point is therefore 6.00 below the instrument, or 2.00—6.00=—4.00 below the datum. The instrument is now moved, and again set up, and the backsight to B, being 3.00, the Ht. Inst. is —4.00+3.00=—1.00, and so on: the Ht. Inst. being always obtained by adding the backsight to the height of the peg on which the rod is held, and the height of the next peg being obtained by subtracting the foresight to the rod held on that peg, from the Ht. Inst.

When the road is level, the "Cutting" or "Filling" at any point is the height of that point above or below the level line. But when, as is generally the case, the road ascends or descends, farther calculation becomes necessary. The following is a form of Grade book, convenient for beginners

1	2	3	4	5	6	7	8	9	10	11
Sta.	Dist	B. S.	Ht. Inst. above datum.	F. S.	Ht. Peg above Datum.	Rise or Fall of Grade.	Ht.grade above Datum.	Ht.Inst. above Grade.	Cut.	Fill.
0					0.000		+4.000			4.000
1	100	9.700	+9.700	3.600	+6.100	+0.300	+4.300	5.400	1.800	
2	100	1.800	+7.900	3.100	+4.800	+0.300	+4.600	3.300	0.200	
3	100	3.480	+8.280	3.170	+5.110	Level.	+4.600	3.680	0.510	
4	100	1.798	+6.908	9.873	—2.965	—0.200	+4.400	2.508		7.365
		16.778		19.743						
				16.778						
				—2.965						

The first six columns are similar to those of the form just given. The 7th column gives the rise or fall of the grade for each distance. The 8th is obtained by a continual addition of the preceding. The 9th is the difference of the 8th and the 4th, and is convenient for the subsequent "Staking out the side slopes." The 10th and 11th are the difference of the 6th and the 8th, as on page 116. On staking out side-slopes, see p. 457.

CHAPTER III.

THE CONSTRUCTION OF ROADS.

"The torrent stops it not; the rugged rock
Opens, and lets it in; and on it runs,
Winning its easy way from clime to clime,
Through glens lock'd up before."
ROGERS.

CONTRACTS.

THE actual "Construction" of a road, after its "Location" has been completed, may be carried on by days' work, under the superintendence of the agents of the company, or town, by which it has been undertaken; but it will be more economically executed by CONTRACT. A "*Specification*" is first to be prepared, containing an exact and minute description of the manner of executing the work in all its details. Copies of it, with maps, profiles, and drawings of the proposed road, &c., are to be submitted to the inspection of the persons desiring to undertake it, who are to be invited by advertisement to hand in sealed tenders of the prices per cubic yard (or other unit of measurement) at which they will agree to perform the work. The proposals are opened on the appointed day, and the lowest are accepted, other things being equal. The "*Contract*," which is to be then signed by the parties, should contain copious and stringent conditions as to the time and manner of performing the work; stipulating when it is to be commenced, how rapidly to progress, in what order of parts, and when to be completed, which of the incidental expenses are to be borne by the contractor, and for which he is to be remunerated; in what cases material carried from excavation into embankment is to be paid for at the united prices of both; what penalties for neglect are to be imposed; when payments for work done are to be made; and so on · always remembering that every thing must be expressed, and nothing left to be inferred.*

The specification is considered to form part of the contract, and a "Bond" is appended, by which the contractor and his sureties are "holden and firmly bound" in a penal sum, "this bond to be null and void, if the said parties shall faithfully execute and fulfil the accompany ing Contract."

Each contract should include such a length of road, called "a section," (usually half a mile or a mile long) that materials for the embankments may be obtained from cuttings within its limits. There should be separate contracts for the mechanical structures required. The works which will need most time for their execution should be commenced first; but no contract should be let, till the land which it includes is secured, or exorbitant demands will be made.

It has been said that the lowest bid is usually accepted, but this is to be taken with great qualifications. The

* In the contracts for the public works of the state of New York, one valuable paragraph comprehends every thing, saying, "To prevent all disputes, it is hereby agreed, that the engineer shall in all cases determine the amount or quantity of the several kinds of work which are to be paid for under this contract, and the amount of compensation at contract prices to be paid therefor; and also that said engineer shall in all cases decide every question, which can or may arise, relating to the execution of this contract on the part of the said contractor, and his estimate and decision shall be final and conclusive."

skill, competency, character, and responsibility of the contractor are as important as the lowness of his prices. A skilful and experienced contractor will often make a profit on a work, which another has abandoned after considerable loss. Bids, less than the actual cost of the work, are often made, both from ignorance and from knavery. In the former case, if the proposals were accepted, the contractor would be ruined, and obliged to leave the work unfinished; in the latter, he would hope to gain something by doing first the easy and profitable parts of the work, and then abandoning it. In both cases the remaining portions would be executed at greatly increased expense. Six contracts in England amounting to $3,000,000 being abandoned, were finished by the company, and cost them $6,000,000. The engineer should therefore ascertain the lowest amount for which the work can be done, and not let it for less.

The work done is usually paid for monthly, according to a measurement made by the inspecting engineer. Five or ten per cent is generally retained till the completion of the contract.

The two main divisions of the operations necessary in the construction of a road, are its *earthwork* and its *mechanical structures*.

1. EARTHWORK.

The term *earthwork* is applied to all the operations in excavation and embankment, whatever the material.

REMOVAL OF THE EARTH.

The problem which is to be solved, both in theory and practice, is, "To remove every portion of earth from the excavation to the embankment by the shortest distance, in the shortest time, and at the least expense."

It must also be deposited so as to form a consolidated mass, and so that not · particle of it will need to be again moved.

The problem is very important in practice, for upon its mode of solution depends a large portion of the cost of the work; and in theory, it requires the aid of the higher Calculus, since, to satisfy its conditions, the sum of the products, arising from multiplying all the elementary volumes of earth into the distances which they are carried, must be a *minimum*.

We have seen, on page 123, that in the simplest case, that in which the whole of one excavation is to be carried into one embankment, we may use the product of the entire mass multiplied by the distance of the centres of gravity of its two positions. But when certain portions of a cutting are to be deposited in spoil-banks; others to form part of an embankment, of which the remainder is to be obtained from side-cuttings; &c., it does not appear *à priori* what arrangement would give a minimum expense. In a few cases the proper course is evident; as, if a hill is to be cut down, and its materials serve only to fill up a valley, and are in excess, the excavation from its summit is clearly the portion to be deposited in spoil-bank; if the materials are insufficient to form the embankment, it is the part most distant from the hill which should be formed from a side-cutting; if the excavation is to be carried in two different directions and is in excess, it is the part of the middle which should be rejected and deposited in spoil-bank.

One general principle of transport may be readily deduced. Let ABCD represent the plan of an excavation

Fig. 70.

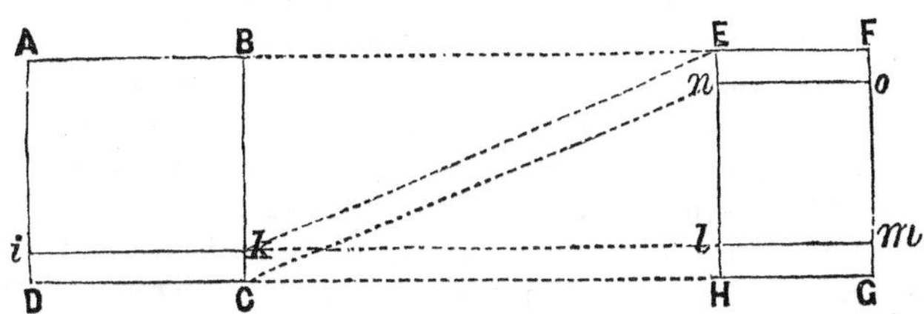

from which the embankment EFGH is to be formed. If the volume CD*ik*, instead of being taken to GH*lm*, should be transported to EF*on*, it follows that the embankment GH*lm* must be obtained from a portion of the excavation beyond the line *ik*, and that the paths of the two volumes will cross each other, which is therefore a disadvantageous disposition, since it increases the distances passed over. Any such crossing of the paths of the volumes transferred, either horizontally or vertically, may be generally avoided by conceiving the solids of excavation and embankment to be intersected by parallel planes, such as DCHG, *ik*, *lm*, &c., and by transferring the partial solids in the manner indicated by the boundaries marked out by these planes.

In many cases the most economical distribution of the earth, can be determined only by a special construction. Thus in the figure, suppose that earth is to be taken from A and B to form embankments at C and D; it is required to know which should form the embankment at C, and which that at D. To bring the case within the application of the principle

Fig. 71.

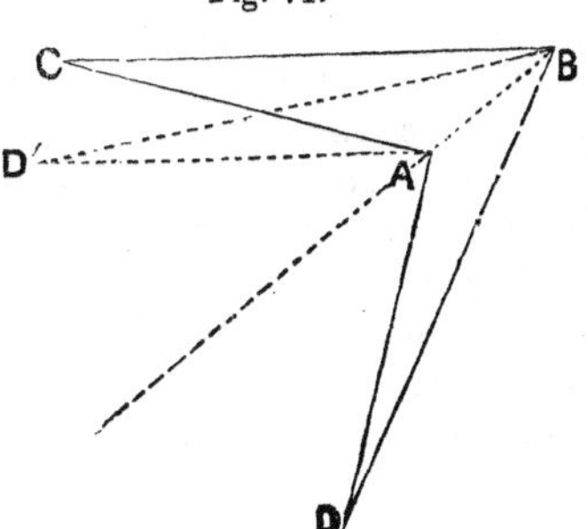

just enunciated, conceive the triangle ABD to turn around the line AB as on a hinge, so that the point D comes to occupy a point D′, symmetrical with its former position.

It is now evident that to avoid the crossing of the paths, the earth from A must be taken to D′, (*i. e.* D) and the earth from B to C; AD′ + BC being less than AC + BD′. If the point D′ had fallen beyond C the reverse would have been proper. If the point D′ had fallen within the triangle ABC, there would be no crossing in either mode of transport, but the proper one would be determined by a similar algebraic condition.*

The choice would be indifferent, if

$$BC - AC = BD - AD,$$

or if $$AD - AC = BD - BC;$$

for then, $$AD + BC = AC + BD.$$

Two points, A and B, Fig. 72, being found which fulfil this condition, other points will be found at the intersection of arcs

Fig. 72.

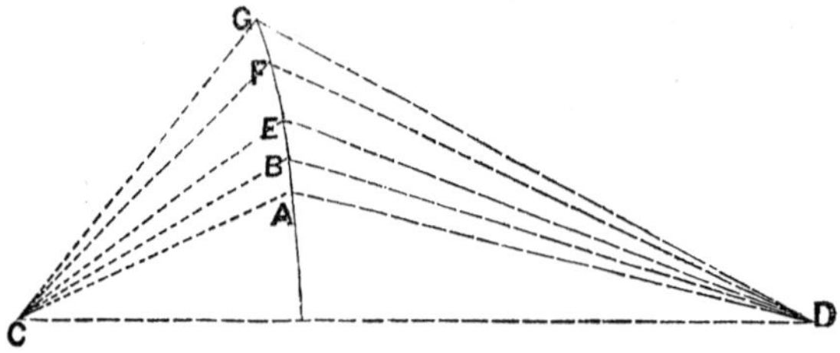

described from C and D as centres, with radii of which the differences are respectively equal to the given difference AD—AC, or BD — BC. If a great number of these points were found, the polygonal line ABEFG would become an hyperbola, possessing the remarkable property of so dividing the transportation, that C should receive all the excavation from one side of it, and D all from the other.

Suppose that embankments at C and D, Fig. 73, are to be made from a mass of earth *mnop*, just equal to them in volume. The minimum of expense will be obtained by finding the curve AG, which shall divide the area *mnop* into two parts equal to

* Gayffier, p. 134.

Fig. 73.

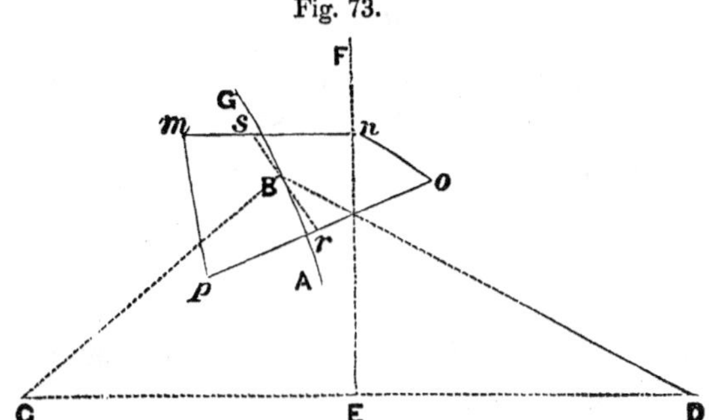

those required at C and D, and which shall also possess the properties enunciated in the preceding paragraph. If the line EF drawn perpendicular to CD, from its middle E, does not cut off a sufficient portion of the area to supply D, this shows that the curve will be concave towards C. Then divide geometrically the area *mnop* in the required proportion, by a straight line *rs*, inclined approximately as the curve would be, and adopt its middle point as a point of the curve. Then will BD — BC be the constant difference of radii required to find the other points of the dividing curve.

If the amount of embankment, which might be deposited at C and at D, was indefinite, and the only requirement was its most economical removal from *mnop*, then the perpendicular EF drawn from the middle of CD, would divide the area into two portions, which should be removed to the points C and D respectively nearest to each of them.

On similar principles may all such problems be resolved. Modifications of them are required, when the paths cannot be taken at will, as when a bridge, or an opening in a wall, is a point through which all the paths must pass. The number of bridges, of openings, of roads, &c., which will be most advantageous, require separate investigations.*

---

* See Gayffier, pp. 137 to 142.

EXCAVATION.

The excavation and removal of earth is performed, according to circumstances, by ploughs, scrapers, barrows, carts, wagons, &c., each of which will be successively considered.

LOOSENING.

Most earth will require to be *loosened* with ploughs, spades, or picks, before being shovelled into the barrow, or cart, in which it is to be removed. The side-hill plough possesses some advantages. The picks should be two feet from point to point, not more than ten or twelve pounds in weight, and very deep and strong in the eye, or socket of the handle. Light and loose soil may however, be at once taken up with the shovel.

When the excavation is deep, the loosening may be facilitated, with a great saving of time and labor, by digging a narrow channel to a depth of five or six feet, and under mining the face of the bank thus formed, letting it fall at once into the barrows, or carts, beneath it. Its disruption is hastened by wedges driven into its upper surface. The concussion of the fall breaks up the mass into small pieces, with great economy, but not without danger to the workmen.

In the ordinary excavation, in which the earth is dug up, the united cohesion and weight must both be overcome; in the method just described, the weight assists in overcoming the cohesion. Representing the force of cohesion by 3, and that of the weight by 2; if both are to be overcome, as in the usual method, their resistance will be $3 + 2 = 5$; while if the weight be made to assist the workman, the resistance will be only $3 - 2 = 1$.

Steam has been applied to excavation, and a machine co structed, which can dig and load 1000 cubic yards per day, in favorable soil at an annual cost, including interest, wear and tear, labor, &c., of $7,500, making the cost per cubic yard, $\frac{\$7,500}{300 \times 1000} = 2\frac{1}{2}$ cents.*

SCRAPER OR SCOOP.

This implement may be used with much advantage, when the earth yields readily to the plough, and is not to be moved more than 100 feet horizontally, nor to be raised to vertical heights of more than 15 feet; though these limits may sometimes be exceeded. The slopes of the banks which it forms, should not be steeper than $1\frac{1}{2}$ to 1. It usually contains $\frac{1}{10}$ of a cubic yard.† The

Fig. 74.

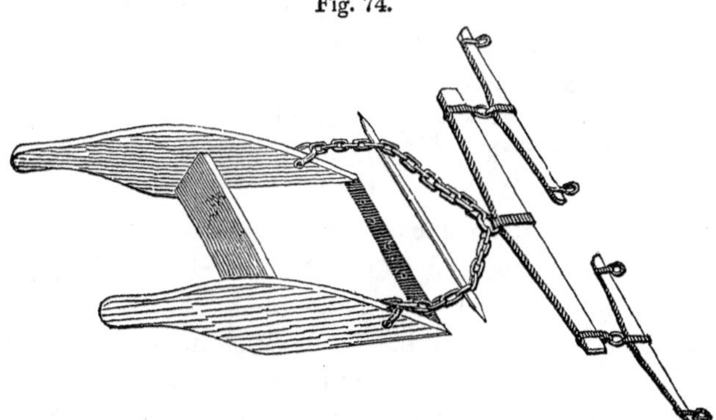

ground, except when soft or sandy, requires to be previ ously ploughed. The scraper is drawn by two horses, managed by a boy. The driver elevates the handles, and the iron-shod edge runs under the loose earth, rising up

* Journal of the Franklin Institute, September, 1843.
† Ibid. October, 1841.

again as soon as the handles are released upon its being filled. It then runs with slight resistance upon two convex iron-shod runners, which project slightly beyond its bottom, and is thus drawn to the place of deposite. At that point the driver raises the handles; its front edge catches in the earth, and its forward motion overturns it, and discharges its load. The horses keep moving; and the scoop is dragged back to the place of excavation, in its inverted position, the handles resting on the tree. It is there loaded, &c., as before.

BARROW-WHEELING.

For excavations of moderate depths, and for distances within certain limits, *barrows* are most conveniently employed. To facilitate emptying their contents, the barrows are made very shallow, with splaying sides, and with a very short axis to the wheel. They contain from $\frac{1}{20}$ to $\frac{1}{10}$ of a cubic yard. They are wheeled on "runs" of plank, (as long and thick as possible) laid on the ground, or supported on trestles, or horses, numerous enough to prevent vibration. When the tracks are inclined, as in ascending from a deep excavation, they should be laid with a slope of one in twelve.* A steeper slope fatigues the workman excessively; a flatter one increases too much the length of his route. The same man does not usually dig, shovel, and wheel, but great advantages are obtained by a division of labor. One man picks, (if that be required) a second shovels into the barrow which stands on the edge of the excavation, and a third wheels the barrow to the place of deposite, or to the next "stage," according to the distance. In the latter case, at the end of

* Dupin. *Applications de la Géometrie.*

the "stage," he meets another wheeler, returning with an empty barrow. The two there exchange their barrows; the second man wheels on the loaded one over another stage, while the first man returns with the empty barrow to the excavation, where he finds a loaded one, which has been filled during his absence; and so the circulation continues.

The length of the "stage" should be such, that the time, taken by the wheeler to travel over it with a loaded barrow, and to return with an empty one, should be just sufficient to enable the shoveller to fill the barrow left at the excavation. It should vary therefore with the nature of the soil; lessening, if this be easily worked, and increasing, if it offer greater resistance. On a level the length of a stage is usually from 60 to 100 feet. On an ascent of 1 in 12, it should be diminished by one-third; on a similar descent it should be increased by the same; for with this slope the labor on an ascent of 60 feet exactly equals a level stage of 90 feet.*

If the distance were not divided into stages, and one man wheeled his barrow the entire length, a number of runs would require to be laid from the excavation. Such an arrangement would be inconvenient, from its blocking up the work, and expensive, from the cost of the plank. At the point where the run terminates in the excavation, two planks are placed, diverging like the letter Y, the full and the empty barrow being wheeled on each alternately. At the meeting of two stages, a double track is laid, to form a turning-out place for the exchange of the barrows. At the place of deposite, several planks should radiate from the main track, so that the earth may be at

* Dupin. *Applications de la Géometrie.*

once evenly distributed, by being emptied from each in turn, thus saving much subsequent levelling.

Barrow-wheeling becomes too expensive after reaching a certain limiting distance of transportation. The frequent neglect of this consideration leads to much waste of labor. When earth is to be conveyed great distances, carts or wagons should be employed. The *limit* is determined by a combination of the cost of filling and of transporting. The table on page 128, makes it 100 feet; the limit in France, with barrows containing $\frac{1}{20}$ of a cubic yard, should be 200 feet; on English works, with barrows holding $\frac{1}{10}$ of a cubic yard, the limit is 300 feet. The limiting distance becomes smaller as the height to which the earth is moved becomes greater.*

CARTS, ETC.

One-horse carts may be advantageously employed for distances exceeding the sphere of barrows. For short distances, the greater proportional loss of time in filling them more than balances their economy while moving. They should be made very light, and their box be balanced on a pivot, so that when loaded they will tend to discharge themselves.† As the distance increases, *wagons*, drawn by two horses, become cheaper, and a temporary railway may often be constructed with profit.

When the length of the *lead*, (*i. e.* the distance from the face of an excavation to the head of an embankment) exceeds $1\frac{1}{2}$ miles, and the amount of earthwork is considerable, a locomotive engine may be advantageously employed to draw trains of wagons upon the rails.

---

* Gayffier, p. 159.

† When horses draw loads out of an excavation, the inclination of their track should not exceed 1 in 20. DUPIN. *Applications de la Géometrie*

"*Casting up by stages*" is a method sometimes employed for removing the earth from deep excavations. A scaffold is prepared with a number of platforms, each five feet above the other, and each successive one receding, like the steps of a staircase. On each platform stands a man with a shovel. The laborer in the excavation throws the earth upon the first platform; the man there stationed throws it up to the second; and so on in succession till it reaches the surface.

Horse-runs are also resorted to in very deep excavations, where the banks are necessarily very high and steep. Upon the slope of the bank are placed two plank "runs," or tracks, reaching from the top to the bottom of the ex cavation. The distance between them must be a little greater than the depth of the excavation. At the top of each is a pulley, over which plays a rope, the ends of which pass down the runs. Each end of the rope is fastened to the front of a barrow, and its length is so adjusted that one barrow will be at the top of one run, while the other barrow is at the bottom of the other run. At the top of the excavation, a horse, attached to the rope, travels horizontally, alternately raising one barrow, which has been filled below, and lowering the other, which has been emptied at the top. A man has hold of each barrow to guide it in its ascent and descent, the weights of the men balancing each other. This method is advantageous for depths exceeding 20 feet.* The use of barrows in such cases, with the proper inclinations for the runs would require too great a distance to be travelled over.

* Gauthey, ii. 197.

SPOIL-BANKS.

The spoil-banks, formed by the deposites of the surplus earth of an excavation, are usually shaped, as in the

Fig. 75.

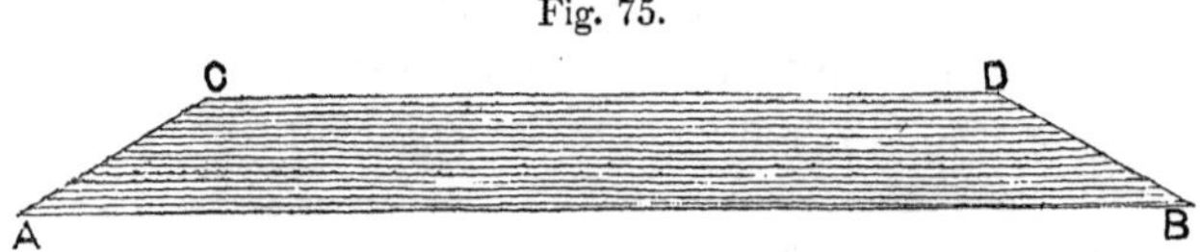

figure, with side-slopes of $1\frac{1}{2}$ to 1. If the land which they occupy be of little value, it will be economical to extend them along the line AB, making them wider and lower within certain limits; since vertical transport costs so much more than horizontal.* The solution of the problem of *minimum* expense shows that for spoil-banks made with barrows, (slopes $1\frac{1}{2}$ to 1, and employing the customary ratio of 18 to 1, for the comparative expense of horizontal and vertical transport) the base AB should be *fifteen* times the height.†

SIDE-SLOPES.

To preserve the slopes of deep excavations from being gullied and washed down into the road, a ditch should be made along the upper edges of the cutting, in order to prevent the surface water of the neighboring land from reaching it. Upon the slopes themselves should be made ditches, called "Catch-water drains," running obliquely downwards, to receive the water of rains, and conduct it into the side ditches.

The side-slopes may be advantageously sown with grass-seed. The roots of the grass will bind the earth

* See page 128 † Gayffier, p. 162

together, and prevent its slipping. A overing of 3 or 4 inches of good soil should be previous.y spread over the side-slopes, but if they are steeper than 1¾ to 1, the soil will not lie upon them. They may also be sodded; the sods being laid on, either with the grass side uppermost, or edgewise, with their faces at right angles to the slope. The latter, "Edge-sodding," is the most efficient, but most expensive.

TUNNELING.

When the excavation exceeds a certain depth, it will be cheaper to make a tunnel as a substitute. The amount of excavation will be much less, but the cost of each yard of it will be much greater. Calculation in each case can alone decide at what depth it would be economical to abandon the open excavation, and to commence the tunnel. Sixty feet is an approximate limit in ordinary earth. The necessity for tunnels seldom occurs, however, in the construction of common roads, though frequent in the great roads of the Alps, and on railroads; in the chapter devoted to which they will therefore be more fully noticed.

BLASTING.

Not only rock, but frozen earth, and sometimes very compact clay, are removed by blasting with powder. The holes are drilled by a long iron bar of the hardest steel, chisel-edged, which is raised and let fall on the desired point, and at each stroke turned partially around, so that the cuts cross each other like the rays of a star *. The holes are made from 1 to 3 inches in diameter, and from 1 to 4 feet deep. One man can drill in a day 18 inches, of one 3 inches in diameter, in rock of average hardness. When water percolates nto the hole, it must be dried with oakum and quicklime, and the powder enclosed in a water-proof cartridge. The proper proportion of powder being introduced by a funnel and copper tube, (so that none may adhere to the side) a wadding, of hay, moss, or dry turf, is placed upon it, and the remainder of the hole is filled with some packing material. This is usually sand, but by far the best, for safety and efficiency, is dried clay, rolled into balls or cylinders, and dried at a smith's forge, as much as can be, without its falling to pieces. The next best material is the chippings and dust of broken brick, moistened slightly while being rammed. An inch or two of the wadding being simply pressed down upon the powder, the filling material is rammed, or "tamped," with a copper wire, till it becomes very compact. Through it passes, from the powder to the surface, some means of ignition. A straw, filled with priming powder, and ignited by a slow match, was formerly employed for this purpose. But of late years this has been generally, and should be universally, superseded by the *safety-fuse*. This has the appearance of a common tarred rope, and is so prepared that the length of it, which will burn any given time, can be exactly known, so that no premature explosion need be feared.

The proper charge of powder, and the direction of the holes, are very important, both for efficiency and economy. The usual charge is one-third of the depth of the hole; but such a rule is evidently irrational, for the amount of a charge so proportioned will vary with the bore. The proper regulator of the charge is the length of "*the line of least resistance*," i. e. the shortest distance from the bulk of the powder to the outside of the rock. Thus in the figure, AB being the hole bored, and B the powder, BC is the "line of least resistance,"

Fig 76.

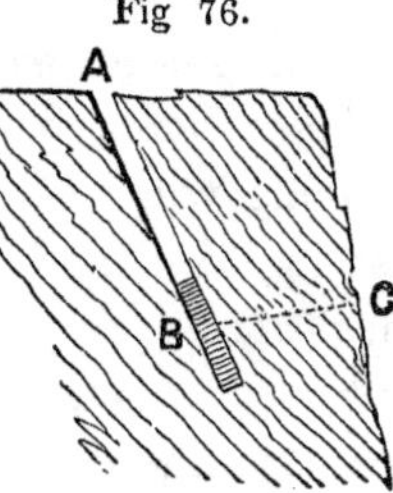

which should not be in the direction of the hole bored. The proper charge depends on it, and not at all on the depth AB. To produce similar proportional results in different blasts, the charges must be as the *cubes* of the respective lines of least resistance. Thus, if four ounces of powder will just suffice to blast a solid rock in which BC is 2 feet, the charge for another in which BC was 3 feet, would be given by the proportion $2^3 : 4 :: 3^3 : 13\frac{1}{2}$ ounces.

On these data the following table is formed.*

Line of least resistance.		Charges of powder.		Line of least resistance.		Charges of powder.	
Feet.	inches.	Lbs.	Oz.	Feet.	Inches.	Lbs.	Oz.
1	0	0	0½	4		2	0
1	6	0	1¾	4	6	2	13½
2	0	0	4	5		3	14½
2	6	0	7¾	6		6	12
3	0	0	13½	7		10	11½
3	6	1	5½	8		16	0

The following table will also be found very convenient

Diameter of the hole.	Powder in one inch of hole.		Powder in one foot of hole.		Depth of hole to contain one lb. of powder.
Inches.	Lbs.	Oz.	Lbs.	Oz.	Inches.
1	0	0.419	0	5.028	38.197
1½	0	0.942	0	11.304	16.976
2	0	1.676	1	4.112	9.549
2½	0	2.618	1	15.416	6.112
3	0	3.770	2	13.240	4.244

* London Mechanics' Magazine, xxxiii. 597, Dec 1840; and professional papers of Royal Military Engineers, vol. 4

When the rock is stratified, having beds and seams, as in the figure, holes bored parallel to the joints will produce much greater effect than the usual vertical ones.

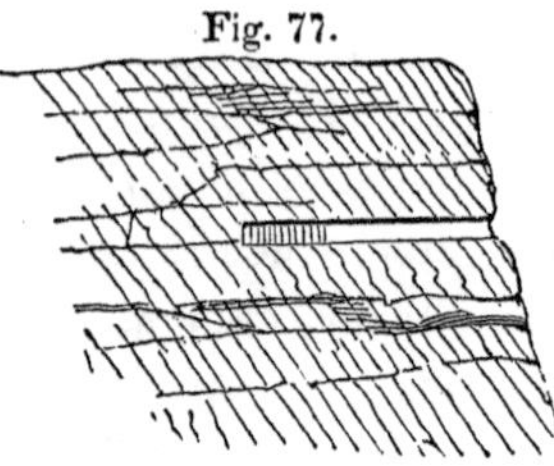
Fig. 77.

When a rocky surface is to be cut down to a line AB, the holes should be oblique, as in the figure. In some cases, a horizontal one, from B towards A would be advantageous.

Fig. 78.
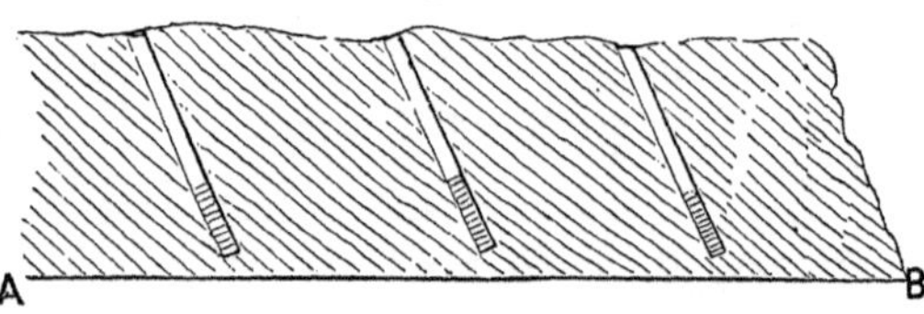

On a high face of rock a system of undermining may be usefully employed, by blowing out a mass below, and removing the remaining overhanging portion by crowbars, wedges, &c.

The *crater*, or cavity formed by an explosion, is assumed to be a truncated cone, which has its inner or smaller diameter equal to half the diameter of the mouth of the crater. It has been found by experiment that the outer diameter of the crater may be increased, in ordinary soils, by excessive charges, to three times the length of the "line of least resistance," but not much beyond this; and that within this limit this diameter increases nearly in the ratio of the square root of the charge.

The most unfavorable situation for a charge is where a re-entering angle is to be blown out, as the rock all around it exerts a powerful resisting pressure. The charge needs to be proportionally increased. This case constantly occurs in blasting out narrow passages.

No loud report should be heard, nor stones be thrown out. The best effect is produced when the report is trifling, but when the mass is lifted, and thoroughly fractured, without the projection of fragments. If the rock be only shaken by a blast, and not moved outwardly, a second charge in the same hole will be very effective.

Any kind of compact brush, such as pine or cedar boughs, laid on rocks about to be blasted, will almost completely prevent the flying of fragments, and thus lessen the danger to persons and buildings in the vicinity.

The safety of blasting operations may be greatly increased by applying galvanism to the ignition of the powder, which can then be effected at any distance. By its aid a row of blasts can be exploded simultaneously, by which their effective power is greatly increased. In this way, a single blast, of nine tons of powder, contained in three cells, removed one million tons of rock from a cliff at Dover, with a saving of $50,000

EMBANKMENTS.

Perfect solidity is the great desideratum in artificial road-making. Every precaution must therefore be employed, in forming a high bank, to lessen its tendency to slip. From the space which the bank is to occupy, all vegetable or perishable matter, and all porous earth and loose stones, should be removed. On this space the earth is then deposited, to form the embankment, which is usually made of full height at its commencement, and is extended by "tipping" earth from the extremity, and so carried out on a level with the top surface. But an embankment thus formed will be deficient in compact ness; for the particles of earth, which are emptied from the top of the bank, will temporarily stop in their descent at the point of the slope at which the friction becomes sufficient to balance their gravity; and when more earth comes upon them, they will give way and slide lower down, causing the portions above them to slip and crack, and thus delaying for a long time the complete consolidation.

This method is, however, cheap and rapid. Its rapidity will be increased by obtaining more "tipping places," which can be effected by forming the bank at first wider

Fig. 79.

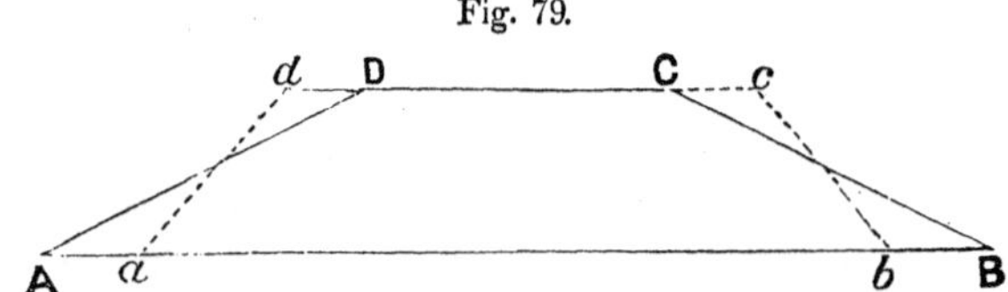

at top, and narrower at bottom, than it is finally to be, (i. e. forming *abcd* instead of ABCD) and subsequently throwing down the superfluous earth from the top to give the proper width at bottom.*

The solidity of embankments, which are made by tipping from the ends may be increased by forming the outside portions of the bank first, and gradually filling up towards the middle, so that the earth may arrange itself with a tendency to move towards the centre, if at all.†

To ensure the stability of embankments, they should, however, be formed by depositing the earth in successive layers or courses, not more than three or four feet thick; and the vehicles, conveying the materials, should be re-

* Laws of Excavation and Embankment on Railways, p. 59

† Mahan, p. 287.

quired to pass over the bank at each trip, so as to compress the earth. If the case warranted the expense, each course might with advantage be well rammed. To lessen the danger of slips, the layers should be made some-

Fig. 80.

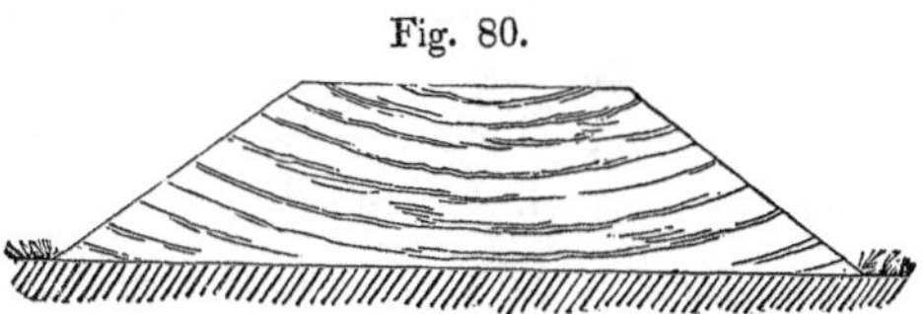

what concave, as in Fig. 80. If made convex, as in the next figure, and as they are apt to become, in the most

Fig. 81.

natural mode of forming them, portions would tend to slip off in the direction of the layers, while the arrangement of concave layers would resist, instead of assisting, any slip. A framework of timber has sometimes been inserted in a bank to bind it more firmly together.

An embankment should always be formed at first of its full width, and not, from a mistaken economy, be at first made narrow, to be subsequently increased by lateral additions; for the new portion will never unite perfectly with the old.

At the foot, or "toe," of the bank, a slight excavation may be made to resist its tendency to spread, or a low but massive stone wall may be there erected.

The slopes, like those of excavations, should be grassed, or sodded. If exposed to the action of water, a row of planks, grooved and tongued, and sharpened at bottom, should be driven at their foot, forming a "*sheet-piling*," and the slopes themselves should be protected with a "*slope-wall*," composed of rough stones, from one to two feet thick, laid without mortar, with their faces at right angles to the slope, and "breaking joints" as perfectly as possible. To prevent their being thrown out of place by the swelling and heaving, which is caused by the freezing of the rain-water retained by the clayey material of which an embankment may be composed, a layer, one or two feet thick, of coarse gravel, should be placed on the slope before laying the stone facing, so that the rain-water can at once pass through this porous coating. At the foot of the slope, an "apron," or mass of loose stones may be deposited.

SWAMPS AND BOGS.

When an embankment is to be made through a swamp bog, marsh, or morass, many precautions are necessary.

If the bog be less than four feet deep, and have a solid bottom, all the soft matter should be removed, and an embankment raised upon the hard bottom.

If it be deeper, but not very soft, the surface may be covered with two rows of swarded turf; the lower being laid with its grassy face downward, the other with that face upward, and the embankment raised upon them.

When the swamp is deep and fluid, thorough draining is the first and most important point. On each side of the road, wide and deep ditches must be cut, to collect the surface water, and to carry it off into the natural water-courses. Numerous smaller ditches must be cut, at short intervals, across the road-way, from one main drain to the other, descending both ways from the centre. This operation will consolidate the surface between the main ditches. The cross-drains may be filled with broken stones, (or bushes, if they will always remain under water, as otherwise they will decay, and cause the road to sink) and on this foundation the embankment may be raised.

In extreme cases, the lower portions of the embankment must be formed of brush-wood, arranged in *fascines* which are a specific remedy against water. They are formed by carefully selecting the long, straight, and slender branches of underwood, and tying them up in bundles, from 9 to 12 inches in diameter, and from 10 to 20 feet long. A layer of these fascines is laid across the road; a second layer in the direction of the road; and so on, to as great a thickness as may be required to raise the road-bed perfectly high and dry. Sharp stakes are driven at intervals to fasten together the layers. Poles, or young trees, may be laid across every other course. Upon this platform of fascines may be laid large flat stones, and upon them a course of earth and gravel.

SIDE-HILL ROADS.

When a road runs along the side of a hill, it will be most cheaply formed, by making it half in excavation and half in embankment. But as the embankment would be

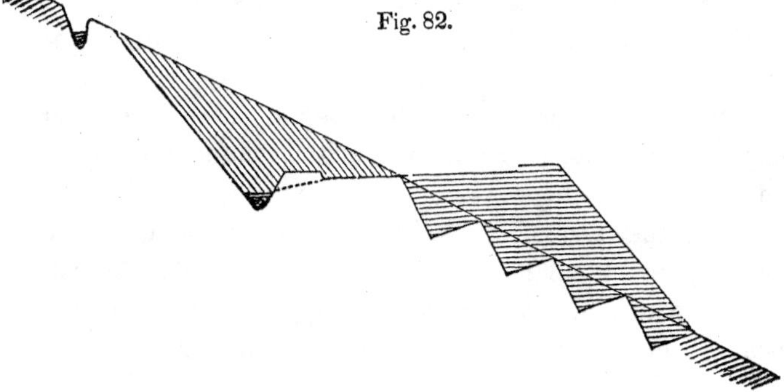

Fig. 82.

liable to slip, if simply deposited on the natural surface of the ground, the latter should be notched into steps, or off sets, in order to retain the earth. In adjusting the height of the made ground, an allowance should be made for its subsequent settling.

If the surface be very much inclined, both the cuttings and fillings will need to be supported by "retaining walls,"

Fig. 83.

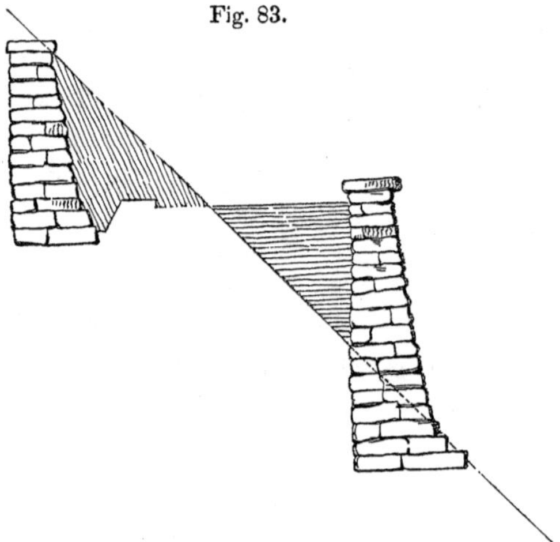

which may be laid dry if composed of large stones, or in mortar. The proper thickness which should be given to them, will be investigated under the head of "Mechanical structures."

If the side hill be of rock, the steep slope at which that material may safely be cut, will enable the upper wall to be dispensed with.

When the road is required to pass along the face of a nearly perpendicular precipice, at a considerable height, (a case which sometimes occurs in passing a projecting point of the rocky bank of a river in a mountainous district) it may rest on a frame-work formed of horizontal beams, deeply let into the face of the precipice, and supported at their outer ends by oblique timbers, the lower ends of which rest in notches formed in the rock.

Fig. 84.

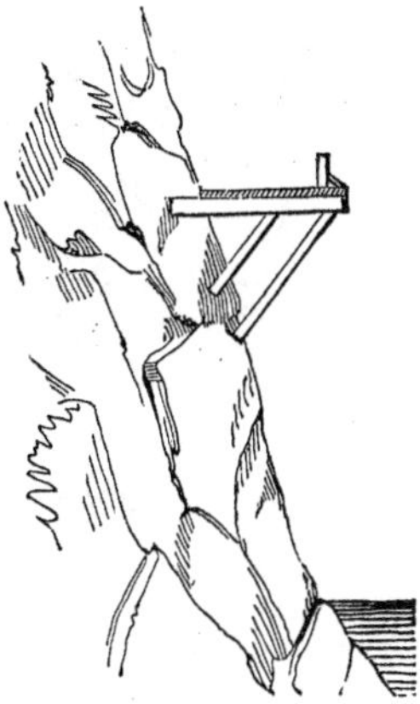

TRIMMING AND SHAPING.

To form the side-slopes with precision, to the proper inclination, a simple bevel, "*bâtir*-level," or "clinometer," may be employed with great advantage. It consists of two strips of board, AB, AC, fastened to each other at right angles and connected by a third one, CB. When the desired slope is 2 to 1, make AB twice the length of AC. Place C, or B, at any known point of the slope; make AC vertical by the plumb-line; and then will BC coincide with the slope desired.

Fig. 85.

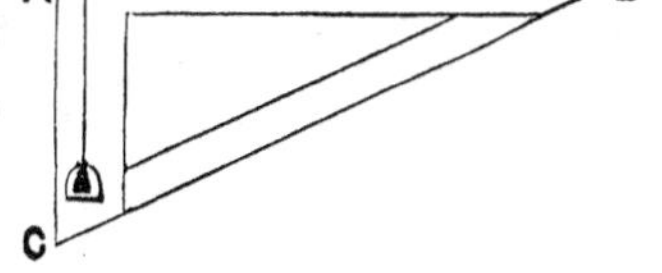

Another implement for the same purpose is formed of a single strip of wood, to which is attached a triangle

Fig. 86.

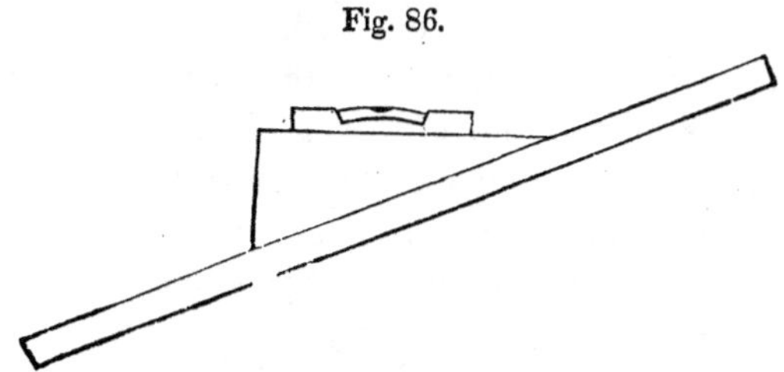

with base and height corresponding to those of the desired slope. When a spirit-level, resting upon the top of this triangle, is horizontal, the inclined strip will coincide with the slope sought.

A more general "Clinometer" is shown in the accompanying figure. It consists of a spirit-level, moveable on

Fig. 87

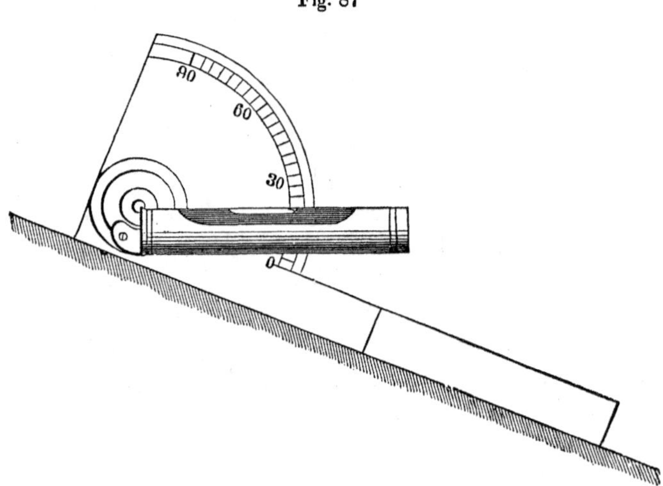

a pivot, which is the centre of a quadrant divided into degrees. To measure a slope, place the bar upon it, and turn the level till the bubble is in its centre. The reading at the top of the level will indicate the inclination of the slope. To increase its portability, the long bar doubles up on a hinge in its middle.*

To shape the tops of the embankments, and the bottoms of the cuttings, in accordance with the desired profile of the road, attach, to the under side of a common

* Simms on Levelling, p. 96

Fig. 88

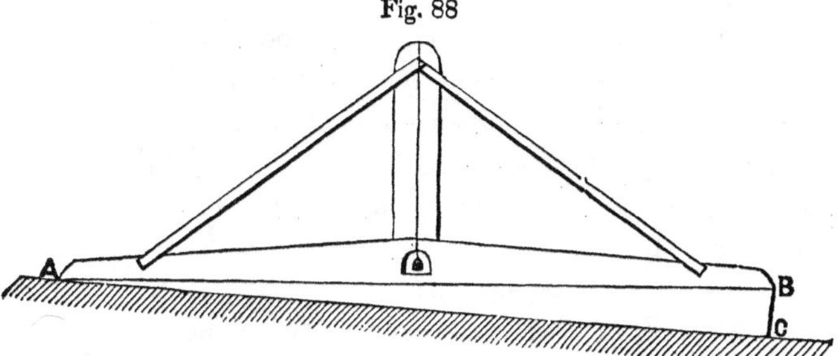

mason's level, a triangle ABC, with its base and height so proportioned as to correspond to the "crowning" of the road ; 1 in 24 for example. Or, instead of the triangle, gauges of different lengths, moveable on thumb-screws, may be made to project below the level, to proper depths.*

2. MECHANICAL STRUCTURES.

Under this head are included the bridges, culverts, and other works of the mason and carpenter, which are required for the purposes of the road.

BRIDGES.

The most simple and natural form of a bridge consists of two timbers, laid across the stream, or opening, which is to be passed over, and covered with plank to form the road-way. Walls should be built to support each end of the timbers, and are named the *abutments*. The width of the opening which they cross is termed the *stretch*, or *bay*. The timbers themselves are the *string-pieces*. Their number and size must of course increase with the stretch. For a stretch of 16 feet, they should be about

* Parnell, p. 261.

15 inches deep by 8 broad, and be placed at intervals of about 2 feet.* The greatest weight which can come upon them is when the surface of the bridge is covered with men standing side by side, and is then equal to 120 lbs. per square foot of surface, independently of the weight of the materials. Recent experiments make this only 70 lbs.

This simple construction is only applicable to short stretches. For spaces of greater width, supports from the bottom of the opening may be placed at proper intervals. They may be piers of masonry, or upright props or shores of timber, properly braced, and supported on piles, if the foundation be insecure. They will divide the long

Fig. 89.

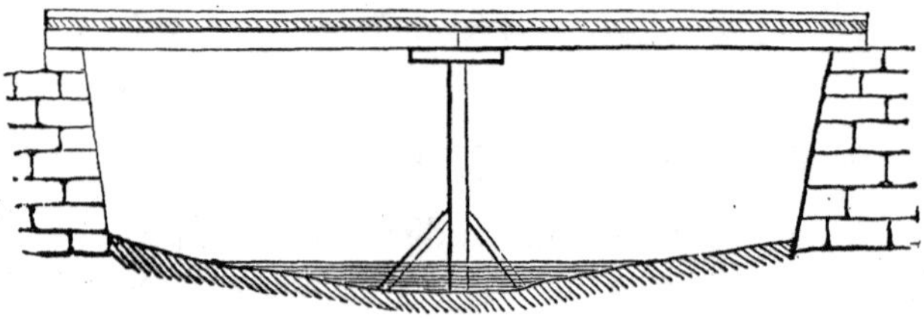

stretch into a number of shorter ones, and support the ends of the timbers by which each of them is spanned.

But if the opening be deep, or occupied by a rapid stream, it is very desirable to avoid the use of any such obstructions. Means must therefore be devised for strengthening the beams, so as to enable them to span larger openings. This may be effected by supports from below, or from above.

Of *supports from below*, the simplest are shorter timbers, (*bolsters*, or *corbels*) placed under the main ones

* Tredgold's Carpentry, p. 148. This gives a great surplus of strength

to which they are firmly bolted, and projecting about one-third of the stretch. This will considerably increase the stiffness.

Fig. 90.

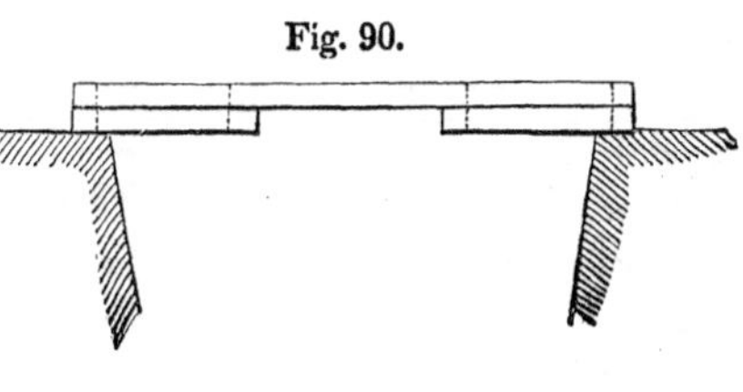

Still more effective are oblique braces, or "struts," supporting the middle of the beam, and resting, at their

Fig. 91

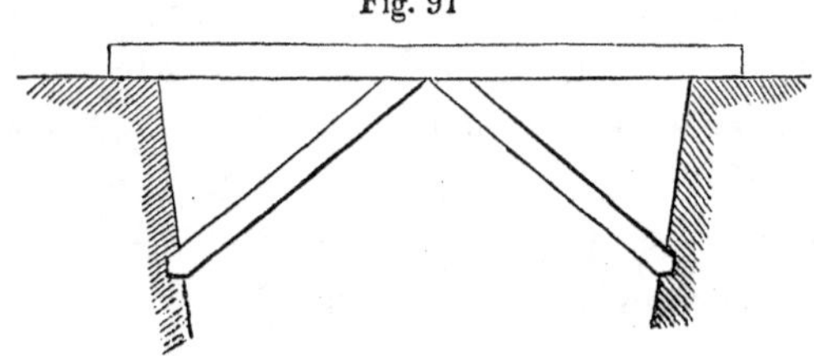

lower ends, in "shoulders," cut into the abutments. Similar braces may be applied to the "bolsters" of Fig. 90.

As the span increased, these braces would become sc

Fig. 92.

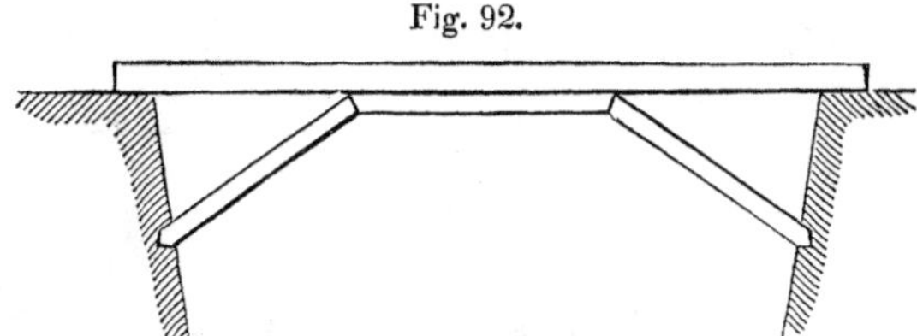

oblique as to lose much of their efficiency. A *straining-piece* is therefore interposed between them. Thirty-five feet may thus be spanned.

For longer stretches, the bolsters, braces, and straining-beams may be combined, as in Fig. 93. The principle ol this method may be extended to very wide openings.

12

Fig. 93.

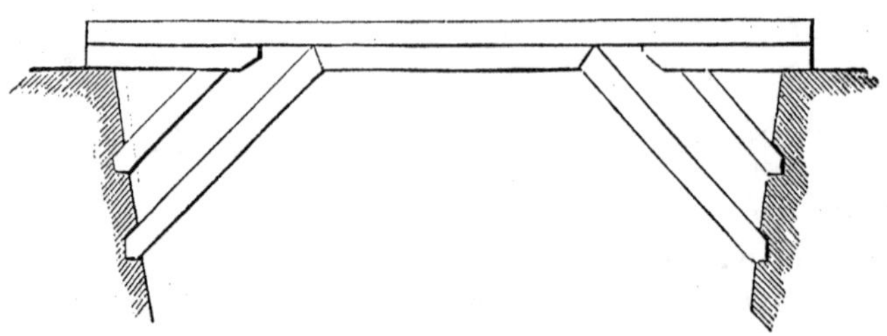

But in many cases supports from below may be objectionable, as exerting too much thrust against the abutments, and being liable to be carried away by freshets, &c. The beams must in such cases be strengthened by *supports from above*

The simplest form of such is shown in Fig. 94, in which the horizontal beam is supported by an upright

Fig. 94.

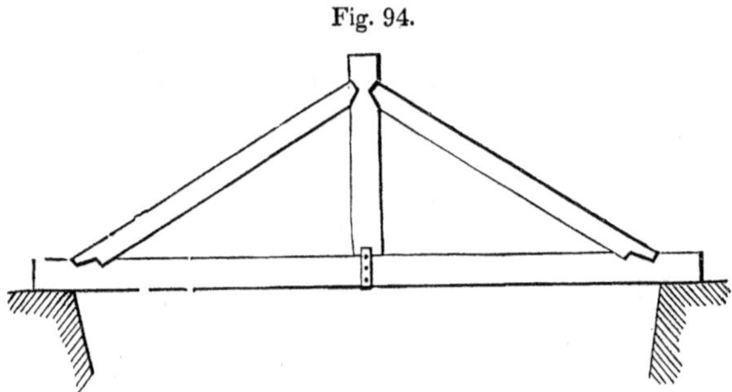

"king-post," to which it is attached by an iron strap, as in the figure, or by the upright "king-post" being formed of two pieces, bolted together, and enclosing the beam between them. The king-post itself is supported by the oblique braces, or "struts," which rest against notches in the horizontal beam.

Since the king-post acts as a suspending tie, an iron

Fig. 95.

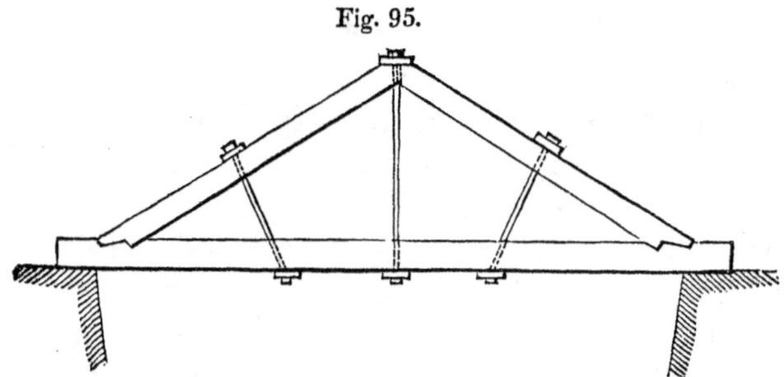

rod may be advantageously substituted for it. The oblique braces may be also stiffened by iron ties, binding them to the main timbers, as in Fig. 95.

For longer stretches, a straining beam may be intro-

Fig. 96.

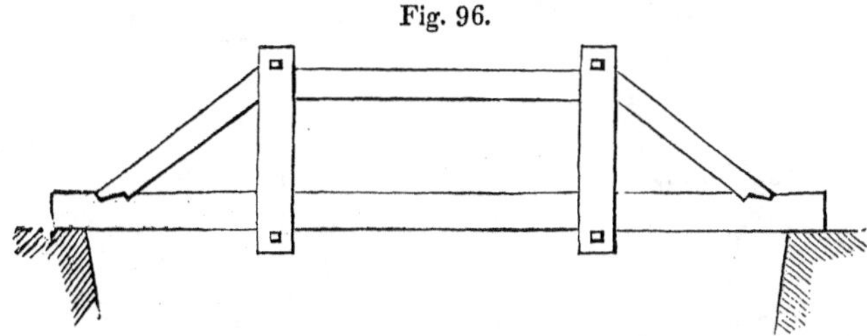

duced between the struts, as in Fig. 96, in which the posts are represented as enclosing the beam.

For bridges of greater span, and more complicated structure, the professional assistance of a civil engineer should be secured. On bridges, see Appendix F. For data and formulas for calculating the strength of beams and trusses, see Gillespie's "Strength of Materials and Stability of Structures."

CULVERTS AND DRAINS.

These structures are necessary for carrying under a road the streams which it intersects. They are also needed to carry the waters of the ditches, from the upper side of a road, to that side on which lie the natural water courses into which they must finally be discharged Their simplest form consists of two walls of stone or brick covered with slabs, and having a foundation, either of wood (if always wet) or of stone, laid in the form of an inverted arch, as shown in cross-section in Figure 97. Their size must be proportioned to the greatest quantity of water which they can ever be required to pass, and should be at least 18 inches square, or large enough to admit a boy to enter to clean them out. Their bottoms should be inclined 1 in 120, or 1 inch in 10 feet. When the road slopes, the inclination of the culvert may be increased, if necessary, by making it cross the road obliquely. At each end flat stones should be sunk vertically, or sheet-piling driven, to guard against the undermining effects of the water. The length of a culvert under an embankment will be equal to the width of the road, increased by the distance on each side, to which the slopes run out, at the depth at which the culvert is placed. At each end of it should be built wing-walls, their tops having an outward and downward slope corresponding to that of the embankment. Their ground plan may be rectangular, trapezoidal, or curved.

Fig. 97.

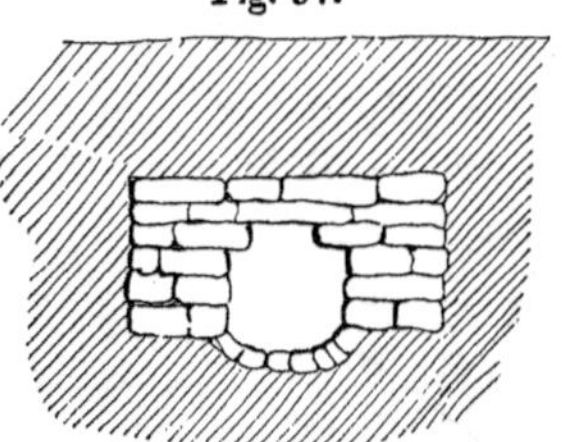

In districts where stone is scarce, a small culvert may

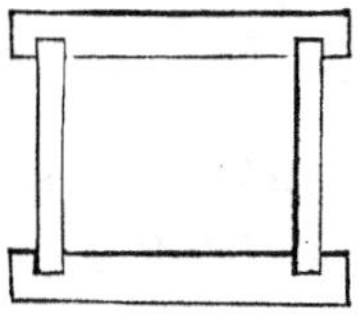

be constructed with four ranges of slabs; grooves being cut in the top and bottom slabs, to receive the upright ones which form the sides.

Fig. 98.

A cheap culvert may be built of brick, with a semicircular arch, of three feet span and 4 inches thick. One thousand bricks will build 26 running feet. If the flow of water be small, the bottom may be merely covered with gravel, over which is then poured grout of hydraulic cement, forming a superficial concrete.

Fig. 99.

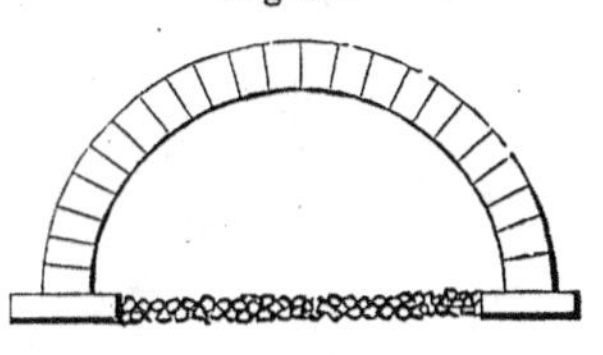

To obtain greater strength, the arch may rest on abutments, sloping inward, and the bottom of the culvert be a flat inverted arch.

Fig. 100.

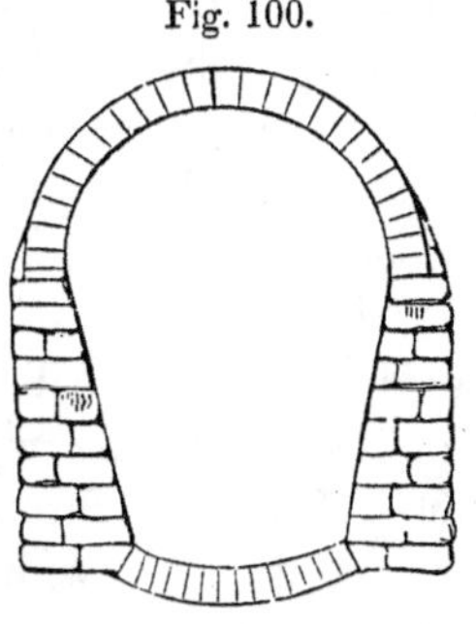

When a road is in excavation, the ditches on either side of it will sometimes require to be covered, to prevent their being filled up by washings from the sides. They may then be formed as in Fig. 97; but spaces of half an inch in width should be left between the covering stones. A layer of brushwood should be placed over these, and the remainder of the ditch filled up to the surface with broken stones, through which the water can filter.*

Similar but smaller drains may be formed at intervals under the road, diverging from its centre like the two

* Parnell, p 95

branches of the letter V, and descending from the angular point to the side-ditches. They are called "mitre drains." In very wet ground, a deep but narrow drain, filled with broken stones, may be carried through the middle of the road.

CATCHWATERS, OR WATER-TABLES.

These are very shallow paved ditches, formed across the road upon a slope, to catch the water which runs down its length, (and which would otherwise furrow up the road-way) and to turn it off into the side-ditches. They are also necessary in the hollows which exist at the points where a descent and ascent meet. They should be so laid that a carriage will not feel any shock in passing over them. Their bottom may be flat, and six feet wide, and for twelve feet on each side they may rise one inch to the foot. The side-slope, down which they discharge their waters, should be also paved. Sometimes for economy they are used as a substitute for a culvert to carry the waters of a small stream across the road; but this is very objectionable, particularly from the ledges of ice which will be there formed in winter. They are sometimes shaped like a V, with the point directed up the ascent, and will then divide the waters. In mountainous situations they should be located obliquely to the axis of the road, and the most advantageous position will evidently be that which has the greatest descent with the least length, and may be geometrically determined.

Let the longitudinal slope of the road descending from A to B, be m to 1; and let its transverse slope from A to C be n to 1; the former being here supposed steeper than the latter. It is required to determine the position of the catchwater AD so that it may have the greatest slope possible.

If a line, BC, be so drawn on the surface of the road as to be horizontal, the desired line of greatest slope, AD, will be perpendicular to it, as explained on page 75. The position of this horizontal line must therefore be first determined. The two points, A and B, which it unites, being on the same level, the descent from A to B equals that from A to C. These descents are expressed respectively by $\frac{AB}{m}$ and $\frac{AC}{n}$, giving the equation, $\frac{AB}{m} = \frac{AC}{n}$; whence $AC = AB \cdot \frac{n}{m}$.

Fig. 101.

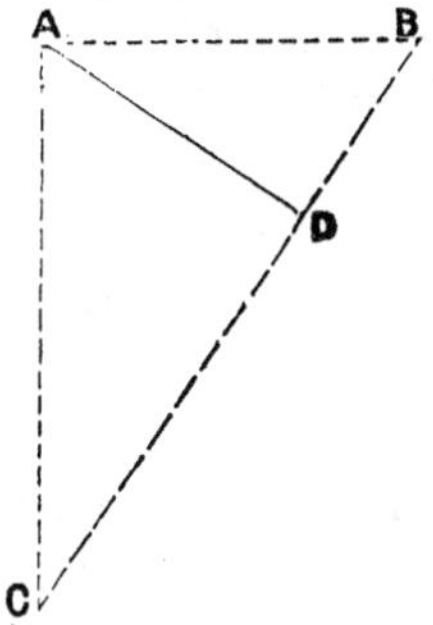

Therefore, to obtain the position AD by a graphical construction, make AB of any length, and set off AC (as given by the equation) at right angles to it; join CB, and from A draw the perpendicular AD, which will be the line required.

If it be required to define the position AD, by the angle BAD, it will be seen that BAD = ACB; and that

$$\text{sin. } ACB = \frac{AB}{CB} = \frac{AB}{\sqrt{(AB^2 + AC^2)}} = \frac{AB}{\sqrt{\left(AB^2 + AB^2 . \frac{n^2}{m^2}\right)}}$$

$$= \frac{1}{\sqrt{\left(1 + \frac{n^2}{m^2}\right)}}.$$

If $m = 20$ and $n = 30$, sin. ACB = .5555, and ACB = 33° 45′.

Care must be taken to avoid placing the catchwater in the direction of the diagonal of the rectangle formed by the four wheels of a carriage; in order to avoid the double shock which would otherwise be caused by two wheels sinking into it at once.

A cheap substitute for a catchwater on a steep slope is a mound of earth, crossing the road obliquely. This will also serve as a resting-place on the ascent. It should be so proportioned, that carriages may pass it without inconvenience.

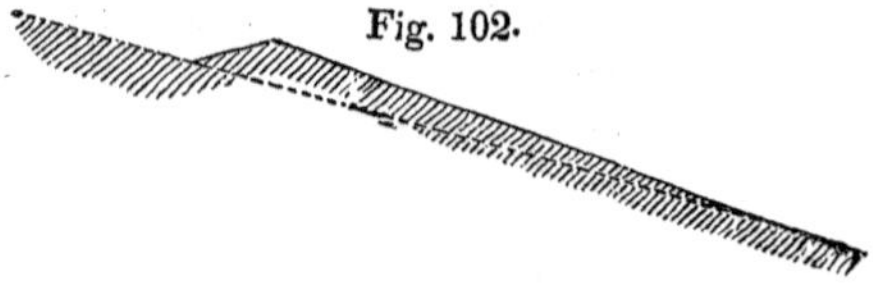

Fig. 102.

RETAINING WALLS.

Retaining, sustaining, revetment, and breast walls, as their various names import, are employed to support masses of earth, and to resist their lateral pressure. Their use, when a road passes along a steep hill-side, has been already explained. In passing through villages also, where land is valuable, a narrower space will suffice for a road in excavation or embankment, if retaining walls be substituted for side-slopes.

The calculation of the necessary thickness for retaining walls, to enable them to resist the thrust of the earth which they are intended to support, is a problem of considerable intricacy of investigation, as well as one of much uncertainty, in consequence of the numerous and greatly varied data required.

When a wall, of which ABCD is a transverse *section*, supports a mass of earth, there is a certain triangular portion, ADE, of the earth, which would slide downward if the wall were removed, and which therefore now presses against

Fig. 103.

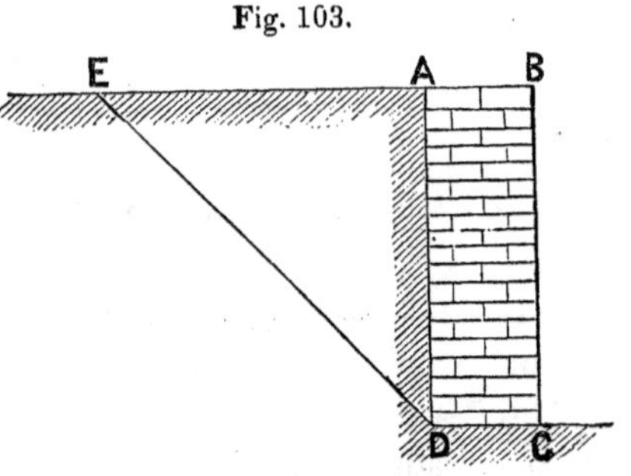

the wall with a force, varying with its height, its specific gravity, and the angle, ADE, at which the earth would stand if unsupported. The wall may yield to its pressure by sliding along its base, or along some horizontal course, or by being overturned and revolving about the exterior edge of one of its horizontal joints. The latter is the only danger to be feared in a well-built wall.

The most complete investigation of the problem of the proper thickness of retaining walls has been made by M. Poncelet in a Memoir,* of which a translation has appeared in the Journal of the Franklin Institute for 1843. It contains valuable tables as well as *formulæ*. Let a denote the angle with the vertical made by the line of the natural slope of the earth, and represented by ADE in the figure. It will vary from 70°, as in the case of very fine dry sand, to 35°, as in the case of heavy clayey earth. Let w denote the weight of any unit of the earth, and w' that of the same unit of the masonry. The specific gravity of the former ranges between 1.4 and 1.9, and that of the latter between 1.7 and 2.5.† The ratio $\frac{w}{w'}$ is therefore usually between $\frac{3}{5}$ and 1. For the simplest case, that in which the embankment does not rise above the wall, the formula‡ for the thickness corresponding to any height H, is

$$\text{Tan.}\ \tfrac{1}{2}\, u \times \tfrac{8}{10} \sqrt{\frac{w}{w'}} \cdot \text{H}.$$

This gives a stability of 1.92 to 1, or nearly double that of a strict equilibrium.

For the usual assumed mean values of $a = 45°$, and $\frac{w}{w'} = \frac{2}{3}$, the formula gives for the required thickness of the wall $\frac{27}{100}$, or a little over a quarter of the height.

* *No.* 13 *du Memorial de l'officier du Génie.* See also PRONY; *Ré cherches sur la Poussée des Terres;* and NAVIER; *Leçons sur l'Appli tion de la Mécanique aux Constructions.*

† Navier. ‡ Poncelet, § 12.

The extreme limits in any case are from $\frac{198}{1000}$, or one-fifth of the height, with compact earth and heavy masonry, to $\frac{452}{1000}$, or not quite half the height, with loose earth and light masonry.* The precise thickness can be calculated by the preceding formula; after noting the slope at which the earth naturally stands, and weighing a certain portion of the masonry, and of the earth previously thoroughly moistened.

When there is an embankment rising above the top of the wall, the proper thickness (in cases in which the height of the superincumbent load does not much exceed the height of the wall) may be approximately obtained by substituting in the same *formula*, instead of the height of the wall, the sum of the heights of the wall and of the earth above it.†

Thus far both faces of the retaining wall have been supposed to be vertical. But the same strength with a less amount of material may be obtained by various modifications of its section.

Fig. 104.

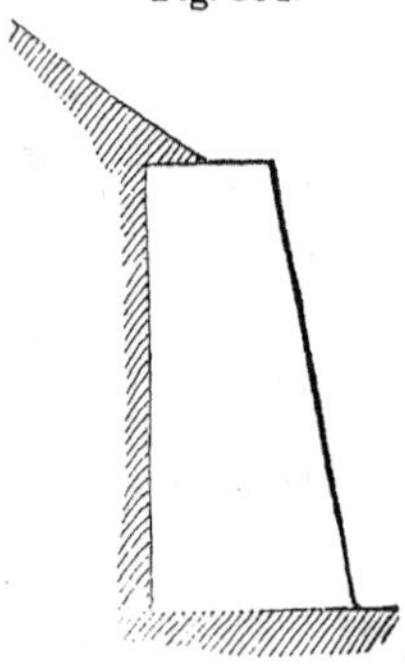

The face of the wall may be advantageously made to slope with a "*bâtir*," varying from $\frac{1}{24}$, or $\frac{1}{2}$ inch horizontal to 1 foot vertical, to $\frac{1}{6}$, or 2 inches to 1 foot.

To find the mean thickness of such a wall, which shall have the same stability as another wall with vertical faces, and of the thickness obtained by the preceding rules, subtract from this given thickness four-tenths of the entire projection of the *bâtir*.‡ Thus, if the given thickness be 4 feet, and the height 24 feet, and the corresponding mean thickness of a wall with

* Poncelet, § 34. † Ibid. § 22 ‡ Ibid. § 72.

a *bâtir* ot $\frac{1}{12}$ be desired, it will be $4 . - \frac{4}{10} \times \frac{24}{12} = 4 . - .8 = 3.2$. The *bâtir* is supposed not to exceed one-fifth of the height From the *mean* thickness, those of the top and bottom are readily deduced, knowing tho height and *bâtir*.

Fig. 105.

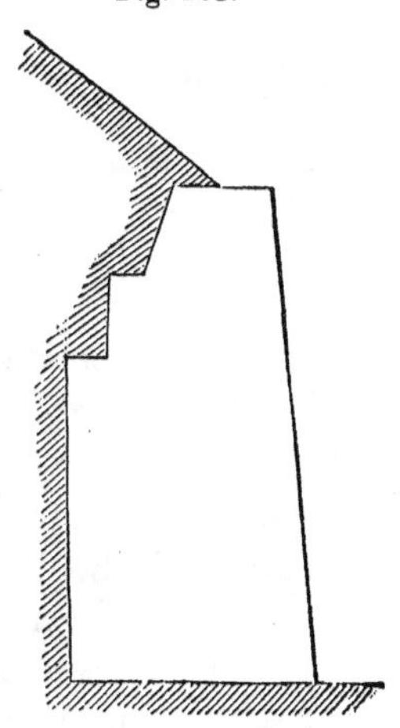

The desired increase of thickness towards the bottom of a wall s often given by offsets at its back. Considerable resistance to the overturning of the wall is offered by the weight of the earth which rests upon these offsets.

Fig. 106

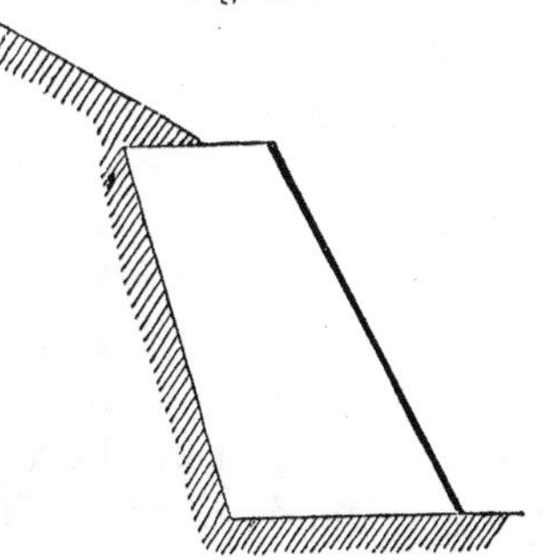

Still more economical of masonry is a leaning retaining wall, in which the back has a *bâtir*, which may advantageously be 1 in 6. In this case strength requires that the perpendicular let fall from the centre of gravity of the section upon the base, should fall so far within the inner edge of the base, that the stone of the bottom course of the foundation may present sufficient surface to bear the pressure upon it.

---

* Mahan, p. 142.

The strength of a wall may be still farther increased by lessening its thickness, and employing the difference of the amount of masonry in buttresses or counter-forts, attached to its back at regular intervals, and firmly banded with it. The trapezoidal section for them is preferred, as giving a broader base of union. Fig. 107 is a ground plan of such an arrangement.

Fig. 107.

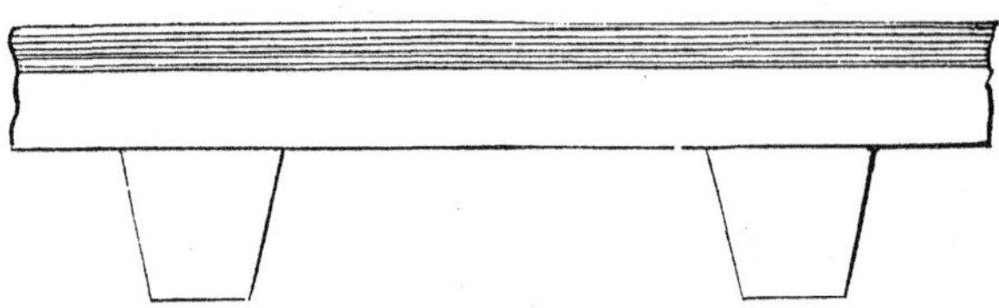

To lessen the pressure of an embankment, that portion of it next the wall should be formed in compact layers, inclining downward from the wall. Through the wall should be left holes (*barbacanes*) six inches high and three wide, disposed, in the *quincunx* form, at distances of six feet horizontally, and four feet vertically, in order to give vent to the water which may filtrate through the bank.

The masonry of a wall which has to sustain great pressures, requires much attention. The following is part of the specification for such walls of rubble masonry on the public works of the state of New York. "The stone shall be sound, well-shaped, and durable, and of not less than 6 inches in thickness, and three feet area of bed. The smoothest and broadest bed shall in all cases be laid down, and if it be rough and uneven, all projecting points shall be hammered off; and the same from the top bed, so as to give the succeeding stone a firm bearing. In all cases the bed shall be properly prepared, by levelling up, before the next stone is laid, but no levellers shall be placed under a stone by raising it from its bed. One-fourth of the wall shall be composed of *headers*, which shall extend *through* the wall, where it is not more than two feet thick, and from 2 to 4 feet back for thicker walls. The whole shall be laid in hydraulic mortar, composed of the best quality of cement, and clean sharp sand; and particular care shall be taken to have each stone surrounded with mortar, and thoroughly bedded in it."

CHAPTER IV.

IMPROVEMENT OF THE SURFACE.

"Next to the general influence of the seasons, there is perhaps no circumstance more interesting to men in a civilized state, than the *perfection* of the means of interior communication."

Committee of House of Commons, 1819.

THE surface of a newly-made road is generally very deficient in the important qualities of hardness and smoothness, and to secure these attributes in their highest attainable degree, it is necessary to cover the earth, which forms the natural surface of the road, with some other material, such as stone, wood, &c. The benefits of such a process are twofold, consisting,

1. In substituting a hard and smooth surface for the soft and uneven earth;

2. In protecting the ground beneath it from the action of the rain-water, which, by penetrating to it, and remaining upon it, would not only impede the progress of vehicles, but render the road too weak to bear their weight.

Such a covering should be regarded, not as an arch to bear the weight of the vehicles, but simply as a roof, to protect the earth beneath it from the weather; not as a substitute for the soil under it, but only as a protection to that soil to enable it to retain its natural strength. Erroneous views on this point have caused very prejudicial practices, particularly in the case of broken stone, or McAdam-roads.

The various surfaces will be considered in the following order; beginning with the most imperfect, that of the unimproved earth, and ending with the most perfect yet attained—that of Railroads.

1. EARTH ROADS.
2. GRAVEL ROADS.
3. BROKEN STONE, OR McADAM ROADS.
4. PAVED ROADS.
5. ROADS OF WOOD.
6. ROADS OF OTHER MATERIALS.
7. ROADS WITH TRACKWAYS.

---

1. EARTH ROADS.

Roads of earth, with the surfaces of the excavations and embankment unimproved by art, are very deficient at all times in the important requisites of smoothness and hardness, and in the spring are almost impassable. But with all their faults, they are almost the only roads in this country, (the scantiness of labor and capital as yet preventing the adoption of better ones) and therefore no pains should be spared to render them as good as their nature will permit.

The faults of surface being so great, it is especially necessary to lessen all other defects, and to make the road in all other respects as nearly as possible "what it ought to be." Its grades should therefore be made, if possible, as easy as 1 in 30,* by winding around the hills, or by cutting them down and filling up the valleys. Its shape should be properly formed with a slope of 1 in 20† each

---

* See page 41. † Page 51

way from the centre. Its drainage should be made very thorough, by deep and capacious ditches, sloping not less than 1 in 125,* in accordance with the minimum road slope. Drainage alone will often change a bad road to a good one, and without it no permanent improvement can be effected. Trees should be removed from the borders of the road, as intercepting the sun and wind from its surface.

If the soil be a loose sand, a coating of six inches of clay carted upon it, will be the most effective and the cheapest way of improving it, if the clay can be obtained within a moderate distance. Only one-half the width need be covered with the clay, thus forming a road for the summer travel, leaving the other sandy portion untouched, to serve for the travel in the rainy season.

If the soil be an adhesive clay, the application of sand in a similar manner will produce equally beneficial results. On a steep hill these improvements will be particularly valuable.

When the road is worn down into hollows, and requires a supply of new material, its selection should be made with great care, so that it may be as gravelly as possible, and entirely free from vegetable earth, muck, or mould No sod or turf should ever be allowed to come upon the road, to fill a hole or rut, or in any other way; for, though at first deceptively tough, they soon decay, and form the softest mud. Nor should the roadmaker run into the other extreme, and fill up the ruts and holes with stones, which will not wear uniformly with the rest of the road, but will produce hard bumps and ridges. The plough and the scraper should never be used in *repairing* a road. Their work is large in quantity, but very bad in quality. The

---

* See page 54.

plough breaks up the compact surface, which time and travel had made tolerable; and the scraper drags upon the road from the side ditches the soft and alluvial matter which the rains had removed, but which this implement obstinately returns to the road.

A very good substitute for the scraper, in levelling the surface of the road, clearing it of stones, and filling up the ruts, consists of a stick of timber, shod with iron, and attached to its tongue or neap obliquely, so that it is drawn over the road "quartering," and throws all obstructions to one side. The stick may be six feet long, a foot wide, and six inches thick, and have secured to its front side a bar of iron descending half an inch below the wood.

Every hole or rut in a road should be at once filled up with good materials, for the wheels fall into them like hammers, deepening them at each stroke, and thus increasing the destructive effect of the next wheel.

EFFECT OF WHEELS ON THE SURFACE.

The effects of broad and narrow wheels upon roads have been much discussed, and many laws enacted to encourage the use of the former. Upon a hard and well-made road, (such as one of broken stone) there is little difference between them, but on a common earth road, narrow wheels, supporting heavy weights, exercise a very destructive cutting and ploughing action. This diminishes as the width of the felloe increases, which it may do to such an extent, that the wheel acts as a roller in improving, instead of injuring, the surface. For these rea sons the New York turnpike law enacts that carriages, having wheels of which the tire or track is six inches wide, shall pay only half the usual tolls; those with wheels nine inches wide, only one-fourth; and that those with twelve inches shall pay none at all. The proportions agree precisely with those deduced from observation by an experienced English roadmaker.* The felloe should have a flat bearing surface and not a rounded one. The benefits of broad wheels are sometimes destroyed by overloading them. To prevent this, when tolls are collected, they should be increased, for each additional horse, more rapidly than the direct proportion; thus, if one horse paid 5 cents, two should pay 11, three 17, &c. Narrow wheels are particularly injurious when in rapid motion, for having less resistance and greater velocity than others, they revolve less perfectly, and *drag* more, thus producing the worst sort of effect. Conical wheels, of which the inner is greater than the outer circumference, tend to move in a curve, and being forced to proceed in a right line, exert a peculiarly destructive grinding action on the road. On McAdam roads, horses' feet exercise a more destructive effect than the wheels of vehicles. It has been calculated† that a set of tires would run 2700 miles in average weather, but that a set of horses' shoes would bear only 200 miles of travel.‡

* Penfold, p. 22. † Gordon on Locomotion.

‡ The imperfect surface of an earth road makes it doubly important to take every precaution to lessen the friction of vehicles upon it. The resistance decreases as the *breadth of the tire* increases, on compressible roads, as earth, sand, gravel, &c.; while on paved and broken-stone roads, the resistance is nearly independent of the breadth of the tire.* *Cylindrical* wheels also cause less friction than conical ones. The *larger* the wheels the less friction have they, and the greater power of leverage in overcoming obstacles. The fore-wheels should be as large as the hind ones, were it not for convenience of turning. *The axles should be straight,* and not bent downward at the end, which increases the friction, though it has the advantage of throwing the mud away from the carriage. The *load* should be placed *on the hind wheels* rather than on the fore ones

* Morin p. 339.

2. GRAVEL ROADS.*

The roundness of the pebbles, which form the chief part of gravel, whether from rivers or pits, prevents them from perfectly consolidating, except under much travel; but still a gravel road, properly made, is far superior to one of common earth. Gravel from the shores of rivers is too clean for this object, and does not contain enough earthy matter to unite and bind together its pebbles, which are too perfectly water-worn, and freed from asperities. On the other hand, gravel dug from the earth contains too much earth, which must be sifted from it before use. Two sieves should be provided, through which the gravel is to be thrown. One should have wires, an inch and a half or two inches apart, so that all pebbles above that size may be rejected. The other should have spaces of three quarters of an inch, and the material which passes through it should be thrown away, or employed for foot-paths. The expense of sifting will be more than repaid by the superior condition of the road formed by the purified material, and the diminution of labor in keeping it in order.

The road-bed should be well shaped and drained. If it is rock, all projecting points should be broken off, and a layer of earth, a foot thick, should be interposed, or the gravel will wear away much more rapidly, and consolidate much more slowly.

Long and pliant *springs* greatly lessen the shock of passing over obstacles, and their advantage has been stated to be equal to one horse in four The *line of draught should ascend* at an angle of 15 degrees, so that when the horse leans forward in pulling, his force will be exerted nearly horizontally

* Parnell, p. 170. Penfold p. 13. Amer. Railroad Journal, vol. ii. p. 4.

A coating of *four inches* of gravel should be spread over the road-bed, and vehicles allowed to pass over it til it becomes tolerably firm, and is nearly, but not entirely, consolidated; men being stationed to continually rake in the ruts, as fast as they appear. A second coating of 3 or 4 inches should then be added and treated like the first; and finally a third coating. A very heavy roller drawn over the road will hasten its consolidation. Wet weather is the most favorable time for adding new materials.

A very erroneous practice is that of putting the larger gravel at the bottom, and the smaller at the surface; for, from the effects of the frost, and of the vibration of carriages, the larger stones will rise to the surface and the smaller ones descend, like the materials in a shaken sieve, and the road will never become firm and smooth.

3. BROKEN-STONE ROADS.

Broken-stone roads have been the subjects of violent partisanship on many disputed points, and the most important of these questions relates to the propriety or necessity of a paved foundation beneath the coating of broken stones. McAdam warmly denies the advantages of this, while Telford supports and practises it. Broken-stone roads may therefore be conveniently divided into *McAdam roads* and *Telford roads*.

McADAM ROADS.

Mr. McAdam, who first brought into general use in England roads of broken stone, and from whom they derive their popular name, is said* to have deduced the

* Millington, p. 234.

leading principles of his improved system from his observation of the passage of a heavy vehicle, such as a loaded stage-coach, over a newly-formed gravel road. The wheels sink in to a considerable depth, and plough up the road, in consequence of the roundness of the pebbles, which renders them easily displaced. Hence ensues great friction against the wheels ; which, moreover, are always in hollows with little hills of pebbles in front of them, which they must roll over or push aside. The evil continues, until at last, after long-repeated passages of heavy vehicles, the pebbles have become broken into angular fragments, which finally form a compact mass.

But since this is so desirable a consummation, the task of breaking the stones ought not to be imposed on the carriages, but should be performed in advance by manual labor, by which it will be executed far more speedily, effectually, and completely.

Hence is deduced the leading principle of the system, viz.: that *the stones should be all broken by hand into angular fragments* before being placed on the road, and that no rounded stones should ever be introduced.

In the next place, whenever a carriage-wheel, or horse's hoof, falls eccentrically on a large stone, it is loosened from its place, and disturbs the smaller ones for a considerable distance around it, thus preventing their consolidation. Therefore *no large stones should be ever employed.*

Small angular stones are the cardinal requisites. When of suitable materials of proper size, and applied in accordance with the directions which will be presently given, they will unite and consolidate into one mass, almost as solid as the original stone, with a smooth, hard, and unelastic surface.

We will examine successively the proper quality of stone to be used; the size to which they should be broken; the manner of breaking them · the thickness of the coating; the best method of applying the stone; of rolling the road; of keeping it in order; and of repairing it when in bad condition.

THE QUALITY OF THE STONE.

The materials employed for a broken-stone road (often called the "Road metal") should be at the same time *hard* and *tough*. "*Hardness* is that disposition of a solid which renders it difficult to displace its parts among themselves; thus, steel is harder than iron, and diamond almost infinitely harder than any other substance in nature. The *toughness* of a solid, or that quality by which it will endure heavy blows without breaking, is again distinct from hardness, though often confounded with it. It consists in a certain yielding of parts with a powerful general cohesion, and is compatible with various degrees of elasticity."*

Some geological knowledge is required to make a proper selection of the materials. The most useful are those which are the most difficult to break up. Such are the *basaltic* and *trap* rocks, particularly those in which the *hornblende* predominates. The *greenstones* are very variable in quality.† *Flint* or *quartz* rocks, and all pure *silicious* materials, are improper for use, since, though hard, they are brittle, and deficient in toughness. *Granite* is generally bad, being composed of three heterogeneous

---

* Sir John Herschel. "Discourse on the study of Natural Philosophy"

† The greenstone of Bergen and Newark mountain (near New York) is good; that of the eastern face of the Palisades above Weehawken is too liable to decomposition. (Renwick, Pract Mechanics, p. 145.)

materials, quartz, felspar, and mica, the first of which is brittle, the second liable to decomposition, and the third laminated. The *sienitic* granites, however, which contain hornblende in the place of felspar, are good, and better in proportion to their darkness of color. *Gneiss* is still inferior to granite, and *mica-slate* wholly inadmissible. The *argillaceous slates* make a smooth road, but one which decays very rapidly when wet. The *sandstones* are too soft. The *limestones* of the carboniferous and transition formations are very good; but other limestones, though they will make a smooth road very quickly, having a peculiar readiness in "binding," are too weak for heavy loads, and wear out very rapidly. In wet weather they are also liable to be slippery. It is generally better economy to bring good materials from a distance than to employ inferior ones obtained close at hand. Excellent materials may be found throughout the primary districts of the United States. In the tide-water regions, south of New York, boulders, or rolled pebbles, must be employed.

As the harder stones cost much more to break than the softer ones, the lower courses of the road may be formed of the latter, and the former reserved for covering the surface, which has to resist the grinding action of the wheels.*

In alluvial countries, where stone is scanty and wood plenty, an artificial stone may be formed by making the clay into balls, and burning them till they are nearly vitrified. The slag, or refuse, of iron furnaces, makes an excellent material. The stony or slaty part of coal may

---

* This is the practice on the avenues of New York; broken gneiss being put below, and covered with broken boulders, which cost three times as much to break

be used near collieries. Cubes of iron have been imbedded among the stones with some advantages.*

SIZE OF THE STONE.

The stone should be broken into pieces, which are as nearly cubical as possible, (rejecting splinters and slices) and the largest of which, in its longest dimensions, can pass through a ring *two and a half inches* in diameter. In reducing them to this size, there will of course be many smaller stones in the mass. These are the proper dimensions, according to *Telford* and *Parnell*.† *Edgeworth* prefers $1\frac{1}{2}$ inches. *Penfold*‡ names two inches for brittle materials. If smaller they would crush too easily; but on the other hand, the less the size of the fragments, the smaller are the interstices exposed to be filled with water and mud. The tougher the stone, the smaller may it be broken. The less its size, the sooner will it make a hard road; and for roads little travelled, and over which only light weights pass, the stones may be reduced to the size of one inch.

Fig. 108

McAdam argues that the size of the stone used on a road must be in due proportion to the space occupied on a smooth level surface, by a wheel of ordinary dimensions; and, as it has about an inch of contact longitudinally, therefore every stone in a road exceeding *one inch* in diameter, is mischievous; for the one-sided bearing of the wheel on a larger stone will tend to turn it over and to loosen the neighboring materials. But this argument proves too much; for however small the stone is, there must be a moment, just as the wheel is leaving it, when the pressure is one-sided, and therefore tends to overturn it. Subsequently McAdam preferred the standard of

* Parnell, p. 245. † Ibid. p. 133 ‡ Pages 14, 15

weight to that of size, and made six ounces the maximum, (corresponding for average materials to cubes of $1\frac{1}{2}$ inches, or $2\frac{1}{2}$ inches in their longest diagonal) directing his overseers to carry a pair of scales and a 6-oz. weight, with which to try the largest stones in a pile. The weight standard has the advantage, that the stones are smaller as they increase in specific gravity, to which the hardness is generally proportional. He subsequently says that he had "not allowed any stone above *three* ounces in weight (equal to cubes of $1\frac{1}{4}$ inches, or 2 inches in their longest diagonal) to be put on the Bath and Bristol roads for the last three years, and found the benefit in the smoothness and durability of the work as well as economy of repairs."* On examining old roads he found that the average size of the stones varied from seven to twenty-seven ounces in weight, and that "the state of disrepair and the amount of expense on the several roads was in a pretty exact proportion to the size of the material used."† The French engineers value *uniformity* of size much less than McAdam, and call it "rather an evil than a good." They therefore use equally all sizes from $1\frac{1}{2}$ inches to dust.‡

Fig. 109.

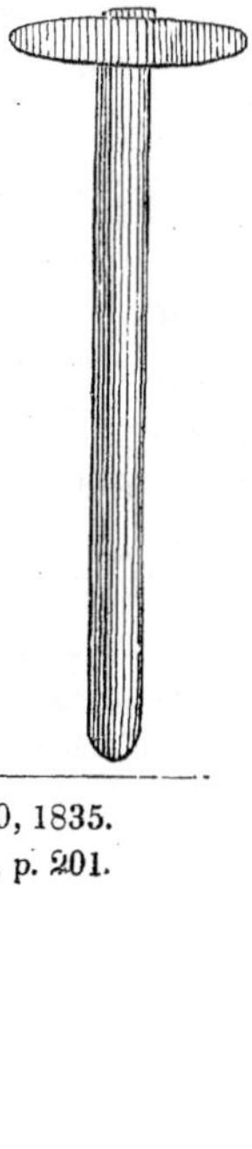

BREAKING THE STONE.

The weight and shape of the hammer, and the manner of using it, are of much importance, making a difference of at least 10 per cent. The head of the hammer should be six inches long, and weigh about one pound; and the handle be tough and flexible, and 3 feet long, if used standing, or 18 inches, if used sitting, which is better. The laborer sits before the pile, and breaks the stones on it, or on a large concave stone as an anvil, on which the stones to be bro-

* Letter of 1834, in Am. Railroad Journal, Jan. 10, 1835.

† System of Roadmaking, 1825.

‡ Gayffier, p. 201.

ken are placed, resting only on their ends, so that, being struck sharply in their middle, they break into angular fragments. Children with smaller hammers can do the lighter work, so that a whole family may be employed. The workmen should not be paid by the day, but at an equitable price per cubic yard. A medium laborer can break in a day from 1½ to 2 yards of gneiss ; but only ½ to ¾ yard of hard boulders, or "cobble-stones."

THICKNESS OF THE COATING.

Twelve inches of well consolidated materials on a good bottom, will be sufficient for roads of the greatest travel, and will resist all usual weights, and frosts. In the climate of France, ten inches is considered enough for the most frequented roads, and six or eight inches for others. The thickness should vary with the soil, the nature of the materials, and the character of the travel over it ; it should be such that the greatest load will not affect more than the surface of the shell ; and it is for this purpose chiefly that thickness is required, in order that the weight which comes on a small part only of the road may be spread over a large portion of the foundation. The severe frosts of our northern states require the maximum of depth.*

McAdam advocates less thickness than the other English constructors. He considers from 7 to 10 inches sufficient, calling the latter depth of "well consolidated materials equal to carry any thing." He adds, "some new roads of six inches in depth were not at all affected by a very severe winter ; and another road having been allowed

* Stone broken into fragments of from 1 to 6 inches occupies twice as much space as in the original solid state ; but the broken stone placed upon the road is reduced by the pressure of the wheels to two-thirds of its former bulk, or more exactly seven-tenths.

to wear down to only three inches, this was found sufficient to prevent the water from penetrating, and thus to escape any injury by frost." He earnestly advocates the principle that the whole science of artificial road-making consists in making a solid dry path on the natural soil, and then keeping it dry by a durable water-proof coating. 'The broken stone is only to preserve the under road from moisture, and not at all to support the vehicles, the weight of which must be really borne by the native soil, which, while preserved *dry*, will carry any weight, and does in fact carry the stone road itself as well as the carriages upon it." . . . "The stone is employed to form a secure, smooth, water-tight flooring, over which vehicles may pass with safety and expedition at all seasons of the year." . . . "Its thickness should be regulated only by the quantity of material necessary to form such a flooring, and not at all by any consideration as to its own independent power of bearing weight." . . . "The erroneous idea that the evils of an undrained wet clayey soil can be remedied by a large quantity of materials, has caused a large part of the costly and unsuccessful expenditures in making broken-stone roads."*

APPLICATION OF THE MATERIALS.

The road-bed, having been thoroughly drained, must be properly shaped and sloped each way from the centre, so as to discharge what water may penetrate to it, and not, as is often practised, be made level, and the crowning given by a greater thickness of stone in the middle. Upon this bed, a coating of *three* inches of the clean broken stones, free from any earthy mixture, is to be spread

* McAdam—" System of Road-making," *passim.*

on a dry day. The travel is then to be admitted on it, men being stationed to rake in the ruts as soon as formed, or a heavy roller used, till it becomes almost consolidated, but not completely so, (the determination of this time being a nice and important practical point) and a second coat of three inches is then to be added during a wet time, as moisture greatly facilitates the union of the two. A third coat is added as was the second, and a fourth if that be required. If the stone be very hard, and the wheeling very difficult, fine clean gravel, free from earth, may be spread over the surface; but it is better for the future solidity of the road to dispense with this, if possible.

If a thick coat be laid on at once, there is a very great destruction of the material before it becomes consolidated, if it ever does so. The stones will not allow one another to be quiet, but are continually elbowing each other, and driving their neighbors to the right and to the left. This constant motion rapidly wears off the angular points, and reduces the stones to a spherical shape, which, in conjunction with the amount of mud and powder produced, destroys the possibility of any firm aggregation, and the road never attains its proper condition of hardness.*

The broken stones need not be spread over a greater width than from 12 to 16 feet, (except near large cities) and "wings" of earth may be left on each side. For a road little used a single track of 8 feet of the "metal" will suffice.†

The perfect *cleanliness* of the stones is strongly insisted on by McAdam. He directs the broken stones to be very carefully kept perfectly free from any mixture of earth, or any matter which will imbibe water, or be affected by frost; since roads

---

* Penfold, p. 15. † See page 47.

made with such a mixture become loose in wet weather, and allow the wheels of carriages to displace the materials, and to cut through to the original soil, thus making the roads rough and rutty, the admission of water being the great evil. He adds that nothing must be laid on the clean stone under the pretence of "binding;" for clean broken stone will combine by its own angles into a smooth solid surface, which cannot be affected by vicissitudes of weather, nor displaced by the action of wheels.

The French engineers consider this cleanliness as unnecessary, since the travelling on the road very soon pulverizes the materials, and fills the interstices with dust and mud; though it might be replied that this took place only on the surface. Some of them, observing the large amount of vacant space in a mass of broken stone,* have even proposed to combine with it in advance a certain proportion of calcareous stone,† or even clay and sand.‡ just sufficient to fill up the existing vacancies. This would doubtless make a road *tolerably* fit for use much sooner than the regular plan, but its permeability to water would entail on it all the evils mentioned in the preceding paragraph.

---

* A cubic *metre* of broken stones, placed in a water-tight box, which they just fill, can receive in the empty spaces between the fragments a volume of water $= \frac{48}{100}$, or nearly one-half of the whole the actual solidity of the stones being therefore only $\frac{52}{100}$. This does not vary for stones from 1 to 8 inches in size. After prolonged travel it increases to $\frac{71}{100}$, leaving a void of only $\frac{29}{100}$. For rolled pebbles and sand the actual solidity may be as much as $\frac{62}{100}$. For perfect spheres, calculation shows that the solidity of a mass of them increases as their diameter decreases. Thus, if a cubic metre be filled with spheres 4 inches in diameter, their solid volume will be $\frac{63}{100}$; if they are 1 inch in diameter their volume is $\frac{70}{100}$; and if only $\frac{1}{25}$ inch it is $\frac{74}{100}$. Pebbles by theory, as well as by the experiment above cited, would be intermediate between broken stones and spheres.—(Gayffier, pp. 204 to 214.)

† M. Polonceau. M Girard de Caudemberg.

ROLLING

The use of a very heavy roller will much facilitate the consolidation of the road. A plan highly recommended is to have a roller made of a hollow cylinder, of cast iron, or covered with iron bands, seven feet in diameter, and five feet long. A strong axle passes through its length. Its ends are closed, and two interior partitions, perpendicular to the axis, divide it into three equal chambers. A longitudinal band of the surface, a foot wide, can be detached, so as to give access to the interior spaces, which are filled with gravel, one or all of them, according to the weight desired. The empty cylinder weighs 7000 lbs.; each compartment filled with gravel adds 4,000 lbs. to the weight; so that the entire weight may be made successively 7,000 lbs., 11,000 lbs., 15,000 lbs., and 19,000 lbs. To compress a new road, ten or twelve strong horses should be attached, on a wet day in summer, to the *empty* roller, and draw it several times over every part of the road, till the materials have been so far compressed as not to form a ridge in front of the roller. Then the middle division is to be filled with gravel, (moistened, to give it solidity) and the rolling resumed till the draught is so much lessened that the end divisions can be filled, the middle one being emptied at first if necessary. There should be an excess of power in the horses, so that they may do ess injury by the violent pressures of their feet. Every part of the road should be passed over from 40 to 100 times. To increase the stability of the compression obtained, an inch of gravel should be spread over the surface and passed over by the roller a few times. If the weathei

---

* Gayffier, p. 212

be dry, the surface should be watered. The season should be summer, that the road-bed may be dry, and the day be wet, to ensure a moist surface, which facilitates the binding of the materials.

When the rolling has finished the compression, the road is still very different from one which has borne the traffic of many years; for although the materials are strongly pressed against one another, and have taken a stable position, they have not acquired the adhesion which takes place after a series of years. The new road, therefore, needs for some time most careful attention. The travel must finish it by being forced to pass over every part of it uniformly, heaps of pebbles being placed very irregularly, so as to direct the vehicles successively on all the points of the road. Every rut, and the slightest hollows and elevations, must be promptly removed by a liberal supply of laborers, whose work will, however have been greatly lessened by the previous rolling. They must rake over every inequality of surface the moment that it is formed.

KEEPING UP A ROAD.

This is a very different thing from "repairing a road," though the two are often confounded. A due attention to the former will greatly lessen the necessity for the latter. The former keeps the road always in good condition; the latter makes it so only occasionally, after intervals of various length, during which it is continually deteriorating in a geometrical ratio, so that *the better the state* in which the road is kept, the less are the injuries to it, and therefore, *the less the expense* of keeping it in this excellent condition.

"Keeping up the road" requires the daily attention of

a permanent corps of laborers. Supposing the road to be already in good condition; that is, in proper shape, and free from holes, ruts, mud, and dust; to keep it so, requires two fundamental operations:

1. The continual removal of the daily wear of the materials, whether in the shape of mud or of dust;

2. The employment of materials to replace those removed.

The first operation requires hoes and brooms. The hoes should be three feet long, and of wood, as iron ones would be more likely to loosen the stones. The lighter dust and more liquid mud must be swept off by birch brooms. The detritus between the little projections of the stones should not be removed by too thorough sweeping, as it protects them from immediate crushing, and preserves their stability. The broom is also necessary to remove every trace of wheels, the moment they have passed, so as to oppose that habit or instinct of horses which leads them to follow in the track of the preceding vehicle, and which would soon convert unremoved tracks into ruts. The broom and hoe have then a double end to be accomplished by the same operation, viz., effacing tracks and removing detritus. Very effective machines have also been constructed for accomplishing these purposes.*

The second operation of applying new materials demands several precautions. To prevent a weak place from being neglected because the materials are not at hand, they should be kept in *dépôts*, never more than a quarter of a mile apart, and carried thence in barrows. They should be applied after a rain, as then they will more easily unite, and no coat, thicker than one stone,

---

* Roads and Railroads, p. 91

should ever be applied at any one time. A cubic yard to a superficial rod will be quite enough at once. They will then soon become incorporated without having their angles worn out by motion, and will be of as much service as double the thickness applied at once. To avoid retarding the travel and increasing the draught too much, a new coat should not be put on any continuous space larger than six or seven square yards. If several depressions are found very near each other, cover the worst, and attend to the next after the first has become solid. The ruts which are formed should not be filled with loose stone, for this would make longitudinal ridges of harder material, but "the laborer should work the rake backwards and forwards on each side of the rut and across it; and if he do it with his eyes shut, he will do more good, than by taking pains to gather all the stones he can find to place in it."*

The number of men required by this system of constant watchfulness may at first seem an objection to it, but the expense will be amply repaid by the advantages obtained. Each laborer should have a certain length of the road assigned to his especial care, and the most intelligent and trustworthy among them should be made inspectors over the others for a certain distance. At times unfavorable for work on the road, they should be employed in breaking stone. The labor of one man will keep in repair three miles of well-made and well-drained road, for the first two years after its formation, and four miles for the next two years, by constantly spreading loose stones in the hollows, raking them from the middle to the sides, opening the ditches, &c. In the fifth year

---

* Penfold, p. 20.

some repairs, "with lifting," may be necessary, as explained under the next head.*

It will be seen by Morin's table, on page 63, that the friction or resistance to draught on a road with deep ruts and thick mud, is four times as great as on one in good order. This shows the importance of very perfectly "keeping up" the road. An incidental advantage is that the prompt removal of the mud after every shower will prevent the annoyance of dust, so general an objection to McAdam roads, but not at all their necessary concomitant.

Where the materials of the road are very brittle stone, they wear away very rapidly in dry weather, and their consumption may be much lessened by *watering* the road judiciously; not so little as to form a crust which adheres to the wheel, nor so much as to make the draught heavy. A moderate use of the watering cart preserves the materials from pulverization, and keeps them settled in their places, at the same time that the comfort of the traveller is greatly enhanced. This is particularly necessary on roads in this country during our hot and dry summers; for after a long drought the crust of the road sometimes becomes so dried out that it ceases to "bind," and permits loose stones to be detached from it, to the great injury of the surface. An excess of moisture must, however, be avoided, since it increases the grinding power of the pulverized stones, as marble is sawn and jewels are cut with their own powder combined with water.

The question may arise, whether the materials thus gradually added to the road, for alimentation rather than reparation, are sufficient to make up for its annual loss,

---

* See Am. Railroad Journal, March 13, 1847.

and diminution of depth, which is too small for direct measurement. Experiments upon this point indicate that the amount of materials annually consumed, and therefore to be replaced, is one cubic yard per mile* for each "collar," or beast of burden passing over it. Others consider it only two-thirds of a cubic yard.†

REPAIRING A ROAD.

A road properly *kept up* by daily attention, needs no *repairs;* but if it be put in order only at intervals, the injuries to it, which have been increasing in geometrical progression, will render very serious repairs necessary. It will be found cut into ruts, deep holes, and irregular projections; and often lower in the middle than at the sides. It must be put into shape, and restored to its proper cross-section, by cutting down the sides, and filling up the middle part. Only a single thin coat of stone should be applied at a time,—not more than a cubic yard to a rod superficial. The surface of the old road may be lightly picked up, or "lifted," (with strong short picks) merely burying the point of the pick one or two inches deep, so that the new materials may be more readily united to the old ones. This is especially necessary on declivities, to prevent the stones rolling down the slope.

When the road to be repaired is one which had been originally formed of large stones, and of superfluous thickness, no new materials should be brought upon it, but the old stones should be loosened with picks, gathered by strong rakes to the side of the road, and there broken to the proper size. The surface of the road having been put in proper shape, the broken stones are to be returned

---

* DUPUIS, *Annales des Ponts et Chausées,* 1842 † Gayffier, p. 232.

to it, being scattered uniformly and thinly over the surface. Only a small space of road should be thus broken up at once, say six or eight feet in length, but the whole width. The old plan of repairing would be to fill up the holes with an additional supply of the same large materials; but the method here recommended makes more work for men and less for horses, and produces a great saving in expense.

The best *season* for repairing broken-stone roads is in the spring or early summer, when the weather is neither very wet nor very dry, for either of these extremes prevents the materials from consolidating, and therefore produces either a heavy or a dusty road. If made at this season, the roads are left in a good state for the summer, and become consolidated and hard, so as to be in a condition to resist the work of the ensuing winter.*

TELFORD ROADS.

This name may be given to the roads of broken stone which rest on a peculiar pavement, as constructed by Telford, on the Holyhead road and elsewhere, and of which he has given the following specification for a width of thirty feet. Fig. 110 is a section of the carriage-way of such a road.

Fig. 110.

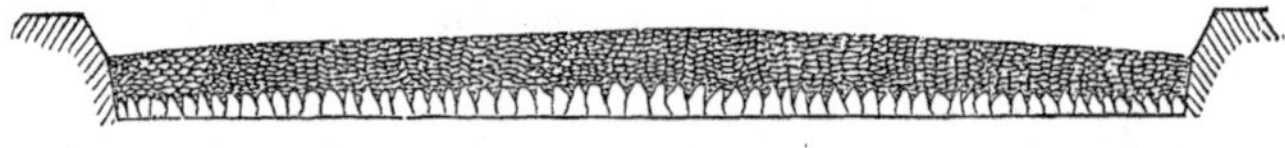

"Upon the level bed† prepared for the road materials

---

* James Walker.

† A bed with the same cross-section as the final road, would certainly be preferable, to ensure drainage. The pavement would then require to be of the same depth at centre and sides.

a *bottom course or layer of stones* is to be set by hand in the form of a close, firm pavement. The stones set in the middle of the road are to be seven inches in depth; at nine feet from the centre, five inches; at twelve from the centre, four inches; and at fifteen feet, three inches.* They are to be set on their *broadest edges and lengthwise across the road*, and the breadth of the upper edge is not to exceed four inches in any case. All the irregularities of the upper part of the said pavement are to be broken off by the hammer, and all the interstices to be filled with stone chips, firmly wedged or packed by hand with a light hammer, so that when the whole pavement is finished, there shall be a convexity of four inches in the breadth of fifteen feet from the centre.

"The middle eighteen feet of pavement is to be coated with hard stones to the depth of six inches. Four of these six inches are to be first put on and worked in by carriages and horses; care being taken to rake in the ruts until the surface becomes firm and consolidated, after which the remaining two inches are to be put on. The whole of this stone is to be broken into pieces as nearly cubical as possible, so that the largest piece, in its longest dimensions, may pass through a ring of two inches and a half inside diameter.

"The paved spaces, on each side of the eighteen middle feet, are to be coated with broken stones, or well cleansed, strong gravel, up to the footpath or other boundary of the road, so as to make the whole convexity of the road six inches from the centre to the sides of it. The whole of the materials are to be covered with a binding of an inch

* The curved section thus obtained, has been shown, on page 50, to be inferior to plane slopes on each side of the centre.

and a half in depth, of good gravel, free from clay of earth."*

The propriety of this foundation, ("*Bottoming*," or "*Pitching*") has been the subject of earnest controversy between the partisans of McAdam and those of Telford. The following are the defects imputed to a road of broken stones, laid on earth, (especially clay) without any foundation.

The weight of vehicles forces the lower stones into the earth, which rises up into the interstices and forms a mixture of earth and stones which will always be loose and open, and never consolidate into a compact mass. In winter the water, which will penetrate, is frozen and breaks up the road. After a thaw and in wet weather, the road is a quagmire, the wheels cut deeply into it, and some times through the entire thickness, so that it resembles a ploughed field. At the best, after a rain the semi-fluid soil will rise up to the surface and form a coat of mud; and after a drought the looseness of the stones will make them rub off their angles and soon wear out. Nor will any thickness of broken stones thoroughly destroy the elasticity of the soil, the evils of which were shown on page 58

McAdam maintains that thorough draining will prevent all these evils, but Telford thinks that they can be removed only by the "bottoming," for which he claims the following advantages.

Roads, being in fact artificial structures, which have to sustain great weights and violent percussion, the first object must be to obtain a permanently firm and stable foundation.

This is effected by the plan of "bottoming;" for by it the pressure of the wheels is distributed over a large

* Parnell, pp. 133–4.

space. Suppose that the wheel touches and presses on a surface of 2 square inches. This pressure is carried tc the foundation stones, which rest at their bottom on a broad surface, averaging 10 by 5 inches, or 50 square inches, so that each square inch of the soil receives only one-twenty-fifth part of the surface pressure, and there is therefore no danger of the pavement stones being pressed into it, nor of the soil being forced to ooze up between them. On a new embankment of soft earth it is best to lay brush or furze, and place the pavement upon this.

The advantages of this system are most striking when the natural soil is retentive of moisture, as when it is clay. The pavement then acts as an under-drain to carry off the water which may find its way through the broken-stone surface. Even on a rock this pavement may be laid with advantage, to form a clear floor.

When the stones are properly set, and wedged with the stone chippings, they will never rise to the surface.* To avoid disturbing them, the carts which bring the broken stone must not be allowed to pass over the foundation.

From the moment that a road thus made begins to be used, it becomes daily harder and smoother. The strength of the resulting surface admits of carriages being drawn over it with the least possible distress to horses. The broken stones being on an immoveable dry bed, do not

* Large stones, placed under a road and not thus wedged down, will invariably work up to the surface. Thus, over Breslington Common, England, the whole of the original soil had been covered at great expense with large flag-stones, and the road-covering laid upon them. Their motion kept the surface in a loose, open state, till, on the road being dug open, they were found almost entirely *turned upon their edges*, having been acting with the force of levers upon the road, which they had made to crack and sink, without the cause at such a depth being suspected.—*McAdam.*

mix with the soil, and become perfectly united together into one solid mass.

The parts of the Holyhead road formed with such a foundation, were unaffected by a series of unusually severe frosts, followed by thaws and heavy rains, while the parts of it differently made, and other roads in the neighborhood, were broken up, and "became as bad as a bog."*

A road thus constructed will in most cases cost less than one entirely of broken stone; for the course of foundation-stones may be of any cheap and inferior stone, as sand-stone, &c., which will bear weight, and not be decomposed by the atmosphere, but which would not be sufficiently hard and tough for the broken-stone covering. The cost of hammering and setting this pavement will be less than that of breaking up an equal mass, and the total amount of stone employed will be no more than would have been required for a road entirely of broken stone.

But even if such a road cost more at first, it would be cheaper in the end; for, beside the saving of draught, stones laid on such a pavement last much longer than those laid on earth, two courses of the former outlasting three of the latter. The expense of scraping is lessened in the same or even a greater proportion.

On the other hand, it is objected that, between the wheel above and the foundation-stone beneath, the broken stone will be in a situation like that of the grain between two millstones, and must therefore be more rapidly ground to powder than if on a soft bottom.† But this will be prevented by using harder stone for the surface than for the foundation.

---

* Telford First Report on Holyhead roads. † Penfold, p 8.

McAdam also maintains that the materials last longer on a soft and elastic bottom than on a hard one; and instances a road in Somersetshire, where a part of it is "over a morass so extremely soft that when you ride in a carriage along the road, you see the water tremble in the ditches on each side," and is succeeded by a bottom of limestone rock, continuing for five or six miles. An exact account of the expenditure on each having been kept, it was found that the cost of keeping up the soft was to that of the hard only as five to seven; *i. e.* five tons of stone on the former would last as long as seven on the latter. But this seems an exceptional case, being contrary to all other experience. Sir John Macneill testifies very strongly that the annual *saving* of a paved bottom will be one-third of the expense in any case, and that if the diminished amount of horse labor were considered, it would be very considerably more than that.*

An artificial *substitute* for a pavement foundation, consisting of a *concrete*, or composition of Roman cement and gravel, has been employed with great success on a wet and elastic soil, where every thing else had failed, and where stones for bottoming would have been very expensive. The locality was the Highgate Archway Road near London, in a deep cutting between two high banks of clay, where the soil was surcharged with water. Many attempts at draining had been made, and a great thickness of broken stone had been used, and subsequently relaid on furze and pieces of waste tin. But the stone mixed with the wet clay, and rapidly wore away, becoming round and smooth, without ever consolidating, and the road was almost impassable. The Parliamentary Commissioners finally took charge of it, and Sir John Macneill succeeded in making a perfect road. Four longitudinal drains were made the whole length of the road, cross drains at every 90 feet, and

* Parnell, p. 163.

intermediate small drains at every 30 feet under the cement.* On the prepared centre, of eighteen feet in width, after it had been properly levelled, was put a layer, six inches thick, of the concrete, formed of one part of Roman cement, one of sand, and eight of stones. The sand and cement were mixed dry in a large shallow trough ; the gravel was added ; as little water as possible was used ; and the whole mixture was then cast upon the ground. Before it had set, a triangular piece of wood was indented into the surface, so as to leave, at every four inches, a triangular groove for the broken stones to lie in and fasten into. These grooves fell three inches from the centre to the sides of the road, in order to carry off any water which might percolate through the broken stones above it. *Six* inches of these were laid upon it when it had sufficiently hardened, (which was in about fifteen minutes) and the sides or wings were filled up with flint gravel. The concrete cost at that place 50 cents per square yard six inches thick. The object was to attain a dry and solid foundation for the broken stone. The result was an excellent road, undisturbed by severe frosts, and on which one horse could draw as much as three in its original state.

4. PAVED ROADS.†

A good pavement should offer little resistance to wheels, but give a firm foothold to horses ; it should be so durable as to seldom require taking up ; it should be as free as possible from noise and dust ; and when it is laid in the streets of a city, it should be susceptible of easy removal and replacement to give access to gas and water pipes.

A common but very inferior pavement, which disgraces the streets of nearly all our cities, is constructed of rounded

---

* See Parnell, pp. 157 and 160, and plates to Simms on Roads.

† Gayffier, pp. 193–8 ; Marlette, pp. 104–8 ; Jullien, pp. 316–18 ; Parnell, pp. 110–123, 348–359 ; Mahan, pp. 292–5 ; Journal of Franklin Institute, Sept. Oct. 1843.

water-worn *pebbles*, or "cobble-stones." The best are of an egg-like shape, from 5 to 10 inches deep, and of a diameter equal to half their depth. They are set with their greatest length upright, and their broadest end uppermost. Under them is a bed of sand or gravel a foot or two deep. They are rammed over three times, and a layer of fine gravel spread over them to fill their interstices.*

The glaring faults of this pavement are that the stones, being supported only by the friction of the very narrow space at which they are in contact, are easily pressed down by heavy loads into the loose bottom, thus forming holes and depressions ; and at best offer great resistance to draught, cause great noise, cannot be easily cleaned, and need very frequent repairs and renewals.†

The pavement which combines most perfectly all desirable requisites, is formed of squared blocks of stone, resting on a stable foundation, and laid diagonally.

We will examine successively the merits of different foundations ; the quality of stone preferable ; their most advantageous size and shape ; their arrangement ; the manner of laying them ; their borders and curbs ; their advantages ; and their comparison with McAdam roads.

---

* The following is part of the specification for the New York pavement : "The paving stones must be heavy and hard, and not less than six inches in depth, nor more than ten inches in any direction. Stones of similar size are to be placed together. They are to be bedded endwise in good clean gravel, twelve inches in depth. They shall all be set perpendicularly and closely paved on their ends, and not be set on their sides or edges in any cases whatever."

† The cost of such a pavement for a new street is in New York from 50 to 75 cents per square yard ; for repairing an old street, about 20 cents

FOUNDATIONS.

The want of a proper foundation is one of the most frequent causes of the failures of pavements. A foundation should be composed of a sufficient thickness of some incompressible material, which will effectually cut off all connection between the subsoil and the bottom of the paving-stones, and should rest upon a well-drained bottom, for which in cities a perfect system of sewerage is indispensable. The principal foundations are those of *sand*, of *broken stone*, of *pebbles*, and of *concrete*.

Foundations of sand.—This material, when it fills an excavation, possesses the valuable properties of incompressibility, and of assuming a new position of equilibrium and stability when any portion of it is disturbed. To secure these qualities in their highest degree, the sand should be very carefully freed from the least admixture of earth or clay, and the largest grains should not exceed one-sixth of an inch in diameter, nor the smallest be less than one-twenty-fifth of an inch. The bed of the road should be excavated to the desired width and depth, and be shaped with a slope each way from the centre, corresponding with that which is to be given to the pavement. This earth bottom should be well rammed, and a layer of sand, four inches thick, be put on, be thoroughly wetted, and be beaten with a rammer weighing forty pounds. Two other layers are to be in like manner added, and the compression will reduce the thickness of twelve inches to eight. The number of layers should be regulated by the character of the subsoil. Two inches of loose sand are to be then added to fill the joints of the stones, which may be now laid. The pressure of loads upon these stones is spread by the incompressible sand over a large surface of the earth beneath. This is the favorite system in France.*

Foundations of broken stone.—A bed is to be excavated, deep enough to allow twelve inches of broken stone to be placed under the pavement. A layer of four inches is first put on, and the street then opened for carriages to pass through it. When it has become firm and consolidated, another layer of four inches is added and worked in as before ; and finally a third layer ; making in fact a complete McAdam road. Upon it the dressed paving-stones are set.† This method, though efficient, is very inconvenient, from the length of time which it occupies, and the difficulty of draught while it is in progress.

Foundations of pebbles.—Such a pebble pavement as is described on page 217, resting itself on sand, gravel, or broken stones, has been recommended to be adopted as the foundation of the dressed block pavement, for streets in which there is a great deal of travel.‡

Foundations of Concrete.—Concrete is a mortar of finely-pulverized quicklime, sand, and gravel, which are mixed dry, and to which water is added to bring the mass to the proper consistence. It must be used immediately. *Béton* (to which the name of Concrete is often improperly given) is a mixture of *hydraulic* mortar with gravel or broken stone ; the mortar being first prepared, fine gravel incorporated with it, the layer of broken stones subsequently added to a layer of it 5 or 6 inches thick, and the whole mass rapidly brought by the hoe and shovel to a homogeneous state. Three parts of sand, one of hydraulic lime, and three of broken stone is a good proportion. A mixture of one part of Roman cement, one of

* Gayffier, p. 126. † Parnell, p. 117.

‡ Committee of Franklin Institute, and Parnell, p 116.

sand, and eight of stone, has also been employed very successfully. *Béton* is much superior to Concrete for moist localities.*

The excavation should be made fourteen inches lower than the *bottom* of the proposed pavement, and filled with that depth of the concrete or *béton*, which sets very rapidly, and becomes a hard, solid mass, on which a pavement may then be laid. This is, perhaps, the most efficient of all the foundations, but also the most costly at first, though this would be balanced by its permanence and saving of repairs. It admits of access to subterraneous pipes with less injury to the neighboring pavement than any other, for the concrete may be broken through at any point without unsettling the foundation for a considerable distance around it, as is the case with foundations of sand or broken stones ; and when the concrete is replaced, the pavement can be at once reset at its proper level, without the uncertain allowance for settling which is necessary in other cases. The blocks set on the concrete are usually laid in mortar. We will examine presently the propriety of this.

QUALITY OF STONE.

The stone should be of a kind which will not wear smooth, but which will always remain rough on the surface. Many varieties of granite are of this character, and are therefore very suitable. The hardest stones are the best, and their specific gravity is a tolerable test of their hardness. The hardest stones will also absorb but $\frac{1}{570}$ of their volume of water; tender ones will absorb $\frac{1}{51}$. The hardest stones also, when struck by a hammer, give a clearer and more ringing sound than soft ones. Tender

* Mahan, p. 40

stones may be made much more durable by plunging them in boiling bitumen, which penetrates their pores and prevents them from absorbing water, which is the most powerful agent in their disintegration.

SIZE AND SHAPE.

The size of the stones should be proportioned to the number and weight of the vehicles which will pass over them, and as each stone is liable to have resting upon it the entire weight borne by one wheel, it should be large enough to sustain this weight without being crushed, or depressed. It should also be no larger than a horse's hoof, so as to prevent any slipping upon its surface, even where unbroken by joints; but the fulfilment of the first condition will generally make this impossible, and the selection of a proper quality of stone will render it unnecessary. If stones of different dimensions are admitted, they should be assorted, and only those of the same size should be used near each other, or the small ones will sink below the rest, and the depressions thus formed will be increased by every passing wheel. It is therefore very desirable that they should be uniform in size. *Cubes of eight inches* in every direction seem to combine most of these requisites. They should be very slightly tapering towards their lower ends, thus making them truncated pyramids.* If they are much larger than this standard, the weight of a wheel coming on one end of one of them, will tend to depress it and to elevate the other end, so

---

* Blocks of this size cost in Philadelphia delivered on the street, $2.75 per square yard of surface. Laying a bed of gravel 15 inches deep, setting the stone, &c., cost 50 cents more, making the entire cost of the pavement $3.25 per square yard.

that such large stones would be less firm than smaller ones.

Hexagonal blocks have been suggested, and would form a more compact mass than those of any other shape; but their superiority in this respect would probably not compensate for the extra cost of cutting them.

ARRANGEMENT.

Fig. 111.

The rectangular stones may be laid in continuous courses across the road, but so as to "break joints" in the direction of its length, as shown in Fig. 111. It has been observed, however, that when stones are laid, as is usual, with their joints parallel and perpendicular to the direction of the road, they wear away most rapidly upon the edges which run across the road, since these receive most directly the shocks of the wheels, and that the stones thus become convex. To prevent this, and to secure equal wear, they should be laid so that the joints cross the road obliquely, making an angle of 45° with the axis of the roadway. One set of joints may be continuous, but the others should break joints, as in Fig. 112.

Fig. 112.

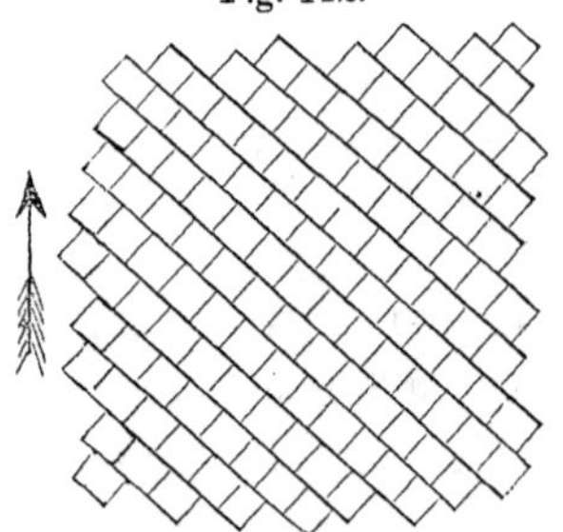

Oblong stones are preferred by the French engineers, with their upper surfaces nine inches by five and a half. They should be laid, (if not diagonally) so that their greatest length is across the street, their narrowest dimension being that passed over by the wheels. They thus offer less resistance to draught than cubical blocks, according to the experiments of Morin.

Fig. 113.

In the steep streets of Genoa the stones are laid in oblique courses, pointing up the ascent, and meeting at an angle in the centre. The continuous joints, which descend to the right and to the left, facilitate the discharge of the rainwater.

MANNER OF LAYING.

The top surface of the foundation (of whatever material it may be) which forms the bed for the paving-stones, is to be shaped, as directed on page 50, sloping each way from the centre, with inclinations ranging from 1 in 50 to 1 in 100, flatter in proportion to the smoothness of the surface. The stones should be so set that the joints between them will not exceed one quarter of an inch. But as they are not cut regularly enough to touch on every part of their surface, some substance must be interposed to fill up the vacancies, and to enable them to support each other. Mortar is used for this purpose on foundations of concrete, and even on those of sand and broken stone. Sometimes gravel is put between them, and a grouting of lime-water poured in. Iron chippings are added to the gravel to increase the adherence. But no adherent compound, such as these, can resist the continual vibrations and play of the pavement. Some other substance should therefore be employed, which will change its position of equilibrium, and never cease to fill up the spaces between the stones, whatever shocks they

15

may receive. Such a substance is pure *sand*. The quality necessary has been indicated on page 218. A coating of an inch should also be spread over the stones When the foundation is any thing but concrete, the paving stones must be rammed, after a certain portion has been laid, with a maul weighing 60 lbs., and those which break under this must be replaced, and those which sink, taken up and reset.

BORDERS AND CURBS.

When the paved road forms the middle portion, or causeway, of a wider road, with wings of earth or broken stone on each side of it, its edges must be supported against the lateral thrust of the stones, by *borders* of larger blocks, 9 or 10 inches wide, 13 to 18 inches long, and 13 inches deep. They are laid as headers and stretchers, so as to form a bond with the pavement. Their outer edge should also have occasional projections into the wings, so that a rut may not be there formed.

When the pavement is a city street, the curb-stones should be long blocks.* There should be no gutter or other channel than that formed, as in the figure, by the meeting of the inclined pavement with the curb-stone, which should rise 6 or 8 inches above the pavement, and be sunk as deep into the ground as possible. The foot pavements should

Fig. 114.

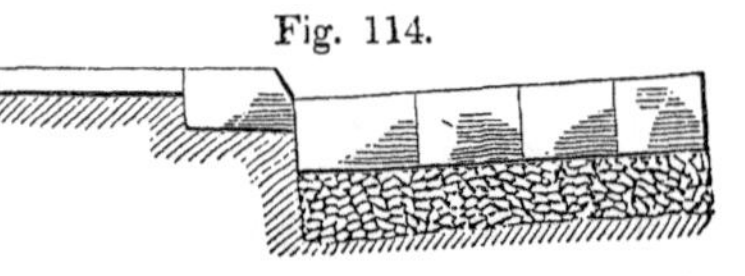

* In the specifications for the New York pavements, the Curb-stones are required to be not less than 3 feet long, 5 inches thick, and 20 inches wide; and the Gutter-stones to be not less than three feet long, 6 inches thick, and 14 inches wide

incline towards the street at the rate of one inch in ten feet, or 1 in 120.*

ADVANTAGES.

The *advantages* of such a pavement are its smoothness and uniformity of surface, enabling vehicles to be drawn over it with ease to the horses, comfort to the passengers, and but little wear and tear of the carriages, which can be therefore made much lighter than at present. At the same time it gives a good foothold to the horses; causes very little noise, yet enough to warn the foot-passengers of the approach of a vehicle, and is very easily cleaned of the dirt which may collect upon it. It is also very durable, thereby rendering unnecessary the frequent stoppage of a street for repairs; and though at first more expensive than cobble-stones, is finally far more economical.

PAVED AND McADAM ROADS COMPARED.

McAdam maintains that his roads are preferable to pavements, even for the streets of cities. He argues that they are cheaper, as requiring no more stone than pavements, admitting an inferior quality, and costing less for repairs; and that they give greater facility of travelling, and cause less annoyance from dust, when properly swept and watered. But experience in the streets of London shows the cost of broken-stone roads to be far greater than pavements, to which they are inferior in every respect.† The result of very full discussions at the Civil Engineers' Institution was, that a whin or granite pavement, of proper form and depth, laid on a sound bottom, is preferable to

* Parnell, p. 120. † Parnell, p. 126.

any other plan for carriageways in the metropolis and other large cities. The objections to the broken-stone roads are that they cannot resist the pressure caused by a very great intercourse, being liable to be thereby crushed and ground into dust, which is easily converted into mud; that this hasty and continual destruction and renewal would, in a great city, prove intolerably troublesome and expensive, while the dust in dry weather, and the mud in wet, would greatly incommode the intercourse in the streets, as well as private dwellings and public shops. The surface of broken stone is also more injurious to the feet of horses than a good pavement, and less easy for their labor; and the expense of making and maintaining the former would be at least fifty per cent. more than the latter.*

ROMAN ROADS.

The ancient Roman roads, which, even at the present day, after the lapse of nearly two thousand years, may be traced for miles, as perfect as when first constructed, were essentially dressed-stone pavements, with foundations of concrete, resting on sub-pavements. The most perfect modern constructions thus appear to be only imperfect and incomplete imitations. The direction and length of the intended road were marked out by two parallel furrows, from the space between which the loose earth was removed. The foundation of the road (*Statumen*) was composed of one or two courses of large flat stones, laid in mortar, a bed of which was first spread over the earth. Next came a course of concrete (*Rudus*) formed of broken stones mixed with quicklime, and

---

* Telford in Parnell, p. 351.

pounded with a rammer. If the stones were freshly broken ones, three parts of them were mixed with one of quicklime ; if they were from old buildings, two parts of lime were used to three of the rubbish. The third course (*Nucleus*) was composed of broken bricks, tiles, and pottery, mixed with lime, which formed one-fourth of the whole. The mixture was spread in a thin layer, and in it were imbedded, so that their top surfaces were perfectly level, the large blocks of stone (*Summa crusta*) which formed the pavement. These stones were irregular polygons, usually with 5, 6, or 7 sides, rough on their under side, but smooth on top, and so perfectly fitted together that the joints were scarcely perceptible. The entire thickness of the four strata was about three feet. When the road passed over marshy ground, the foundation stones rested on a framework of timber, (made of a species of oak not subject to warp or shrink) and to protect this from the lime, it was covered with a bed of rushes or reeds, and sometimes of straw. On each side of the road were paved footpaths, and parapets ; with stones at regular intervals for mounting on horseback Milestones marked the distances to all parts of the empire from the *Milliarium aureum*, a gilt column in the Forum of Rome.

The *Russ Pavement*, (named from its introducer,) in New York, is constructed thus :—The street (Broadway) is graded with a crown of 7 inches= $\frac{1}{36}$. Granite chips are spread over this, and rammed down flush with the earth. A concrete foundation, 6 inches thick, is formed in rectangular sections. It contains 1 part of Rosendale cement, 2½ parts of clean coarse sand, 2½ of broken stone, and 2 of gravel. On it rest rectangular blocks of sienitic granite, 10 inches deep, 10 to 18 long, and 5 to 12 wide. They are laid diagonally, at angles of 45° with the line of the street, and so as to form lozenge-shaped compartments. Lewis holes in certain blocks, and iron plates under them, give easy access to water and gas pipes, permitting excavations 4 feet long, and 3½ wide. The contract price in 1849 was $5.50 per square yard of pavement.

5. ROADS OF WOOD.

The abundance, and consequent cheapness, of wood in our new country, renders its employment in Road-making of great value. It has been used in the form of logs, of charcoal, of planks, and of blocks.

LOG ROADS.

Fig. 115.

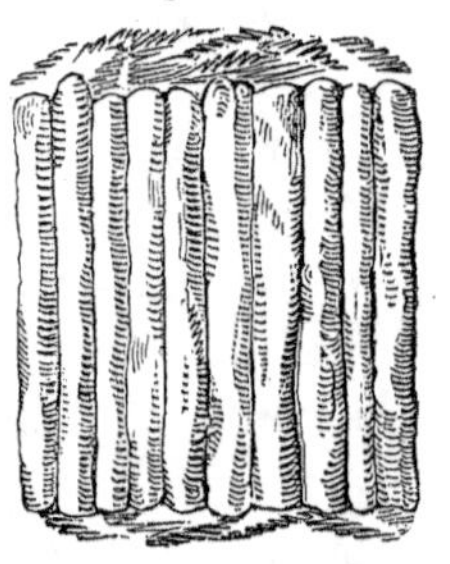

When a road passes over soft swampy ground, always kept moist by springs, which cannot be drained without too much expense, and which is surrounded by a forest, it may be cheaply and rapidly made passable, by felling a sufficient number of young trees, as straight and as uniform in size as possible, and laying them side by side across the road at right angles to its length. This arrangement is well known under the name of a "*Corduroy*" road, of which the figure gives a top and end view. Though its successive hills and hollows offer great resistance to draught, and are very unpleasant to persons riding over it, it is nevertheless a very valuable substitute for a swamp, which in its natural state would at times be utterly impassable. But necessary and desirable as these roads may be to accomplish such an end in the infancy of a settlement, their retention upon a great thoroughfare is a disgraceful proof of indolence and want of enterprise in those who habitually travel over them; though several such instances might be specified.

CHARCOAL ROADS.

A very good road has been lately made through a swampy forest, by felling and burning the timber, and covering the surface with the charcoal thus prepared.

"Timber from six to eighteen inches through is cut twenty-four feet long, and piled up lengthwise in the centre of the road about five feet high, being nine feet wide at the bottom and two at the top, and then covered with straw and earth in the manner of coal-pits. The earth required to cover the pile, taken from either side, leaves two good-sized ditches, and the timber, although not split, is easily charred; and, when charred, the earth is removed to the side of the ditches, the coal raked down, to a width of fifteen feet, leaving it two feet thick at the centre and one at the sides, and the road is completed."

A road thus made in Michigan cost $660 per mile, and is said to be very compact and free from mud or dust. At a season when the mud on the adjoining earth road was half axletree deep, "on the coal road, there was not the least standing, and the impress of the feet of a horse passing rapidly over it was like that made on hard washed sand, as the surf recedes, on the shore of the lake. The water was not drained from the ditches, and yet there were no ruts or inequalities in the surface of the coal road, except what was produced by more compact packing on the line of travel. It is probable that coal will fully compensate for the deficiency of limestone and gravel in many sections of the west, and, where a road is to be constructed through forest land, that coal may be used at a fourth of the expense of limestone."

Two such roads in Wisconsin were let by contract at $1.56 and $ .62½ per rod, or $499 and $520 per mile.

PLANK ROADS.

Plan and Cross Section of a Plank Road.

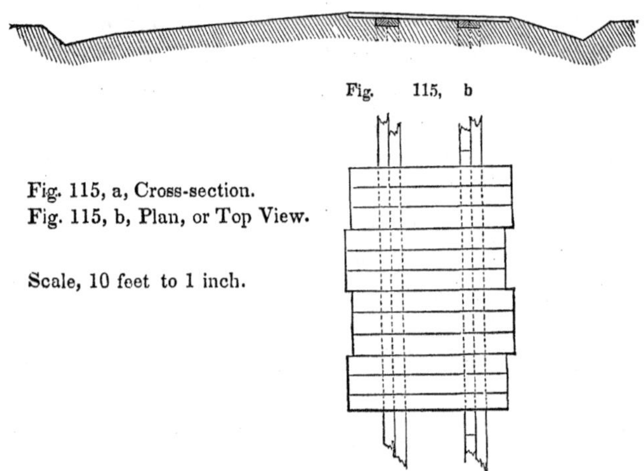

The most valuable improvement since McAdam's, and one superior to his in many localities, is the recent invention of covering roads with planks. The first plank road on this continent was constructed in Upper Canada in 1836. A short piece, laid down experimentally, gave so much satisfaction, as to ease of travelling, and cheapness of keeping in repair, that a mile of it was constructed the next year at a cost of $2100. Its success caused it to be continued. Since then 500 miles have been constructed in Canada, and more than 2000 registered in the State of New-York; and probably several thousands more in the other states of the Union from Maine to Texas and Wisconsin.

In the most generally approved system, two parallel rows of small sticks of timber (called indifferently *sleepers*, *stringers*, or *sills*) are imbedded in the road, three or four feet apart. Planks, eight feet long and three or four inches thick, are laid upon these sticks, across them, at right angles to their direction. A side track of earth, to turn out upon, is carefully graded. Deep ditches are dug on each side, to ensure perfect drainage; and thus is formed a Plank Road.

The benefits of covering the earth with some better material have been indicated on page 188, and the peculiar advantages of this plank covering will be more fully made known, when we shall have discussed in order the various details of construction.*

LAYING THEM OUT.

The waste of labor caused by unnecessary ascents in a road, has been pointed out in the early part of this volume, (pages 32–36.) It was also shown (page 28) that it is profitable to the traveller to go two or three thousand feet around to avoid ascending a hill a hundred feet high; though the cost of constructing the additional length of road *partially* counterbalances this consideration. It was also proved that the smoother the surface of the road was made, the more injurious proportionally were such ascents. They are therefore especially objectionable on plank roads, which hold an intermediate place between common roads and railroads. Some distinguished engi-

* Hon. Philo White's report to the Council of Wisconsin, February, 1848, embodies a very extended and systematic collection of information on this subject. To it, and to the valuable published and obliging private communications of Hon. George Geddes, C. E., (who first introduced and naturalized this improvement in the United States,) the author is much indebted, as also to many other recent sources.

neers have been led astray on this point. Their arguments, if carried out to their full extent, would lead to the construction of railroads also with similarly steep grades. It is true, as they state, that a given load can be drawn up a much steeper hill on a plank road than on a common one, the friction on the former being so much less, but (as proven on pages 34 and 35, which see) this will lessen in an equally increased ratio the advantages of the level portions of the road. Let us assume the resistance of friction, or "stick-tion," (as Professor Whewell calls it,) on a plank road to be one-third of that on a good earth road. It will therefore be one-sixtieth of the weight carried, if that of the earth be one-twentieth. If, now, a horse can draw one ton on the level earth road, the total resistance will be doubled when he comes to a hill which rises one foot in going twenty, (1 in 20,) and he will be able to draw only half a ton up this hill, and therefore his load on the level parts of the road would be but half a ton; for it would be useless for him to take more *to* the hill than he could drag *up* it. Now suppose the same road to be planked, and this hill to remain untouched. On the level portions the same horse can now draw *three* tons, by our hypothesis. But the hill, rising 1 in 20, will offer a resistance three times as great as does the "stiction" of the plank road, and the whole resistance in going up it will therefore be *four times* as great as on a level. The horse can therefore draw only one-fourth of his former load, or only three-quarters of a ton, which is consequently the limit of his load on the level. Thus then this hill has brought down the *gain* of the plank road over the earth to only a quarter of a ton, instead of two tons, which it would be, were the hill removed. Therefore, in laying out a plank road, it is indispensable, in order to secure all the benefits which can be derived from it, to avoid or cut down all steep ascents.

A very *short* rise, of even considerable steepness, may, however, be allowed to remain, to save expense; since a horse can, for a short time, put forth extra exertion to overcome such an increased resistance; and the danger of slipping is avoided by descending upon the earthen track.*

A plank road, lately laid out, under the supervision of Mr. Geddes, between Cazenovia and Chittenango, N. Y., is an excellent exemplification of the true principles of roadmaking. Both these villages are situated on the "Chittenango creek," the former being 800 feet higher than the latter. The most level common road between these villages rises, however, more than 1,200 feet in going from Chittenango to Cazenovia, and rises more than 400 feet in going from Cazenovia to Chittenango, in spite of this latter place being 800 feet lower. It thus adds one-half to the ascent and labor, going in one direction, and in the other direction it goes up hill one-half the height, which should have been a continuous descent. The line of the *plank road*, however, by following the creek, (crossing it five times,) ascends only the necessary 800 feet in one direction, and has no ascents in the other, with two or three trifling exceptions, of a few feet in all, admitted in order to save expense. There is a nearly perpendicular fall in the creek of 140 feet. To overcome this, it was necessary to commence, far below the falls, to climb up the steep hill-side, following up the sides of the lateral ravines, until they were narrow enough to bridge, and then turning and following back the opposite sides till the main valley was again reached. The *extreme* rise is at the rate of one foot to the rod, (1 in $16\frac{1}{2}$;) and this only

* The steeper the grade, the more rapid is the wear of the planks, in a very remarkable degree; a foot in a rod doubling the wear on a level.

for short distances, and in only three instances. with a much less grade, or a level, intervening. The line passes through a dense forest, which supplied its material, being cut into plank by sawmills erected in a gulf never before approached by a wheeled carriage.

WIDTH.

A single track of plank, eight feet wide, with an earthen turn-out track beside it, of twelve feet, will in almost all cases be sufficient. This gives twenty feet for the least width necessary between the inside top lines of the ditches, the width of which is to be added, making about two rods on level ground. If extra cuttings or fillings be required, the width occupied by their slopes must be added to this. An earthen road of eight feet wide on each side of the plank track, has sometimes been adopted. The New York general plank road law fixed four rods (66 feet) as the least permissible width that plank roads might be laid out. This provision has since been repealed.

Wider plank tracks were at first employed. In Canada *single* tracks were made from 9 to 12 feet wide. But it was found, on the 12-feet Toronto road, after seven years' use, that the planks were worn only in the middle seven or eight feet, and that the remaining four or five feet of the surface had not even lost the marks of the saw. One-third of the planking was therefore useless, and one-third of the expenditure wasted.

A *double* plank track will rarely be necessary. No one without experience in the matter can credit the amount of travel which one such track can accommodate. Over a single track near Syracuse, 161,000 teams passed in two years, averaging over 220 teams per day, and during three days 720 passed daily. The earthen turn-out track must, however, be kept in good order, and this is easy, if it slope off properly to the ditch, for it is not cut with any continuous lengthwise ruts, but is only passed over by the wheels of the wagons which turn off from the track, and return to it. They thus move in curves, which would very rarely exactly hit each other, and this travel, being spread nearly uniformly over the earth, tends to keep it in shape rather than to disturb it.

If, however, there is so much travel that the earth track will not remain in good order, then this travel will pay for the double track which it requires. But this should be made in *two separate eight-feet tracks*, and not in one wide one of 16 or 24 feet, as was at first the practice. On a wide track the travel will generally be near its middle, and will thus wear out the planks very unequally, besides depressing them in their centre, and making the ends spring up, and when it passes near one end that will tilt up, and loosen the other. Besides, when a light vehicle wishes to pass a loaded one moving in the centre, as it naturally will, the former will be greatly delayed in waiting for the other to turn aside, or else will have one wheel crowded off into the ditch. But where there are two separate tracks, the whole width of one is at the service of the light vehicle. On a sixteen-feet track near Toronto, the planks, having become loose and unsettled, were sawn in two in the centre, and this imperfect double track, even without any turn-out path between, worked better than in ts original state. An experienced constructor states that if he were desired to build a road fifty feet wide, he would make it in separate eight-feet tracks.

The wide track of 16 feet plank has sometimes been divided into two of eight feet, by spiking down scantling

20 feet long, and six inches square, along the middle of the road, at intervals of 100 feet in the clear, between each scantling. This, however, only partially remedies the objections adduced.

When the ground is of such a very unsettled and yielding nature, such as loose sand, marsh, &c., that a solid turn-out track of earth cannot be made, planks, sixteen feet long, may be used, resting on three, four, or five sleepers, crowning in the middle three or four inches, and the ends sprung down, and pinned to the outer sleepers

GRADING.

The importance of elevating a road-bed above the level of the adjoining fields, and digging deep ditches on each side, has been already urged, (pages 53, 54,) and this is a fundamental requisite in making a good plank road. Employ the earth from the ditches, if good material, rejecting the sods, to raise the road-bed. Give the ditches free outlets, cut their bottoms with true slopes, make under-drains, of cobble-stones and brush, across the road in wet places, and use every precaution to ensure thorough and complete drainage. This will be more difficult in a flat than in a hilly country. If it be effected, however, the plank will last much longer, and the road be always in better condition.*

The "cross-section" of the road-bed, or its shape cross

* The ditches and side slopes of the road-bed, after being ploughed up, may be most rapidly shaped by the use of a scraper of this form, >, composed of two planks hinged together in front, and kept apart in the rear by an adjustable cross-piece. The team is attached to the *outer* plank at such a distance from the point as to keep the inner plank in the direction of the road, so that it forms the straight edge of the bank, while the skew of the outer plank throws the earth to one side in the manner of a snow-plough. A man with a long lever inserted in the outer side regulates this more exactly

wise, between the ditches, must be carefully adjusted so as to freely carry off the rain which may fall on it. First decide on which side of the road the plank track is to be laid. It should generally be on the right-hand side coming from the country into a town, so that the farmers' wagons may keep upon it, when they bring in their heavy produce, and that the turning out may be done by those which are going back light.* The twelve feet width intended for the earth track should be heavily rolled or beaten, to make it firm and hard. It should slope down from the centre three-quarters of an inch to the foot, (1 in 16,) and the eight feet of plank should fall off three inches, or 1 in 32. From each side of the 20 feet thus graded, the bank should slope down to the bottom of the ditches at the rate of three inches to the foot, or 1 in 4. (See Fig. 115, a; page 230.)

The proper shape may be most easily and accurately given by the use of a common mason's level, having a tapering piece of wood under it, (as shown in Fig. 88, page 173,) or having one leg so much longer than the other, as will give the slope required. If the plank be laid on an old roadway, no more of it should be broken up than is absolutely necessary for imbedding the sleepers, as it is very desirable to preserve as solid a foundation as possible.

SLEEPERS, SILLS, OR STRINGERS.

Material.—Pine, hemlock tamarack, oak, and walnut, have been used in Canada. Hemlock has been mostly used in New York, from its abundance and cheapness Pine would be more durable.

Number and size.—At first, five or six, each six inches square, were placed under 16 feet plank. The Canada

* But, in ascending a long hill in *either* direction, it should be on the right hand side.

Board of Works' Specification, 1845, directs four to be put under a 16-feet road, and three under a 10-feet road; the outer ones to be five inches square, and the inner ones to be six inches wide, and two inches thick, laid flatwise. On the New York roads of eight feet planks, two sleepers, four inches square, have been generally employed. They have, however, been found insufficient, and the experienced engineer of the original Syracuse road, strongly recommends sleepers 12 by 3, laid on their flat sides, and for an important road would make them 12 by 4, or even 12 by 6.* They should be large and strong enough to hold up the plank road in case of a soft place for a few feet. Others argue, however, that they should be small enough to sink down with the earth as it settles under the planks, so that these may continue to bear upon the ground; as otherwise the planks would be rapidly worn out by the springing thus caused, and would be soon rotted by the confined air under them. They also assert that the only use of the sleepers is to keep the road in shape when first laid down. Indeed, a road three miles long has been laid in Canada, without any sleepers at all under the planks and it worked quite well. Its advocates say that sleepers form a trench in which water collects, and is by them prevented from running off. It therefore floats the planks, or washes out mud from under them, and thus forms a cavity, which produces the bad effects above mentioned. This consideration would make light sleepers appear to be worse than none. The conclusion seems to be that large sleepers should be used for an important road; and that for a poor one, which expects to receive only light loads, and which runs over a hard bottom, sleepers might perhaps be altogether dispensed with.

* The *lower* sleeper may be 14 inches wide, and the other 10, as the former acts as a bridge over the channels made under it to let off the water; and also sustains a somewhat larger share of the weight.

Length.—The sleepers used should be as straight and true as possible. On the Syracuse road none less than 13 feet long were admitted. On the Canada roads they are required to be not less than 16 feet, nor more than 20 feet long.

Laying.—Their distance apart, centre to centre, should be such that the wheels of loaded wagons may pass directly over their middle ; or somewhat nearer to their outer than their inner sides. This distance will therefore vary in different sections of the country, according to the usual "track" of wagons.* If this principle be varied from, it should be by bringing the sleepers nearer the middle than the ends of the planks, to prevent any de pression in the centre. The foot-wide sleepers in the figure are drawn three feet apart in the clear, or four feet centre to centre.

They should be well bedded in the earth, in trenches cut to receive them, with their top surface barely in sight. They should bear firmly and evenly throughout their whole length, and the earth between them be well rammed down, and made firm, solid, and even.† The sleeper nearer the ditch is to be laid so much lower than the inner one, as to give the proper slope to the road, which is so important for carrying off the rainwater.

Joints.—At the joints, where two sleepers come together, end to end, they are liable to sink under passing loads. To prevent this, various means may be employed.

* The common track of wagons, measured "from inside to outside," which is the same as from centre to centre, is four feet eight inches in the state of New York. In New Jersey and the Southern States, it is five feet. In Connecticut it varies from three feet eight inches for light wagons, to five feet two inches for heavy ones. In Wisconsin, it is five feet four inches.

† A wooden *roller*, weighing two tons, has been very successfully used for settling the sleepers and the earth between them, being drawn over them several times before they are planked.

The broad sleepers (12 by 3) may be sawn in two length-wise, so as to be each 6 by 3, and laid side by side, so as to "break joints;" the joints of one set being opposite the middle of the adjoining pieces, which form the other set. This arrangement is shown in Fig. 115, b, page 230. The sawmills charge no more for the sleepers in two pieces, each 6 by 3, than in one 12 by 3. A second remedy is to lay a short board under the joints of the sleepers, as shown in Fig. 115, c. A third is to connect the ends by a mortice and tenon, two inches long, as in Fig. 115, d. A fourth is to unite them by a bevel scarfing, three inches in length, reversed on each half, as shown in Fig. 115, e, in which, for distinctness, the two sleepers are represented as separated. In every case the joint on one side of the road ought to be opposite the middle of the sleeper on the other side.

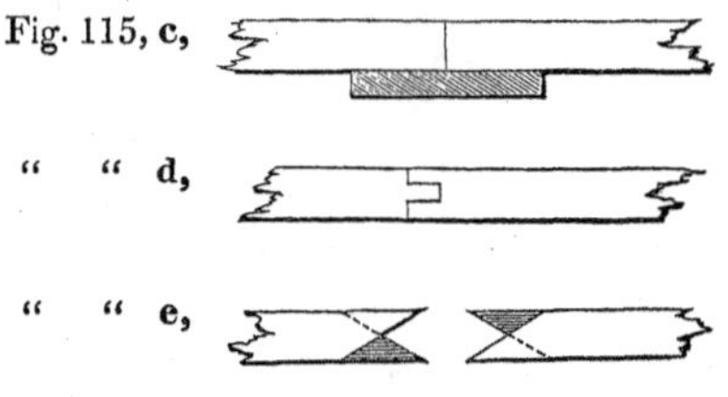

PLANK.

Material.—In Canada, pine, hemlock, tamarack, oak, and walnut, have been employed. In this State, hemlock alone has been used, being the cheapest material to be obtained. Its defects are its perishable nature, and its numerous knots, which soon make the road rough, when the softer portions of the planks have worn away. Pine, oak, maple, or beach, would be preferable. In Wiscon sin, &c., white and burr oak are abundant, and would therefore be advantageously used. Oak would make the most permanent road, from its superior capabilities of resisting both wear and decay. From its greater weight it would cost a little more for hauling and handling. The slipperiness of hardwood has been made an objection to it, but the sand with which the road should be covered, would obviate this. Whatever sort of timber is employed, it should be sound, and free from sap, bad knots, shakes, wanes, or any other imperfections. The plank should be full on the edges, and not less than nine nor more than sixteen inches wide, if of soft wood, or not more than twelve, if of hard wood.

Thickness.—The planks are usually either three or four inches thick; but the builders of the later roads prefer giving less strength to the plank, and more to the sleepers, which are more durable; and therefore recommend three-inch plank, with sleepers a foot wide. With hemlock plank, any thickness beyond three inches is wasted, for when two inches have been worn down, the projecting knots will make the road too rough to travel on, and it will require renewal. One inch more will be sufficient to hold the knots in, so that we get three inches as the proper thickness.* With less knotty timber, thicker plank may be used, *provided* there will be travel enough to wear out the whole thickness from above, before it unprofitably rots out from below. When two tracks are laid, that which would be travelled by the loaded wagons going to market may be laid with four-inch plank, and the other track, for the light wagons, with three-inch plank.

Laying.—The planks should be laid directly *across* the road, at right angles, or "square," to its line, as shown in Fig. 115, b, on page 230. The ends of the planks are not laid evenly to a line, but project three or four inches on each side alternately, so as to prevent a rut being formed by the side of the plank track, and to make it easier for loaded wagons to get upon it; as the wheels,

* The knots may, however, be cheaply dubbed down with an adze.

instead of scraping along the ends of the planks, when coming towards the track obliquely after turning off, will on coming square against the edge of one of these projecting planks, rise directly upon it. On the Canada roads, every three planks project three inches on each side of the road alternately, as shown in Fig. 115, b.

The planks were laid *lengthwise* of the road, on the first one running from Quebec, it being supposed that they would wear better, and could be more easily taken up and replaced. But it was found that loaded horses slipped upon them, (the longitudinal direction of the grain giving no hold to the feet,) that ruts were soon worn in them, and that they did not keep their places. This arrangement is therefore now abandoned.

The planks have also been laid *obliquely*, diagonally, or "skewing;" so as to make an angle of 45 degrees with the line of the road, twelve feet plank making an eight-feet wide road. This plan is adopted on the Longeuil and Chambly road near Montreal. Its advantages are, that the edges of the plank are not worn down so soon as when the wheels strike them directly, (as was shown in reference to pavements, on page 222;) that the zigzag ends of the plank facilitate the getting on the track; and that there is less loss on the rejected, or "cull" planks of 12 feet, than on those of 8 feet. But when a wagon-wheel comes upon one end of a plank laid thus obliquely, the other end, having no load to keep it down, will spring up, if not fastened to the sleeper; and if it is, the spikes or pins will finally be loosened. Each end of each plank undergoes this action in turn, and thus the road is injured and broken up. The first method of laying the planks—at right angles to the direction of the road—is much to be preferred.

The planks must be laid so as to bear equally on the sleepers, and on the ground between them, depending chiefly on the latter for their support. The earth must be well up to and touching the planks at every point, for if any space of confined air be left, dry rot soon follows If any water be allowed to get under the planks, it forms a soft mud, which is pressed up between them, and deposited on their surface, thus excavating a cavity under them, and rendering them liable to move under passing loads in a manner which soon wears them out. They must also be laid to close joints.*

Fastening.—On the Canada roads the planks have generally been spiked or pinned down to the sleepers. The specification of the Board of Public Works directs them to be spiked "with one spike at each end for planks 12 inches wide or less, and two at each end for planks of a greater width. The spikes are to be of the description called 'pressed' spikes, made of the best English or Canadian iron. They are to be $6\frac{1}{2}$ inches long, $\frac{3}{8}$ inch square, with chisel-shaped edges, and good broad heads, and are to weigh five to a pound. They are to be driven with the chisel-edge across the fibres of the wood."

On the New York roads this has been considered an unnecessary expense, since the loads come equally upon both ends of the transverse planks, and thus tend to keep them down in their places, their own weight assisting in this. But in wet, and badly-drained places, a new consideration intervenes. If the planks are not fastened down, they will float as soon as an inch of water gets under them. The wheels of a loaded wagon pressing down each plank in turn, drive the water before them, till it finally attains force enough to throw up a plank, and thus break up the road. On the other hand, when the planks

* Never allow the earth on the sides of the track to rise above the ends of the plank

are fastened down, the whole road is floated, and the vibrations produced by the passing loads drive the water out on the sides and top of the road, and excavate cavities, which ought to be immediately filled up, an operation which is made difficult by the fastening down of the planks to the sleepers. It is therefore thought better to leave the plank free, and allow them to be thrown out of place, and thus at once give free passage to the water, and prevent further mischief; a repairer being kept constantly at work upon the road, and required in rainy weather to pass over every portion of it once or twice a day. It might be well, as a compromise, to spike down planks at short intervals, say every fifth or tenth plank, the rest being well driven home against these.

Covering.—The planks having been properly laid, as has been directed, should be covered over one inch in thickness, with very fine gravel, or coarse sand, from which all stones, or pebbles, are to be raked, so as to leave nothing upon the surface of the road, that could be forced into and injure the fibres of the planks. The grit of the sand soon penetrates into the grain of the wood, and combines with the fibres, and the droppings upon the road, to form a hard and tough covering, like felt, which greatly protects the wood from the wheels and horses' shoes. Sawdust and tan-bark have also been used.

The road is now ready for use.

COST.

The chief items in the cost of a plank road are the timber and the earth-work. The price of the former will vary greatly in different localities and at different times. The cost of the latter, as well as of bridges, culverts, &c., will generally be different on every mile of road. The cost of plank roads in general, therefore, cannot be definitely stated. The following estimate gives the extremes.

On the plan recommended, the *planking* will require, per mile, 8 × 3 × 5280 = 126,720 feet; and the *sleepers* (2) × 1 × 3 × 5280 = 31,680 feet; in all 158,400 feet; or, say, 160,000 feet, board measure. *Shaping* the road-bed, and *laying* the sleepers and planking, costs from 30 cents to $1 per rod, according as the line is new, or on an old bed, and the soil easy or hard to work. The number of *gate-houses* will be governed by the opposing considerations of making them many, so that no one can travel far on the road without paying therefor; and few, so that the expenses of collection may be small. By the New York Plank Road law, the toll-gates are not to be within three miles of each other. The item of *contingencies* will not bear any relation to the varying cost of the plank, and therefore should not be estimated by a percentage, as is usually done These points being premised, we arrive at the following estimate of *Cost per mile :*

Plank: 160 M.; $4 to $10 per M.; .	$640	to	$1600
Shaping and Laying; 30 cents to $1 per rod,	96	"	320
Gate-houses: per mile, ...	50	"	150
Engineering and superintendence, . .	100	"	100
Contingencies, ...	100	"	200
	$986	to	$2370

We thus see that the cost per mile will range from, say, $1000 to $2400, *exclusive* of extra earth-work, bridges, culverts, &c. From 10 to 15 cents per cubic yard may be estimated as the cost of the *excavation*, including putting it into *embankment*, except when carried over one or two hundred feet, (see page 132;) and it should be stipulated that no cutting of less than two feet depth, should be counted, or paid for, as "excavation;" but be considered as included in the general price for laying. In making a new road through a forest, the *clearing* and *grubbing* will be a new item of expense. Add ten per cent. upon the cost of these items for contingencies incident to them. The land is supposed to be given.

The *Syracuse and Central Square* plank road, 16 miles, cost $1487 per mile, with lumber at $5.20 per M. It has a single eight-feet track, except over a few spots of yielding sand. The *Rome and Oswego* road, 62 miles, cost $80,000, or about $1300 per mile; lumber costing from $4 to $5 per M. It is of eight feet hemlock plank, three to four inches thick; with grades cut down to 1 in 20 near Rome, and at the western end, where it is more hilly, to 1 in 16½. The *Utica* northern road, 22 miles, cost $42,000, (besides $8000 for right of way over a turnpike,) being nearly $2000 per mile, five miles being a new line cut through woods, at an extra cost for clearing, of $500 per mile. Deduct this, and the average cost would be about $1800 per mile. A short road near *Detroit*, eight feet wide, laid on a travelled roadway, cost, with lumber at $6 per M., $1500 per mile.

The first New York road (Syracuse and Central Square) was not built by contract, but by days' work, so as to ensure the perfect bedding of the timbers. It was also found that the work was done at a less cost than the bids of contractors, who made such offers as would secure them against loss in a work then new and untried. In a road where there was much earthwork, that at least should be let by contract. The road should also be divided into quarter-mile sections, and the lumber for each be contracted for, to be equally distributed along the line, when delivered The actual laving upon the graded bed could then be done by days' work. All the operations should be under the charge of an intelligent and efficient engineer.

DURABILITY.

A plank road may require renewal, either because it has been worn out at top by the travel upon it, or because it has been destroyed at bottom by rot. But, if the road have travel enough to make it profitable to its builders, it will wear out first; and if it does so, it will have earned abundantly enough to replace it twice over, as we shall see presently. The liability to decay is therefore a secondary consideration on roads of importance.

Wear.—The actual wear is of course proportioned to the amount of travel. The most definite results have been obtained on the first New York road, that from Syracuse to Central Square. In its first two years, ending July, 1848, more than 160,000 teams passed over its first eight miles. This travel wore its *hemlock* plank down *one inch*, where they had not been floated. Another inch could be worn down before the projections of the knots would make it necessary to relay the road, so that it would have borne the passage of 320,000 teams. But this is an under-estimate, inasmuch as the wear and tear of the first year is more than that of several following; since the first travel upon the road tears off the outer splinters and fibres cross-cut by the saw, while the coating subsequently formed protects the plank from wear. Upon a Canada *pine* road, travelled over by at least 150 two-horse teams per day, (50,000 per year.) the road had worn down in two years only one-quarter of an inch; and this too was attributed chiefly to its exposure the first year without sanding. It was estimated that sanded plank on this road would wear at least ten years. *Oak* would of course wear longer.

Decay.—As to natural decay, no *hemlock* road has as yet been in use long enough to determine how long the plank can be preserved from rot. Seven years is perhaps a fair average. Different species of hemlock vary greatly; and upland timber is always more durable than that from low and wet localities. The *pine* roads in Canada generally last about eight years, varying from seven to twelve. The original Toronto road was used chiefly by teams hauling steamboat wood, and at the end of five years, began to break through in places, and, not being repaired, was principally gone at the end of ten years. Having been poorly built, badly drained, not sanded, and no care bestowed upon it, it indicates the minimum of durability. *Oak* plank cross-walks in Detroit, the plank being laid flat on the ground, have lasted two or three times as long as those of pine. It is believed that oak plank, well laid, would last at least 12 or 15 years. One set of sleepers will outlast two plankings; several Canada roads have been relaid upon the old sleepers, thus much lessening the cost of renewal.

A Canadian engineer thinks that $20 per mile would be required the first year; to restore the grade where it had settled, to fasten loose plank, &c. For the next five years, $10 per mile, and then there would be some planks to be replaced. The repairs would then increase so as to amount to a renewal of the surface at the end of four years more, making ten for the age of the road.

ADVANTAGES.

Plank roads are the *Farmer's Railroads*. He profits most by their construction, though all classes of the community are benefited by any such improvement, as has been fully shown in the "Introduction" to this volume. The peculiar merit of plank roads is, that the great diminution of friction upon them makes them more akin to railroads than to common roads, with the advantage over railroads, that every one can drive his own wagon upon them. Their advantages naturally divide themselves into two classes; their utility to the community at large, and their profits to the stockholders who build them.

1. *To the community*. A horse can draw on a plank road from *two to three times* as much as he can on an ordinary Macadam or good common road. On the latter roads one ton is a fair load for a single horse, and 3000 lbs. the utmost allowance. But upon a plank road, a two-horse team has drawn *six tons* of iron; another has drawn, for several days in succession, over two cords of green beech and maple wood, estimated at six tons also, and could draw four or five tons, thirty miles a day continuously. These results of experience agree with the calculations founded on the data of p. 62, taking the friction on a Macadam road at $\frac{1}{36}$, (the average of the two values there given,) and that on planks at $\frac{1}{98}$. The resulting ratio is $2\frac{3}{4}$ to 1.

A great degree of speed can also be obtained upon plank roads with much less injury to the vehicles and to the horses feet than on a Macadam road, though contrary impressions have sometimes been caused by the excessive speed with which their light draught often causes horses to be driven, without the driver being aware of it. Eight feet of a Canadian McAdamized road, disrupted by frost, was *taken up*, and planked over; and the horses when reined from the plank to the stone, in turning out, would of their own accord, if not prevented, immediately turn back upon the plank.

But the peculiar advantage to the community of plank roads is their continuing in perfect order, and affording undiminished facilities for travel, at all seasons, while common roads are rendered impassable by the continued rains of autumn, the occasional thaws in mid-winter, or the "breaking up" in spring. They thus enable the farmer to carry his produce to market at seasons and in weather when he would otherwise be imprisoned at home, and could not there work to advantage. His farm will therefore be made more valuable to him; and it has accordingly been found that the value and price of lands contiguous to those roads have been enhanced by their operation to such a degree as to excite the envy and complaints of those living off their line. The lessened "stiction" will also enable him to carry his former load to a more distant market, if desired, or to carry to his former market a larger load, and therefore at less cost per bushel, hundred-weight, or cord. He can therefore sell cheaper, and yet gain more. The consumer of his produce, wood, &c., gets a better supply of all articles, and at lower prices. The shopkeepers carry on an active trade with their country customers, at times when, were it not for these roads, they would have nothing to do. It is one of those few business arrangements by which all parties gain, and which, therefore, in the words of Clinton, actually "augment the public wealth."

2. *To the stockholders.* The annual profits of a plank road will of course be governed by the two elements of its first cost, and the amount of travel upon it. The latter should be approximately determined in advance, as directed on page 66. One important point has, however, been determined with considerable accuracy, viz.: how much a road will earn before it is worn out. Upon the first eight miles of the Syracuse and Central Square plank road, the tolls during its first two years, ending July, 1848, amounted to $12,900, and the expenses for salaries and repairs to $1,500; leaving $11,400 for dividends and rebuilding. This amount of travel had worn the plank down one inch. Another inch could be worn down before a renewal would be necessary, and the road would then have earned $22,800 above expenses, or $2,850 per mile. This experience indicates that hemlock plank before being worn out, will earn two or three times their original cost. The surplus above the cost of renewal will therefore be payable in dividends, amounting in gross to between 100 and 200 per cent. upon the first cost of the *plank*, (that of the *whole road* bearing no constant ratio to this ;) the amount of *each* annual dividend being of course greater the more rapidly this wearing out, with its concomitant and proportional earning, takes place.

This calculation is predicated on the tolls established by the New York Plank Road law, which are as follows: For any vehicle drawn by two horses, &c., 1½ cents per mile, and ½ cent for each additional animal; for vehicles drawn by one horse ¾ cent per mile; for a horse and rider, or led horse, ½ cent; for every score of sheep, swine, or neat cattle, one cent per mile. But the companies are not to charge more than will enable them to pay annual dividends of 10 per cent. upon the stock actually paid in and expended on the road, after keeping the road in repair, and setting aside 10 per cent. for its reconstruction. This restriction has since been repealed.*

The great *objection* to plank roads in the eyes of an engineer is their perishable nature, and consequent final destruction. But this fault is one not peculiar to *plank* roads, but common to all in a greater or less degree. Thus in the case of broken-stone, or McAdam roads, usually cited as contrasting models of durability, we find that they wear away so rapidly as to require not only constant repairs, but, when well kept up, an actual addition to their substance of one cubic yard per mile for each beast of burden passing over them, (see page 209;) and the 80,000 teams per year which passed over the Syracuse road, would have required an amount of broken stone, to replace their wear, enough to renew it many times over. A Canadian report to the Board of Public Works shows that the cost of one mile of McAdam road will there make and maintain nearly four miles of plank road ; and on one road the substitution of plank for broken stone effected a saving of an amount sufficient to replank the road every three years, if that had been necessary. The New York Senate report states that a plank road over the same line with a McAdam one can often be built and maintained for less than the interest on the cost of a McAdam one, added to the expense of its

* The New York Laws relating to Plank Roads, are –1847, chapters 210, 287, 398 : 1848, chapt. 360 : 1849, chapt. 250.

necessary annual repairs. But even if a plank road was still more perishable than it is, and was worn out in one year, still, if in that time it had repaid its cost two or three fold, (as we have seen it would do,) it would be so much the more profitable investment; and this is the final object of all private engineering constructions.

It should not be forgotten by the engineer engaged in laying out a road for a private company, that their interests, and those of the public who are to use the road, are not identical. The public wish the road to be so laid out that they can carry over it the greatest possible loads at the least possible cost. The stockholders generally wish only to secure to themselves the largest possible amount of tolls in return for the smallest possible investment. These two interests conflict. The steep ascents, so injurious to the travelling public, as shown on pp. 231–3, are advantageous to the company who plank the road, since they prevent large loads being carried, and thus produce a twofold gain—the amount of tolls being proportioned to the number of the loads, and not (as they should be) to their weight; and the carriage of such excessive ones as would break defective plank being thus prevented. The engineer of the company must therefore sacrifice the absolute perfection of his road to this requisition of policy, and may leave steep ascents untouched, thus saving the first cost of cutting them down, as well as increasing the subsequent receipts. But, on the other hand, if the grades of the road be not sufficiently improved, it may not attract the expected amount of travel. A prudent compromise must therefore be made between these opposing interests

WOODEN PAVEMENTS.

Fig. 116.

Pavements formed of wooden blocks, usually hexagonal in shape, possess many advantages. They cause little resistance to draught; are almost entirely free from noise; are easily kept clean; are easy to a horse's hoof; lessen very much the wear and tear of vehicles; are pleasant to travellers; admit of great speed, and are cheaper in their firs cost than granite blocks.

To counterbalance these recommendations, they are slippery and therefore dangerous in wet weather; and are very perishable, both from wear and from decay. The slipperiness has been obviated by grooving and striating their surface, but this lessens their ease of draught and noiselessness, and increases their cost.* The rapidity of their wear may be lessened by setting them on a foundation of broken stone, or of concrete, so shaped as to rapidly drain the water from their bottoms; and by covering their surfaces with a mixture of boiling tar and clean gravel. Their decay may be prevented by various chemical preservatives, of which the principal are, Kyan's, who saturates the wood with a solution of bichloride of mercury or corrosive sublimate (one pound to five gallons of water); Burnett's, who uses a solution of chloride of zinc, (one pound to ten gallons of water) absorbed in a vacuum; Renwick's, with coal tar; and Boucherie's, with the impure pyrolignite of iron, absorbed by the vital action of the sap vessels.

---

* A description of various forms proposed for wooden pavements may be found in the N. Y. American Repository, vol. iii.; and in London Mechanics' Magazine, and Repertory of Patent Inventions, *passim*.

6. ROADS OF OTHER MATERIALS.

BRICK.

Roads are made in Holland of hard burnt bricks, or "clinkers," laid on a firm foundation, and set on edge, with their longest dimension across the road. A better bond would be obtained by such an arrangement as is shown in the figure. But the pressure of heavy loads and the blows of horses' feet are too powerful for bricks, which should therefore be reserved for foot-pave ments only.

Fig. 117.

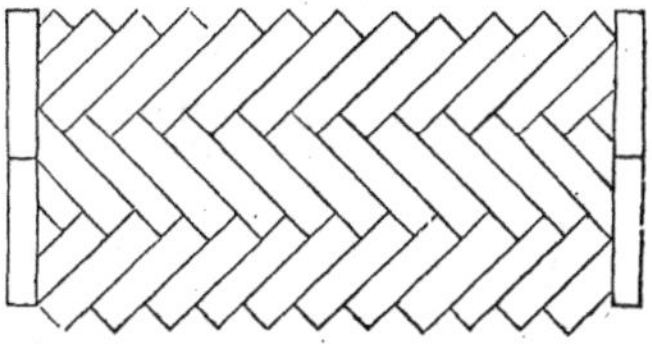

CONCRETE.

Roads of concrete, or *béton*, six to eight inches thick, (such as has been described as the best foundation for granite blocks) have been warmly advocated in France, particularly for the use of steam carriages, in the place of the more costly, though more perfect railroads. Concrete will sustain great weights, carried on wheels, with little injury, but has been found (on the towpath of a canal aqueduct) to be rapidly destroyed by the feet of horses.

CAST IRON.

This material has been tried several times, but abandoned in consequence of its wearing so smooth as to cause horses to slip

17

ASPHALTUM.

This name has been given to a bituminous mastic, of which the principal localities are Seyssel in France, and Val-de-travers in Switzerland. A limestone is also there found, which contains from 3 to 15 per cent. of bitumen. The stone is broken into fragments of the size of an egg, and ground to powder. A certain proportion, usually from 6 to 10 per cent., of mineral tar (obtained by boiling in water the bituminous sandstone of the same place) is combined with the limestone, by heating the former in iron boilers, and gradually adding and stirring in the powdered stone. In this state it is poured upon a level surface, and forms smooth cakes, over which gravel is spread. It is too weak for carriage-ways, and in this climate too soft in summer, and too brittle in winter, for even foot-pavements; but in Paris the asphaltum side-walks of the Boulevards are most perfect specimens of pavements. The asphaltum is melted on the spot in large caldrons, and poured within a moveable frame to the desired thickness. The edges of these slabs are united with the same material, and the pavement before an entire block of houses is thus made one smooth level surface, unbroken by a single joint.

CAOUTCHOUC.

A pavement, formed by mixing gravel with melted caoutchouc, or gum elastic, has been tried in London. A specimen in the court-yard of the Admiralty, in 1844, was very pleasant to walk upon, but showed permanent depressions where heavily loaded vehicles had passed over it.

7. ROADS WITH TRACKWAYS.

When wheeled carriages are drawn by horses, the wheels should move on the smoothest and hardest surface possible, while the horses require one rough enough to give them a secure foothold, and soft enough to be easy to their feet. These two opposite requirements are united only in *Roads with Trackways*, on which two parallel tracks of suitable materials are provided to receive the wheels, while the space between the tracks is filled with a different material, on which the horses travel. The wheel-tracks are usually of stone, of wood, or of iron.

STONE TRACKWAYS.

The Egyptians seem to have first discovered the value of stone trackways in moving great weights, for traces of such contrivances have been found in the quarries which supplied the enormous stones of their Pyramids. In modern times they reappeared in the streets of Pisa, and are now general in those of Milan. They have of late years been used with great advantage in London, upon a road over which 250,000 tons annually passed, in wagons carrying each five tons. The repairs of this road for thirteen years cost less than one hundred dollars. The friction upon this stone trackway was so much reduced (being only $\frac{1}{180}$ of the weight) that a small horse (weighing 4½ cwt.) could draw on a level 15 tons; and a powerful horse (weighing 14 cwt.) 30½ tons, at the rate of 4 miles per hour. On this road the tracks were blocks of granite 5 or 6 feet long, 16 inches wide, and 12 inches deep. The space between them was paved.*

---

* Parnell, p. 106

A similar trackway of stone has been used with grea advantage to facilitate the ascent of a steep hill, as a sub stitute for reducing the inclination. Upon the Holyhead road, two hills, each a mile in length, had an inclination of 1 in 20. To reduce this to 1 in 24 would have cost $100,000. Nearly the same advantage, in diminishing the tractive force required, was obtained by moderate cutting and embankment, and making stone trackways, at a total expense of less than half the former amount. To draw one ton over the original hills required a power of 294 lbs.; to draw it over the trackways laid on the same inclinations required only 132 lbs.; so that the tractive force was reduced more than one-half by this improve ment; and the effect was the same as if the hill had been cut down to a level, its surface remaining unchanged The arrangement of this trackway is shown in Fig. 118

Fig. 118.

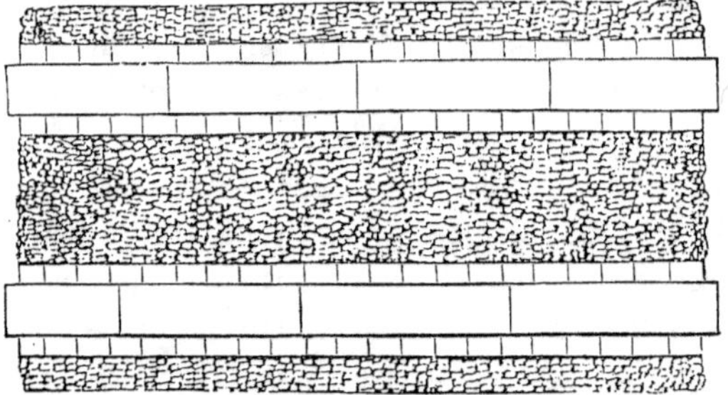

The blocks were of granite, twelve inches deep, fourteen inches wide, and not less than four feet long. A foundation for them was prepared by making an excavation, 8 feet wide and 25 inches deep. On its levelled bottom was laid a rough pavement (like that described for the "Telford road," page 210) eight inches deep. The joints were also filled with gravel. Upon this pavement were laid three inches of broken stones, none exceeding one and a half inches in their longest dimensions. On them was a layer of two inches of the best gravel, over which a heavy roller was passed. Upon this the stone blocks or "trams" were laid to a very accurate level. The spaces between and outside of them were filled up to a depth of six inches with broken limestone. On each side of the blocks was placed a row of paving-stones of granite, six inches deep, five inches wide, and nine inches long. The remaining space was filled up with hard broken stone, and the whole covered with a top dressing of an inch of good gravel.*

WOODEN TRACKWAYS.

In districts where timber abounds, it may be substituted for stone in forming tracks, on which the wheels of ordinary vehicles may run. Projections on the sides of the tracks may be employed to retain the wheels upon them, but the moisture retained in the joints would cause rapid decay, and if any such precaution be thought necessary, a furrow or gutter in one of the tracks would be preferable.† It would of course be necessary in this case, that the road should everywhere have sufficient inclination to carry off the water, which would otherwise fill the furrow.

Fig. 119.

The road-bed should first be properly shaped, with an inclination each way from the centre, and the timbers be completely imbedded in it. Two tracks should be laid, for the travel in the two directions. A faster vehicle, overtaking a slower one, could easily leave the track and

* Parnell, p. 109.

† See report of Mr Jno. S. Williams, American Mechanics' Magazine, p. 210

re-enter it after passing. The outside timber of each track should be smooth on its upper surface, and the inner one have hollowed in it a furrow, about 3 inches deep, 4 inches wide at bottom, and twice that at top. The flat timber should be wide enough to allow for the usual variation in the widths of vehicles. The rise of the road between the two timbers should just equal the depth of the furrow, so that the two wheels may be on the same level. The distance between the centres of the timbers should be about 5 feet; between the two tracks a space of four feet should be left; and on the outside of each, nine and a half feet for a summer road, making a total width of 33 feet, or two rods.

The railroad from Clifton to the Adirondack mines, New York, is made of wooden rails. They are of hard maple, 6 by 4 inches, and 14 feet long. They are set on edge into notches in the ties, and are fastened by wooden wedges.

IRON TRACKWAYS.

The wooden tracks, adopted more than two centuries ago in the coal-mines of England, were before long covered with thin plates of iron to increase their durability and to lessen their friction, and subsequently replaced by tracks entirely of iron. While a flange on their sides was used to keep carriages upon them, they were "tram-roads," but when the flange was transferred from the road to the wheel, the trackway became a RAILWAY. The extent of this topic demands for it a separate chapter.

Fig. 120

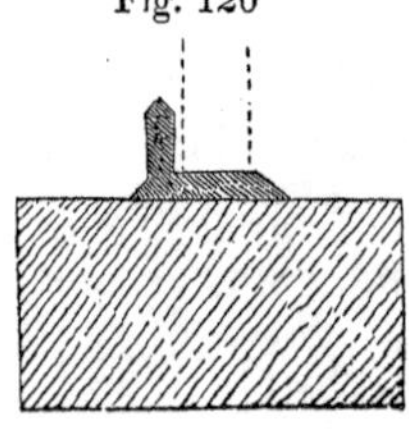

CHAPTER V.

RAIL-ROADS.

"Nothing can do more harm to the adoption of railroads, than the promulgation of such *nonsense* as that we shall see locomotive engines travelling at the rate of 12, 16, 18, and 20 miles per hour!"

WOOD, on Railroads, 1825.

"An express train on the Great Western Railway, drawing 59 tons, has travelled, for three hours, at the rate of 63 miles per hour!"

RITCHIE, on Railways, 1846.

THE great success and rapid extension of railroads, are due to that appreciation of the value of time, which is the characteristic of the present age. The speed obtained upon them virtually and practically shortens distances in the precise ratio in which it abridges the time occupied in travelling over them.

The rapidity of motion and power of traction, which are attainable on railroads, depend on their diminution of friction. This is the chief element in the improvement of the surface of all roads, and in the preceding chapter we have considered, in the order of their progressive merits, the various means which may be employed for that object. In railroads we have arrived at their climax.

The essential attributes of a railroad are two smooth surfaces, usually of iron, for the wheels to run upon. These surfaces must be made as narrow as possible, to lessen their cost, and some contrivance to keep the wheels upon them is then rendered necessary; the usual one at present being a projection, or "flange," on the inner rim of the wheel.

Since the peculiar wheels, which are the chief source of the superiority of railroads, prevent the vehicles which are adapted to run upon them, from being used on ordinary roads, railroads pass out of the practical scope of the present treatise; for the details of their construction no longer belong to the community at large, but demand the highest professional skill of the Civil Engineer. The general interest, however, in the subject of railroads seems to demand some explanation of the leading principles which should govern those engaged in their establishment, and some account of the ingenious contrivances which have been adopted to overcome the difficulties, which have, one after another, risen up in vain efforts to stop the progress of the giant. A brief popular view of these topics (without the minute practical details with which the subject of roads in general has been treated) will accordingly be given in the present chapter.*

Wooden railways were employed as a substitute for common roads, in the collieries of England, soon after the year 1600.† The earliest record of their existence is in the life of the Lord Keeper North, wherein it appears that about the year 1670, they were used at Newcastle-on-Tyne, for transporting coal from the mines to the river, and enabled one horse to draw four or five chaldrons.

---

* The principal authorities consulted have been *Lecount*, "Treatise on Railways," from the seventh edition of the Encyclopedia Britannica; *Ritchie*, "On Railways;" Professor *Vignoles'* Lectures; and the reports and discussions in the "Civil Engineer and Architect's Journal;" "Journal of the Franklin Institute;" "American Railroad Journal," &c.

† Ritchie on Railways, p. 19.

Subsequently these wooden rails were covered with plates of iron ; but the introduction of rails wholly of iron seems not to have taken place till 1767.* A projection, or flange, on the outer side of the rails, kept the wheels of carriages upon them. They were then called "Tram-roads." The objections to them were the broad surface of the plate, which collected obstructions upon it, and the great friction of the wheels against the side-flange.

In 1789, was constructed the first public railway in England, at Loughborough, by Mr. William Jessop, and he introduced cast iron edge-rails, and wheels with the flanges cast upon them instead of on the rail. "Tram-roads" were, however, still in use in 1808.†

In 1803, malleable iron rails were first tried, but not approved of. In 1808, they were introduced into some coal works in Cumberland, and used with complete success.‡ Since that time they have become almost universal, and have been formed into a great variety of shapes, the best of which will be noticed in the section on "Construction." The progress from the use of horse power to locomotives of the present power and speed, will be included in the examination of "Motive powers."

In our condensed sketch of the extensive subject of Railroads, divisions and subdivisions, analogous to those of the previously examined topic of roads in general, will be employed, and thus the coincidences, and the differences, of the principles appropriate to each, will be made more prominent and striking.

The following is an outline of the proposed arrangement:

---

* Hornblower's Report to House of Commons in 1811.

† Ritchie, p 22.

‡ R. Stevenson's Report, 1818.

I. What Railroads ought to be.

To determine "What Railroads ought to be," it is first necessary to ascertain what are the *Resistances* to motion upon them which we must seek to overcome or diminish. The nature and amount of these resistances upon a *straight* and *level* road will be first examined, and then their increase on *curves*,* and on *ascents*.†

RESISTANCES ON A STRAIGHT AND LEVEL ROAD.

The amount of these resistances has been usually taken at 8 lbs. to a gross ton of 2240 lbs; or 1 to 280; i. e. it was assumed that a weight of eight pounds suspended from a cord passing over a pulley, and allowed to descend by its own gravity, (as down a well) would draw, on a

* See page 273. † See page 276.

straight and level railroad, a car attached to the other end of the cord, and weighing one ton; or that 1 pound would thus draw 280 lbs. But later experiments have shown that the resistance varies with the velocity; that it is 10 lbs. per ton at a speed of 12 miles per hour, and over 50 lbs. at 60 miles per hour. The most satisfactory analysis of it, is given by the following empirical formula, deduced by Mr. Scott Russell (and communicated to the British association in 1846) from experiments on five different railroads, mostly of the narrow gauge.

The resistance has three principal elements; Friction, Atmosphere, and Concussion.

The first resistance is that of the *Friction* proper of the wheels and axles. It is constant at all velocities, and amounts, in the best-constructed carriages, to 6 lbs. per ton weight of train.*

The second resistance is that of the *Air*. It is considered to be proportional to the surface of the front of the train, and to the square of the velocity. It equals the weight of a column of air, whose base is the frontage of the train, and whose length is the height due to the velocity. This weight, for each square foot of frontage, and for a velocity of one mile per hour, equals 0.0027 lb., or $\frac{1}{400}$ lb. For the usual frontage of 80 square feet, it is therefore one-fifth of a pound at one mile per hour.

The third, or *residual* resistance, is probably due to the unavoidable *Concussions*, oscillations, flexures, imbedding of wheels in rail, friction of air against sides, &c. It may be hereafter decomposed into various elements, but is now taken as proportional to the weight of the train and the velocity, and as being equal to $\frac{1}{3}$ lb. for each ton of train, at one mile per hour.

---

* The tons here used are all gross tons of 2240 lbs.

We are now prepared to find the resistance (in lbs.) of a straight and level railroad to the motion of a train of cars, whose weight (in tons), velocity (in miles per hour), and frontage (in square feet), are given, by the following

RULE.

1. Multiply the weight by 6,—for friction.

2. Multiply the weight by the velocity, and divide by 3,—for concussion.

3. Square the velocity, and multiply this square by the frontage, and divide this product by 400,—for the atmosphere.

4. Add these three results, and the sum is the total resistance. Divide this by the weight, and the quotient is the resistance per ton.

Example 1. A freight train of 100 tons is to be drawn 12 miles per hour. Its frontage is 80 square feet. What is the resistance to be overcome by the motive force?

$$\text{Friction} = 100 \times 6 = 600 \text{ lbs.}$$

$$\text{Concussion} = \frac{100 \times 12}{3} = 400 \text{ "}$$

$$\text{Atmosphere} = \frac{12 \times 12 \times 80}{400} = 29 \text{ "}$$

$$\text{Total resistance} = 1029 \text{ lbs.}$$

$$\text{Resistance per ton} = \frac{1029}{100} = 10\tfrac{1}{4} \text{ lbs.}$$

Example 2. A passenger train of 50 tons is to be drawn 35 miles per hour. Its frontage is 80 square feet. Required its resistance.

$$\text{Friction} = 50 \times 6 = 300 \text{ lbs.}$$

$$\text{Concussion} = \frac{50 \times 35}{3} = 583 \text{ "}$$

$$\text{Atmosphere} = \frac{35 \times 35 \times 80}{400} = 245 \text{ "}$$

$$\text{Total resistance} = 1128 \text{ lbs.}$$

$$\text{Resistance per ton} = \frac{1128}{50} = 22\tfrac{1}{2} \text{ lbs}$$

Example 3. A train of 25 tons, at 60 miles per hour, would meet a resistance (by both theory and experiment) of 55 lbs. per ton. This rapid increase of resistance with velocity, is very striking, though it has been disputed by some experimenters.

The above formula has been tested by Mr. Scott Russell, and Mr. Wyndham Harding, chiefly for passenger trains of from 20 to 64 tons, and at speeds from 30 to 60 miles per hour. At lower velocities, its results somewhat exceed those of the experiments. When the railroad or carriages are in bad repair, or side-winds prevail, the resistances will be greater than are here given. For head-winds, the velocity of the wind should be added to that of the train.

The following Table shows the Resistances to Trains of different weights, and at different velocities, as given both by actual experiments and by the above formula: the frontage being 60 square feet.

Velocity.	Weight.	Resistance by Exper.	Resist. by Formula.	Velocity.	Weight.	Resistance by Exper.	Resist. by Formula.
mi. per hr.	*tons.*	*lbs. per ton.*	*lbs. per ton.*	*mi. per hr.*	*tons.*	*lbs. per ton.*	*lbs. per ton.*
14	9	12.6	13.9	34	30½	25.0	23.1
16	20½	8.5	13.2	34	18	23.4	27.2
19	40¾	8.5	12.9	35	21½	22.5	26.1
21	18	12.6	16.7	39	24	30.0	31.0
25	40¾	12.6	16.6	47	31¾	33.7	33.1
27	40¾	12.6	17.7	50	30	32.9	35.3
31	15½	23.4	25.4	53	25	41.7	42.1
32	14½	22.5	27.2	61	21½	52.6	54.8

When the motive power is a Locomotive *Engine*, its own resistance must also be taken into account. The friction on its machinery, or working parts, may be taken at 7 lbs. per ton of its weight; and its friction considered as a carriage at 8 lbs. per ton. To this should be added according to Pambour, 1 lb. for each ton of the load drawn by it. Its atmospheric resistance is already taken into account, since, if again calculated and attributed to the engine, it should be deducted from the train of cars, which the engine in front of them shields from it.

The usual mode of recording the resistance as so many lbs. "per ton," does not give a satisfactory standard of comparison; one of the resistances (that of the atmosphere) being independent of the weight of the train. An increase of this weight (which is the divisor of the whole) would therefore lessen the resistance *per ton*, while it increased the *total* resistance.

On the other hand, this atmospheric resistance no doubt varies somewhat with the length of the train, and the consequent increased friction of the air against the sides of the carriages. Dr. Lardner (in his report of 1841 to the British Association) considers "the resistance due to the air to proceed from the effect due to the entire volume of the train, and not to depend in any sensible degree on the form of the foremost car." Sharp fronts did not diminish it, nor did an increased frontage (as formed by boards projecting on each side) much increase it. Barlow, in a paper read before the Royal Society in 1836, considers the resistance of the air to increase in a ratio, not as the square, but, not much higher than the simple velocity.

A *new formula*, which assumes this resistance to be directly proportional to the *bulk* of the train, and which also more minutely analyzes the resistances of the engine, has been deduced by Mr. *D. Gooch*, from experiments made in 1848 on a "broad gauge" road. His results have been much disputed. The following is an analysis of them:—

For the Cars, the *Frictional* resistance is taken at 6 lbs. per ton, as before.

The *Atmospheric* resistance is assumed as equal to the square of the velocity, multiplied by the bulk of the train in cubic feet, and that product by $\frac{2}{1000000}$. Each ton weight of the train is supposed to correspond to 180 cubic feet. The atmospheric resistance obtained by this formula would equal that given by Russell, in the case of a load of $55\frac{1}{2}$ tons. For a greater load, this formula makes this resistance proportionally greater than Russell's, and for a less load proportionally less.

The *residual* or oscillatory resistance is taken at only $\frac{1}{15}$ the product of the velocity by the weight, instead of $\frac{1}{3}$, as in the former formula. Mr. Gooch considers this "oscillatory" resistance to be mainly the increased friction of the axle bearing upon its collars, in consequence of the transverse vibrations at high velocities, while Mr. Russell makes it include all the resistances remaining, after "friction" and "atmosphere" are deducted from the total amount.

Example 4. Let weight of train = 100 tons; velocity = 50 miles per hour; required the resistance to the motion of the cars.

$$\text{Friction} = 100 \times 6 \quad - \; - \; - \; - \quad = 600 \text{ lbs.}$$

$$\text{Oscillation} = \frac{50 \times 100}{15} \quad - \; - \; - \; - \quad = 333 \text{ "}$$

$$\text{Atmosphere} = 50 \times 50 \times 100 \times 180 \times \frac{2}{1000000} = 900 \text{ "}$$

$$\text{Total resistance of cars} = 1833 \text{ lbs.}$$

For the Engine *and tender*, the resistance is separated into two parts. That caused by the friction of axles and machinery, is (in pounds per ton of their weight) equal to 5, plus one half the velocity in miles per hour. That due to atmosphere and load equals $\frac{4}{100000}$ of the square of the velocity multiplied by the weight of the train. These resistances would of course be different for each different engine.

Example 5. With weight of train = 100 tons; velocity = 50 miles; and Engine and tender = 50 tons, required resistance of the Engine and tender.

$$5. + \tfrac{1}{2} \times 50 = 30 \text{ lbs. per ton of their weight.}$$

$$\frac{4}{1000000} \times 50 \times 50 \times 100 = \underline{10} \quad \text{“} \quad \text{“} \quad \text{“}$$

Or, the total resistance of Engine and tender $= 40 \times 50 = 2000$ lbs.
Total resistance of Train and Engine $= 1833 + 2000 = 3833$ lbs.,

$$\text{or } \frac{3833}{100 + 50} = 25.5 \text{ lbs. per ton.}$$

The discrepancies in the results obtained by various experimenters and theorizers, show the great deficiencies which exist in the data of the experiments and in the application of the theoretical principles involved.

Assuming for the present Mr. Scott Russell's formula to be approximately correct, we are next to examine the increased resistances which occur on CURVES, and on ASCENTS. This will be done under the heads of "What Railroads ought to be," as to their *Directions*, and as to their *Grades*.

1. WHAT RAILROADS OUGHT TO BE AS TO THEIR DIRECTION.

Straightness of direction is much more important on railroads than on common roads, for two reasons; the economy of straightness, and the resistances and dangers of curves.

ECONOMY OF STRAIGHTNESS.

From the great cost of the superstructure of a railroad, and the continually increasing expense of keeping it in repair, it is highly desirable that it should be as straight, and consequently as short, as possible.

As the earthwork of a railroad costs almost nothing for repairs, while those of its perishable superstructure are very great, and proportioned to its length, as is also the cost, in fuel, wages, and wear and tear of the engines, of running the road, it will often be advantageous to make large expenditures for the former element of cost, in order to lessen the length of the road, and consequently the annual expenditures for the latter.*

Suppose the total cost of a railroad to be $30,000 per mile, the interest of which is $1800; the annual repairs of the superstructure $1000 per mile; and the expenses of engines also $1000 per mile. The total annual expense will then be $3800, which is the interest of $63,000, which sum might profitably be expended to shorten the road one mile, or $12 to shorten it one foot of length. If this single foot gained was the only result of a day's labor of a locating party, it would be a satisfactory equivalent for the expenses of such a day's work.

On these grounds, a *short* route, which has the faults of steep grades and curves of small radius, may profitably receive an outlay of capital upon it, for the purpose of lessening these defects, equivalent to the cost of the difference of distance between it and a *longer* line, which has better grades and curves.

From these considerations it is also seen that a line ought not to diverge from the direct course between its extremities, and thus increase its distance, for the sake of the trade of a small town, for whose benefit the time and fare of all the passengers and freight on the whole line would thus be taxed. It would be preferable to make a branch track to the town.

EVILS OF CURVES.

Curves are necessary evils on most routes, enabling them to pass around obstacles, such as projecting hills, deep hollows, houses too valuable to be removed, &c.

* See Amer. Railroad Journal, August, 1839, for an able development of this position by W. B. Casey, C E.

The greatest economy in curving is found when the line is located in a narrow and sinuous valley, with rocky banks, whose windings can be cheaply followed by suitably adjusted curves. When the line crosses a series of ridges transversely, and nearly at right angles to their general direction, there would be little economy in lateral deviation and curvature.

The *evils* of curves are the resistances which they offer to the motion of cars, and the dangers to which they expose them.

The following are the four principal causes of the resistances on curves :*

1. The obliquity of the direction of the moving power; *i. e.* the angle which the line of traction, drawn from the engine to each car, makes with the tangent to the curve at the middle of each car, in the direction of which the cars tend to move.

2. The pressure and consequent friction of the flanges of the wheels against the outer rail, due to the centrifugal force.

This is partially obviated by elevating the outer rail, as will be hereafter explained.

3. The pressure and consequent friction of the flanges, due to the parallelism of the axles ; for the directions of the tangents at the points of contact of each pair of wheels are different, and therefore if one pair of wheels be perpendicular to its corresponding tangent, the other pair will be oblique to *its* tangent.

This resistance is partly remedied by allowing a "play" of an inch or less between the wheels and the rails. It diminishes as the axles are placed nearer to each other.

* Report of E. F. Johnson, C. E., 1842

and is therefore much lessened by supporting the cars on two trucks, each resting on four wheels, the two axles of which are very near to each other.

4. The fastening of each pair of wheels to the same axle, with which they turn.* The wheel on the outer side of a curve must revolve farther, and therefore faster, than the inner one, which must slide (if both are of the same diameter) by an amount equal to the difference between the lengths of the inner and outer rails of the curve.

To lessen this resistance, the wheels are made conical, with their inner diameters greater than the outer, so that on curves, the outer wheels run on their greater diameter and the inner ones on the less. This cone may be so adjusted, that the wheel can run in a circle of 595 feet diameter without the flanges touching the rail. It was at first 1 in 7, but has of late been reduced to $\frac{1}{20}$ and $\frac{1}{40}$.

Without these arrangements, the resistance of a curve of even a mile in radius, at a speed of 25 miles per hour, would equal that of an ascending grade of $9\frac{1}{2}$ feet per mile ; and one of 700 feet radius, a grade of 77 feet, &c.

The actual resistance has been very imperfectly ascertained.

Grave objections have been urged by eminent engineers against "coning" the wheels. One is, that if the rail is canted so as to fit the coning, on the straight line portions of the road there will be a constant grinding action, owing to the dragging of the outer and smaller portions of the wheels. It is also stated that much of the concussions of the flange of the wheel against the rail is due to the coning, and that where cylindrical

* If they turned on the axle, as in ordinary carriages, they would not have sufficient steadiness to run truly at high velocities.—LECOUNT, p 134.

surfaces have been used for the wheels (instead of the conical ones, as is usual), the trains have run much steadier.

The actual resistance on curves has been very imperfectly ascertained. See note on page 454.

The amount of mechanical power absorbed in passing around a curve is altogether independent of the radius of the curve, and depends only on the amount of the entire angular change in the direction of the line. When the curve has been run by "Angles of deflection," its length in chains, multiplied by its angle of deflection, equals the entire angular change. Thus, a curve of 1°, 30 chains long, offers the same resistance as one of 3°, 10 chains long.* Sharp curves are therefore not objectionable on the score of loss of power, though highly so from their wear and tear of engines and cars, displacement of rails, danger, &c.

The *danger* of running off the track is much increased by curves, even of large radius, especially at high velocities. The momentum of the cars impels them onward in a straight line, and they are kept within the rails only by the flanges of the wheels and the firmness of the outer rail, the resistance of which gradually makes them follow the curvature of the road. If the momentum should exceed the resisting force, the cars must obey the former and leave the track. Curves at the foot of inclinations are therefore especially objectionable, since the cars will come upon them with excessive velocity. The rocking and twisting motion thus given to the cars indicates the dangerous tendency which they thus acquire.

* The angle of deflection of any curve may be found by dividing 5730 by its radius in feet.

When sharp curves are unavoidable, they should, if possible, be located near stopping-places. They should not be placed on a steep slope, on account of the double resistance which would then be caused to trains ascending, and the increased danger of running off to trains rapidly descending. But if such location on a long slope be unavoidable, the grade should be flattened along the curve, and the difference applied to the straight portions. Curves should not be in deep cutting, where the impossibility of seeing far ahead might cause collisions, but on the parts in embankment, or on the surface.

The increased velocities of the more recent railroads have greatly lessened the permissible smallness of the radii of curves. For the usual speeds employed on the English railways, it is recommended, that the minimum radius should be one mile. On the Baltimore and Ohio railroad, however, one of the earliest in the United States, there are several curves of 400 feet radius, ($14\frac{1}{4}°$) and one of 318 feet, (18°) over which locomotives pass without difficulty at a speed of 15 miles per hour.

The minimum in France, allowed by "*L'Administration des Ponts et Chaussées,*" is 2700 feet; or about 2°.

The minimum curve upon the Hudson River railroad has a radius of 2062 feet$=2\frac{3}{4}°$.

By the Parliamentary "Standing Orders" of 1846, a Railroad Company cannot diminish the radius of any curve to less than half a mile (2640 feet) without the special permission of Parliament.

2. WHAT RAILROADS OUGHT TO BE AS TO THEIR GRADES.

The question of the steepest grade admissible on a railroad is not one of practicability, as is often supposed, but only one of comparative economy. Locomotive engines can be made to ascend grades of almost unlimited steepness, by a proportionate increase of their power and adhesion, but their ascent becomes less and less useful in proportion as the grades become more and more steep. On an ascent of 19 feet to the mile, an engine can draw only about one-half its load on a level; at 38 feet to the mile, only one-third, and so on, (adopting the usual, though insufficient, ratio of 8 lbs. to the ton, or 1 to 280, as the resistance on a level) since, on this supposition, if the railroad rises 1 foot in 280, an *additional* force of 8 lbs. will be required to draw one ton up this ascent, (see page 32) and therefore *double* the former force will be needed to draw the former load. Only half the load, therefore, could be drawn by the *same* force; or that amount of power which could draw a load a mile on a level, would be exhausted in drawing it half a mile up this ascent.*

---

* The precise ratio between the total resistance on a level road, and that on any ascent, and therefore between the comparative loads which can be carried on each, may be found by the proportion which will now be investigated.

The loads on a level, and on an ascent, are in the inverse ratio of the resistance thereon: i. e.

The load on the level *is to* the load up the ascent, *as* the total resistance on the ascent *is to* the resistance on the level.

The resistance on the ascent is compounded of that of friction, &c. on the level, and that of gravity, which is such a part of the whole load, as the height of the ascent is of its length, as shown on page 32.

Adopting the more correct ratio of $10\frac{1}{4}$ lbs. per ton, or 1 to 218, as the resistance at the usual freight speed of 12 miles per hour, (see page 266) it would require an ascent of 24 feet per mile to double it, 48 feet to triple it, and so on. When the resistance is increased to 20 lbs. per ton, or 1 to 112, (as in the case at high velocities) an ascent of 47 feet per mile is required to double it; and a resistance of 30 lbs. per ton corresponds to an ascent of 70 feet.

These results show that heavy grades are *proportionally* less injurious on a road where great speed is employed, with correspondingly great resistances, though the *absolute* loss of power caused by them remains the same. The late discovery, that the resistances at even slow rates of travel are greater than had been supposed, lessens greatly the objections to heavy grades, and shows them to be *relatively* much less injurious than had been imagined, seeing that so much greater an ascent is required to *double* the resistance. Besides, a small diminution in the

---

Let then f = Resistance (in lbs. per ton) on a level.

h = Ascent in feet per mile; and $\frac{h}{5280}$ = Inclination.

$$\frac{h}{5280} \times 2240 = \frac{14h}{33} = \text{Resistance per ton of Gravity.}$$

$$f + \frac{14h}{33} = \text{Total resistance on the inclination.}$$

The above proportion then becomes,

$$\text{Load on level} : \text{Load up ascent} :: f + \frac{14h}{33} : f. \qquad \text{Whence,}$$

$$\text{Load up ascent} = \text{Load on level} \times \frac{f}{f + \frac{14h}{33}} = \frac{\text{Load on level}}{1 + \frac{14h}{33f}}$$

When the motive power is a Locomotive Engine, as is usual, its weight must be included in the "Load on level," used in the calculation, and finally subtracted from the resulting "Load up ascent."

Example.—Let the weight of the cars drawn on a level, at 12 miles per hour, be 447 tons; the engine 20, and the tender 14 tons: required

velocity of the train would compensate for the increased resistance of quite a steep grade.

The cost of draught on a railroad is nearly as the power employed, so that it will cost nearly twice as much to carry a load on a railroad with an ascending grade of 24 feet to the mile, as to carry it on a level route. This consideration will therefore justify large expenditures upon the excavations, embankments, &c., of a railroad, with a view of reducing its grades. The propriety of such expenditures is to be determined by comparing the annual interest of the amount with the annual saving of power ever after, in drawing the expected loads over the flattened road.

But, on the other hand, this principle may be carried to excess. These great expenses for graduation should be incurred only when maximum loads are to be constantly carried at high speeds, as on important leading lines of great traffic. Much steeper grades, than would be otherwise allowable, may be adopted on roads on which maximum loads are not often carried, and on which the trains are required for public convenience to go often, and will therefore generally go light. The engine may be able to draw 400 tons on a level, and may seldom have more than 100 to draw. In such cases the true economy is,

---

the load which the same power can draw up an ascent of 10 feet per mile.

Here $f = 10\frac{1}{4}$, and $h = 10$. By the above formula,

$$\text{Load up ascent} = \frac{(447 + 20 + 14)}{1 + \dfrac{14 \times 10}{33 \times 10\frac{1}{4}}} = \frac{481}{1 + 0{,}41} = 341.1$$

$341.1 - (20 + 14) = 307{,}1 =$ The load up the ascent.

For the method of calculating the tractive power of locomotives, see page 325.

not to go to great expense in order to reduce the grades below such a degree of steepness as would permit the engines to draw up their usual small loads ; nor to attempt to make a very level road, on which the engines could do a great deal, but would have very little to do. The same reasoning applies to railroads between places furnishing but a moderate amount of travel, such as the thinly settled parts of this country. Should the travel subsequently greatly increase, in an unanticipated degree, more frequent light trains could be sent. The enormous expenditures sometimes made in such situations to make a perfect road, have been too great for the scanty travel to pay interest upon, and have discouraged the proper construction of such as would have been really profitable.

A great reduction of the first cost of a railroad may often be made, without much increasing its subsequent expenses ; inasmuch as the capital expended in the graduation of a road has averaged, in England, fifteen times the cost of the locomotive power, and as the daily cost of transit, due to this last, is also very small. Locomotive power forms only about one-third of the whole working expenses of a road ; and only a part of this, say one-half, is likely to be affected by the grades ; so that there is only one-sixth of the whole working expenses, which can be saved by making a road theoretically perfect in grades ; a small consideration for the interest of the extra capital, unless the traffic is likely to be continued, regular, and very heavy.

In brief, first determine precisely what is wanted. If the best possible road would be justified by the importance of the traffic, make it as perfect (*i. e.* as straight, level, and unyielding) as possible, so that it can accomplish the greatest amoun of labor in the least time and with the smallest expenditure of power. If a cheap though inferior road will accommodate the traffic expected, let such a one be made.

In comparing two roads between the same points, one of which is level and the other has a summit, reached by an ascending grade, succeeded by a descending one, it must not be overlooked that there is a certain degree of compensating power in the descent. As to how much of the power lost in the ascent, is gained by the assistance of gravity in the descent, there is great difference of opinion. It was formerly supposed that on descents steeper than the angle of repose, 1 in 280, or 19 feet to the mile, the cars would be accelerated by the force of gravity, (which is just balanced by friction at that inclination) and that the brake would then need to be applied, so that beyond that limit no more assistance could be derived from gravity. But it has been found by recent experiments that the resistance of the air to the motion of cars is far greater, and increases with the speed much faster, than had been imagined. This resistance, therefore, opposes the accelerating tendency of gravity with a force increasing with the velocity, so that trains of cars may safely descend inclinations of 60 feet to the mile. On planes of 53 feet to the mile, trains have commenced the descent at a speed of 40 miles per hour, but instead of this velocity being increased, it was reduced to 30 miles per hour. Railroads may therefore be laid out with grades of nearly 60 feet to the mile, with little or no loss of power in the descent; and there is little practical loss of power in the ascent, if the loads are such as do not task the engines to their full power on the level portions of the road. In England it has been found that cheap lines with steep grades have not cost much more to work them than some which had cost two to three hundred thousand dollars per mile. We may therefore conclude that Navier's maxim, that "The amount of power required to effect the transit of a line of railroad, depends entirely on the length of the line and the difference of level of its two extremities," is true, if none of the inclinations upon it exceeds 60 feet to the mile, and if the engine is not obliged to carry its maximum load on a level.

This principle of compensation on descents was carried to such excess a few years ago, that it was sanguinely recommended to make all railroads *undulating*, carefully avoiding all levels, and establishing a continual succession of ascents and descents. It was argued that the momentum which the cars acquired in descending one slope, would carry them up the next, just as a pendulum swings as far to the one side as to the other; and that having received an impelling force at one end of the road, they would reach the other end, down one of these slopes and up the next in turn, by the assistance of gravity alone. Volumes have been written in attack and defence of this theory; but the most fatal objection to it, even supposing the undulations all properly arranged, is, that the velocity which a train must have acquired when it had reached the foot of one slope, to be sufficient to carry it up the next, would be too great for safety, and that the irregularities of speed would be destructive to the cars and to the road.

3. WHAT RAILROADS OUGHT TO BE AS TO THEIR CROSS-SECTION.

The *width* of a railroad is the first element of its cross section to be considered, and it depends upon the width between the inner sides of the rails, which is called its "*Gauge*."

THE BROAD AND NARROW GAUGE QUESTION.

The customary gauge is 4 feet $8\frac{1}{2}$ inches; varying from 4"8 to 4"9, according to the space deemed necessary for the play of the flanges of the wheels. This is called the "narrow gauge." The "broad gauge," first introduced by Mr. Brunel, on the Great Western Railway in England, is 7 feet. Between these two gauges is still going on the fiercest contest of the many which have arisen on the various doubtful points in the construction of railroads.

The original railroads were made of the same width as the tram-roads, on which ran common wagons. This width happened to be 4 feet $8\frac{1}{2}$ inches. The new railroads adopted the same width, for the convenience of using upon them the same cars, and thus this width became almost universal. Our American roads, using at first English engines, were necessarily formed with an identical gauge. Other gauges have also been employed. Four feet 10 inches is the New Jersey and Ohio gauge. Five feet is the gauge of Virginia, East Tennessee, and the north of Georgia. Five feet 6 inches is the gauge in Maine, (Atlantic and St. Lawrence Road) in Canada, (by general law) and in Missouri, (by law of 1835). Six feet is the gauge of the Erie Railroad, and of its connecting roads.

ADVANTAGES OF THE BROAD GAUGE.

The track being wider, the cars have a broader base; so that if the frost, or any other cause, raises or lowers one side of the road a certain amount, say one inch, it will cause an angular inclination of only 1 in 84 on the wide track, but 1 in 56½ on the narrow one.

The breadth of base being greater, the centre of gravity, with equal loads, is lower; so that there is less danger of the cars running off the track. They have also less lateral motion and greater steadiness, and thus add much to the comfort of the traveller. This steadiness may also be increased by placing the wheels outside of the cars.

The broader base permits the wheels of the cars to be proportionally increased in size, and thus is obtained greater leverage for overcoming the friction at the axles.

Instead of letting the cars remain of the same width as now, in order to increase the steadiness, their width may be increased to correspond with that of the track, (making it 10 or 11 feet instead of the present 8 or 9) and then they will be as steady as at present, but be much more commodious for passengers, (giving space to sleep and eat) and more convenient for packing bulky freight, as hay, cotton, lumber, barrels, cattle on the hoof, &c.

The preceding are the advantages belonging to the *cars* · those gained by the *engines* are still greater.

The narrow track does not give width enough to make sufficiently large, and to arrange to the greatest advantage, the various parts of the engine. With their usual construction, the highest profitable speed for maximum loads (at the average working pressure of steam) is about 10 miles per hour. To carry the same load at twice the speed, it would be necessary to double the quantity of steam generated by the boiler, and therefore to double either its length or its diameter. The length of its flues cannot be advantageously increased; therefore the enlargement must be that of its breadth. To effect this, more space between the wheels is needed, and to get it, a wider track is required.

Even if it be not required to carry great loads at high speeds, the surface of the boiler, being larger, may be less intensely heated, and will therefore last longer.

As larger driving wheels may be used on the wide track, their adoption will enable greater speed to be attained without increasing the rate of motion of the piston. The expansive force of steam may therefore be employed.

The larger and more powerful engines will do more work, with no more men, than smaller ones. In them there is therefore the same economy as in large ships.

OBJECTIONS TO THE BROAD GAUGE.

More ground is required; and the excavations and embankments are wider, and therefore more expensive.

The axles must be heavier to have the same strength as before.

There is an increased resistance on the curves, in consequence of the increased sliding of the inner wheels, which is equal (as was seen on page 273) to the difference between the lengths of the outer and inner rails, and therefore proportional to the difference of the respective radii of the curves.

The larger engines of the broad gauge roads have more power than is generally needed, and therefore part of it is practically wasted.

But on the whole, for a great road, the advantages of the broad gauge would indisputably overpower the objec tions to it, if it were not for the evils of "The break of gauge."

THE BREAK OF GAUGE.

This is the name given to the interruption which occurs whenever a road of broad gauge meets one of narrow gauge, and which renders necessary the change of passengers, baggage, and freight, from one set of cars to another, and prevents the same cars being run through without transhipment, or "breaking bulk." Passengers thus suffer much delay, confusion, and discomfort; and merchandise is exposed to damage and risk of loss, in being thus changed from one car to another midway in its route, besides incurring much unnecessary expense. The speedy conveyance of troops is also an important consideration; for railroads are one of the most powerful means of national defence, enabling an army to be concentrated rapidly at any point attacked; but their value for this purpose would be greatly lessened if it were necessary, at some "break of gauge" on the route, to stop and lose the time necessary for transferring the troops, with their artillery, stores, &c., from one set of cars to another.

Most roads belong to the narrow gauge class. In England the proportion is as 7 to 1; there being in operation, in 1846, 1901 miles of the narrow gauge, and only 274 of the broad. Every new road of broad gauge, connecting with a narrow one, therefore increases the evils of the break of gauge. The importance of lessening them has given rise to various contrivances for that purpose. The following are the four principal remedies proposed.

1. *Telescopic axles.* The axles have been so arranged that one portion slides in the other, like the joints of a

telescope, so that the distance between the wheels can be so adjusted as to suit either the broad or the narrow gauge. To lessen their gauge, the catch which fastens them is loosened, and the carriage is pushed along a pair of rails, the space between which gradually narrows from 7 feet to 4 feet 8½ inches, and thus the wheels of the carriage are gradually forced nearer to each other. To widen their gauge the operation is reversed. But, besides the expense of the alteration, there is a resulting unsteadiness, and consequent liability to danger.

2. *Low trucks* on the broad gauge roads may have rails laid on them 4 feet 8½ inches apart, upon which the narrow gauge cars may be run, and thus be carried on the broad roads. But this contrivance raises the centre of gravity, making the whole top-heavy; and adds so much extra dead-weight to the load. Besides, it does not provide for conveying the broad gauge cars on the narrow roads.

3. *Shifting car-bodies* for passengers have been provided, which could be swung, by powerful cranes, from one set of wheels to another; and *Moveable boxes*, to receive merchandise, have been made of such a size, that one should be carried on a narrow gauge track, and two on a broad one.

4. *Extra rails* have been laid, so that the same road could be used for both classes of cars; a pair of narrow gauge rails being laid within the broad ones, or only a single rail being laid, so as to be 4 feet 8½ inches from one of the broad gauge rails. But, besides the expense of these arrangements, there would be increased danger at the crossings.

All these remedies are imperfect; and the "break of gauge" seems to be an evil for which there is no cure, except in destroying its cause.

The Royal Commission, appointed by the British Parliament, in 1845, to investigate this subject, made an elaborate report in 1846, and sum up as follows:

"1. As regards the *safety, accommodation, and convenience* of the passengers, no decided preference is due to either gauge; but on the broad gauge the motion is generally more easy at high velocities.

"2. In respect of *speed*, we consider the advantages are with the broad gauge; but we think the public safety would be endangered in employing the greater capabilities of the broad gauge much beyond their present use, except on roads more consolidated, and more substantially and perfectly formed, than those of the existing lines.

"3. In the commercial case of *the transport of goods*, we believe the narrow gauge to possess the greater convenience, and to be the more suited to the general traffie of the country.

"4. The broad gauge involves *the greater outlay*; and we have not been able to discover, either in the maintenance of way, in the cost of locomotive power, or in the other annual expenses, any adequate reduction to compensate for the additional first cost."

They recommend "that the gauge of four feet eight inches and a half, be declared by the legislature to be the gauge to be used in all public railways now under construction, or hereafter to be constructed, in Great Britain." They add, that "great commercial convenience would be obtained by reducing the gauge of the present broad gauge lines to the narrow gauge;" and "think it desirable that some equitable means should be found of producing such entire uniformity of gauge, or of adopting such other course as would admit of the narrow gauge carriages passing, without interruption or danger, along the broad gauge lines."

The final conclusion seems to be that, if all railroads were now to be constructed anew, a gauge of five and a half, or six feet, would be considered most desirable ; but that the evils of a "break of gauge" are so great, in the present preponderance of narrow gauge roads, as to overbalance the disadvantages of the narrow gauge ; which should therefore be adopted by all future railroads which are to connect with others.

WIDTH OF ROAD-BED.

When the gauge has been decided upon, the necessary width of the roadway can be determined. When the road has a double track, the middle space between the two pairs of rails, for convenience and safety, should not be less than six feet. The side-spaces, outside of the rails, should, for safety, be a little more than the width of the track, particularly on embankments ; so that if the engine gets off the track, it may still remain upon the bank. This width also gives greater stability to the embankment and to the rails laid upon it, diminishing their liability to be disturbed by slips. These side-spaces are from 5 to 8 feet on different railways. They should be greatest on roads where great velocity is adopted ; on high embankments ; and on the outside of curves. They will of course be less in tunnels, viaducts, bridges, &c.

The total width of the road-bed of a railroad, with narrow gauge and double track, will therefore be $6 + 2\ (4\frac{3}{4}) + 2 \times 6 = 27\frac{1}{2}$ feet. In excavations, the widths of the ditches on each side must be added.

The total width of a double track railroad is usually about 28 feet in excavations and 24 feet in embankments Single track railroads are about ten feet less, say 18 and 14 feet. These dimensions are used when great economy is required, and are increased for wet cuts and high banks.

If it be proposed to lay only a single track at first, and subsequently to add a second one, the cuttings and fillings should always be made *at first* of the full width for a double track; for the extra expense of the additional width is but a small proportion of the whole, and a narrow cut or bank, is, from the want of room for the carts, &c. to pass, worked much more disadvantageously, and therefore much more expensively, than a wide one. If an embankment be subsequently widened, the new portion will not adhere to the side of the old one without forming the latter into steps; and in widening a rock excavation, a single blast might render the road impassable for many hours.

The other subjects properly belonging to the "Cross-section," such as the elevation of the outer rail on a curve, &c., will be more advantageously examined under the head of "Superstructure."

Very narrow gauge railroads have been worked with great success, and their use is increasing. The best known of these is the Festiniog Railway in North Wales. The gauge is only 23½ inches. A heavy traffic is carried on over this road, and the profits yield a large percentage on the original outlay. A 3½ feet gauge has been successfully used in Norway, Sweden, Belgium, and other places.

It seems advisable to make 4 ft. 8½ in. the maximum width, and to adopt a much narrower gauge where the traffic is limited. On the Great Western Railway of England the famous 7 ft. gauge, introduced by Brunel, is being abandoned; a large part of it having already been replaced by the 4 ft. 8½ in. gauge. Several wide gauge roads in the United States have been changed to a 4 ft. 8½ in. gauge.

II. THE LOCATION OF RAILROADS.

The location of railroads is guided by the same principles as that of common roads, and made in similar manner, but the greater importance to railroads of straight lines and easy grades, as has been shown in the preceding section, justifies and requires a much greater expenditure in the surveys which seek the attainment of these, and in the excavations, embankments, and bridges by which they are secured. The minor undulations of the country are disregarded, for they can be readily overcome by the cuttings and fillings which will be demanded by any traffic which is important enough to need a railroad for its accommodation ; and straightness is the first object : where a common road should go around a hill, a railroad should cut through it. For this reason, the compass, or some other angular instrument, usually takes the lead in the location, and is followed by the level. Upon the rough plot of the survey, *curves* are pencilled in, their centres and radii are determined, and then they are laid out on the ground, being corrected, if necessary, by calculation or by trial, till they pass through the desired points. The important calculations for excavation and embankment are identical with those of common roads ; but the *estimate* must include the new items of superstructure, of engines cars, &c., which are to be presently examined.

In examining the comparative merits of two rival routes, the relative importance of distance and grade or shortness and steepness, must be determined by the considerations given on pages 276—281. To determine which is the least objectionable in amount of curvature, calculate the angular deflection of each curve, as indicated on p. 274. The sum of all of these on each line will be its total deflection, and the proper standard for comparing it with others.

III. THE CONSTRUCTION OF RAILROADS.

The two principal divisions of this part of the subject are—"Forming the Road-bed," (which corresponds to the general "Construction" of common roads); and the "Superstructure," which includes the Rails, and their supports, ties, &c.

1. FORMING THE ROAD-BED.

EXCAVATIONS.

The *Excavations* on railroads are often of much greater depths than are ever necessary on common roads, the extra expense being amply repaid by the advantages of the easier grades and straighter lines thereby attained. There is an excavation (in sand) on one English railway, 110 feet deep; and on another, 16,000,000 cubic yards of material were removed. The thorough drainage of these excavations by ditches, cross-drains, &c., is of the highest importance. Their sides often need to be supported by retaining walls, in order to make steeper slopes possible, and thus to lessen their top width, when they pass through valuable ground. Sometimes these retaining walls are supported by iron beams, or flat arches, extending across the railway at a sufficient height to clear the engines. In one remarkable cutting, 60 feet deep, the upper portion of it was rock, but the lower looser matter. If the whole cutting had been extended upwards, with such side slopes as the looser and lower portion required, it would have been more than 200 feet wide at its top, and would have involved very great expense in the removal of so large an amount of rock. The sides of the cutting were therefore made nearly perpendicular, and the loose strata at the bottom were supported by retaining walls, carried up till they reached the solid rock. Such are some of the ingenious expedients rendered necessary by the gigantic constructions of modern railroads.

The side slopes will depend upon the material through which the cutting is made. For common earth they are usually 1½ to 1, although in deep cuts they should be 2 to 1. It is sometimes economical to make the slopes at first as steep as they will stand, and remove the surplus earth after the track is laid. Clay slopes are very troublesome. Some that have stood at first at 2 to 1, have finally slipped until they were 6 to 1. Slope walls have been used in such soil, as being more economical than taking out the requisite amount of earth to secure stability to the banks. It has been observed that the mass of earth which slides out of a clay slope leaves the bank with a concave face. It will then be better to make the original excavation, so as to leave the slopes with this curved batir.

In making long, deep cuts, the bottom should be left so as to slope up from both ends toward the centre, to secure drainage while the work of excavating is going on. The bottom can be taken out, down to the desired grade of the finished work, after the main portion of the earth has been removed. When the cut is very deep, the work can be pushed forward more rapidly by working in tiers or stories, keeping the upper one in advance of the lower one.

Ditches should always be dug on the surface, a few feet back from the edge of the cutting, to prevent the surface water from running down the side slopes.

TUNNELS.

The depth of an excavation frequently renders a *Tunnel* more economical. In constructing one, the centre line of the road must be set out with very great accuracy upon the surface of the ground, (by a Transit instrument) and "shafts" sunk at proper intervals along this line. The excavations are made by "headings," or "drifts," from shaft to shaft, and to the open ends of the tunnel. The material excavated is raised through these shafts, which, after the completion of the tunnel, serve as ventilators. Their distances apart should be from 500 to 1000 feet. If the material be earth, or stratified rock, the crown of the tunnel, and its sides, must be supported by a brick arch, and the excavation kept only a few feet in advance of the completed arch.

The height of tunnels, in the clear, varies on the English railways from 17 to 30 feet, and the width, for a double track, from 22 to 30 feet. The average sectional area in the clear, is 450 square feet: when an arch is required, the excavation would contain about 700 square feet. The cost per lineal foot of the English railway tunnels, has ranged from $30 to $150. If sufficient time had been allowed, they could generally have been executed for $60 per lineal foot. A number in England are over a mile in length, and one is more than a mile and three quarters.

The greatest work of the kind is under the Alps, at Mount Cenis, connecting Modane, in France, with Bardonneche, in Italy, and is seven miles and one thousand and forty-four yards in length. No shafts could be used, as the mountain rose rapidly from both sides, and there were no depressions. The tunnel is one mile below the summit of the mountain.

The expense of construction was borne by the French and Italian governments, each paying in proportion to the length of tunnel lying in their territory.

EMBANKMENTS.

The embankments of railroads demand the use of every possible precaution to ensure their solidity; not only on account of their size, but because the vibrations imparted to them by the passing trains, greatly increase their tendency to slip. The expense and time required to form them in layers, as recommended on page 166, often forbid the adoption of that method. They are usually constructed by raising them to their full height at one end, and so carrying them onward. Temporary rails are laid along the bank, and extended with it, and on them wagons, containing each about 3 cubic yards, are drawn by horses, or by locomotive engines, if the distance, or "lead," be great.

It has been ascertained that, contrary to the usual theory and practice, the quantity of work which can be done on an embankment so made, and, consequently, the time which will be required for its completion, does not depend on the area of the face of the cutting which supplies it, or on the number of wagons which can be filled in it together; but on their rate of speed, and on the number of them which can be emptied in a given time over the head of the embankment, to the top width of which this element is proportional. The number of wagons drawn together in a "set," should increase or decrease with the length of the "lead," and the breadth of the end of the bank; and the number of "sets" should be increased at certain exact periods in the progress of the work, which are susceptible of mathematical determination.*

---

* These points are very clearly and fully examined in "Laws of Excavation and Embankment on Railways," London, 1840.

When the road crosses a swamp, the banks may be formed as explained on pp. 168, 9. A railroad is carried across the Montezuma swamp, N. Y., by spreading the pressure over a large surface, by means of a wooden platform. On another railroad a spreading base was secured, to carry a bank across a swamp, by placing trees and brush under the bank, at right angles to the line of the road, and extending for some distance beyond the foot of the side slopes. Water should never be allowed to stand within three feet of the top of the bank.

When a railroad passes through a wooded swamp, where no materials for embankment are at hand a cheap and efficient substitute will be formed by a series of timber trusses. Piles of 15 inches diameter, not sharpened, are driven so as to form two lines, at a distance from each other equal to the width of the railroad. Transverse ties are fastened across their tops, which are braced by inclined struts, the lower ends of which abut against short piles. Longitudinal timbers are laid on the heads of the piles to carry the rails. Various combinations of the trusses are employed, according to the height of the superstructure above the surface of the ground. After the railroad has been thus constructed, it may be gradually banked up to the level of the rails, by taking advantage of its facilities of transportation, to bring earth from a distance to the places where it is needed.

The side-slopes of both the excavations and the embankments should be sown with grass seed, or sodded, as directed in the construction of roads. Some deep cuttings on the English railways, have been planted with flowers, shrubs, and trees; an improvement as delightful to the passenger and therefore profitable to the proprietors of the road, as it is beneficial to the permanence of the slopes.

BALLASTING.

The tops of the embankments, and the bottoms of the excavations, are brought to a height called the "Formation level," about two feet below the intended level of the rails, and there shaped with a fall from the middle to each side, as in common roads, in order to drain off the water which falls upon them. The remaining space of two feet (more or less, according to circumstances) is filled up with "ballast."

There are four objects in using ballast:

1. To spread the bearing of the sleepers over a large surface of the ground.
2. To keep the track in place.
3. To secure drainage.
4. To give a medium elasticity.

It is composed of some porous material, such as gravel, broken stones, quarry rubbish, cinders, etc., through which the water of rains can readily pass. Burnt clay has been used in alluvium countries. It should be laid on rock as well as earth, and should extend one or two feet beyond the ends of the ties in cuttings, and be of the full width of the embankments.

Upon this "ballast" are laid the supports of the rails. Without this precaution, the water absorbed by the earthy materials of the road-bed would render it soft and spongy in ordinary weather, and by freezing in winter would disturb the position and the levels of the rails. On many American railroads the neglect of this safeguard against the effects of our Northern winters renders them very unsafe at high velocities in the early spring, when the frost is coming out of the ground. Ships' ballast was first used for this purpose on the early railroad at New-castle, and from this circumstance the substitutes have retained the original name.

2. THE SUPERSTRUCTURE.

Under this head will be considered the best forms and weights of *rails*, whether supported at intervals (in chairs, on stone blocks or wooden cross-sleepers) or on continuous bearings for their whole length; and their proper arrangement, inclination, elevation, &c., when laid.

RAILS SUPPORTED AT INTERVALS.

When rails are supported only at intervals, on props, like a bridge on piers, they are liable to be depressed between these supports by the heavy loads which pass over them. It is therefore very important to give them such a shape as will secure the greatest strength with the least quantity of material. The form indicated by theory, and originally adopted in practice, is that called "fishbellied," from the rounded profile of its under side. A rail of such

Fig. 121.

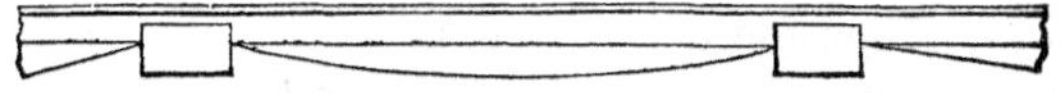

a form, will have more power to resist deflection than a straight one of the same weight, in the proportion of 11 to 9.* But whatever the theoretical advantages of this form, its inconvenience in practice, owing to its requiring a higher support, which is therefore less steady, has caused it to be generally discarded.

The forms now used are all varieties of the *parallel* or *straight* rail, in which the top and bottom are parallel, and which has the same cross-section at all parts of

---

* Lecount, p. 110. Barlow doubts this.

its length. Usually the rail is thinner through its middle than at its top and base. The various forms are named the T rail, the H rail, the hour-glass rail, &c. from the shape of their cross-section. The popular division of rails is into the "Plate rail," and the "Edge rail;" the latter including all the varieties just mentioned.

Fig. 122.

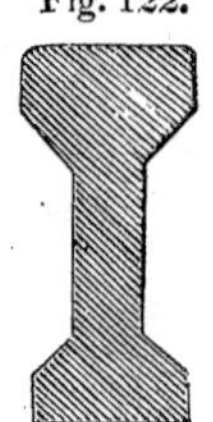

The best form of the parallel rail was investigated by Professor Barlow, in behalf of the London and Birmingham Railway Company, and Fig. 122 shows the section of the rail which he found to possess the greatest strength with the least material, the bottom web being much smaller than the head.

Fig. 123

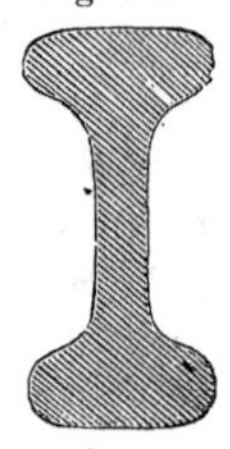

But a double headed, or H rail, as shown in Fig. 123, with its top and base of the same size and shape, is now generally preferred in England. Professor Barlow considers this shape to be inferior in strength and convenience in fixing, its broader bearing to be of no advantage, and the proposed plan of turning it over, when the upper table is worn down, to be impracticable; but still it is found preferable in practice, as enabling the best side to be selected, as being more easily keyed in its chair, and as having a broader bearing.

Fig. 124.

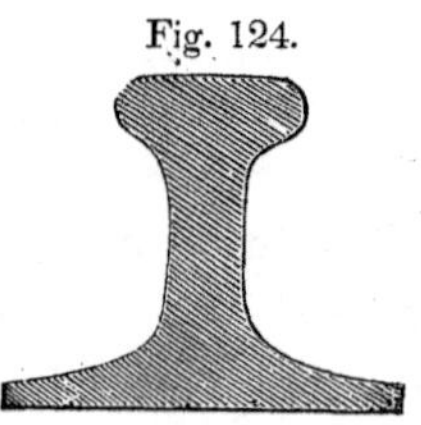

The favorite form in this country is shown in Fig. 124. The following dimensions are recommended: height, 4 in., width of bottom, 4 in., width of head, $2\frac{3}{8}$ in., thickness of vertical web, $\frac{9}{16}$ in.

Steel rails (usually Bessemer steel) are coming into almost universal use, owing to their greater durability and strength, although the first cost is much greater.

The *form* of the rail being decided upon, its *weight*, on which its strength depends, is next to be determined. The weight is expressed by the number of pounds in a lineal yard. Its *minimum* may be determined thus. A certain breadth is necessary for the bearing surface of the rail, that the wheel may run upon it without being grooved. $2\frac{1}{2}$ inches seems the minimum for this. The minimum breadth is desired, in order that as much as possible of the material may be put in the depth, the strength being as the simple breadth, but as the *square* of the depth. The minimum depth, to resist abrasion and exfoliation, is $1\frac{1}{2}$ inches. This gives a sectional area of $2\frac{1}{2} \times 1\frac{1}{2}$ inches $= 3.75$, or, say 4 inches, which corresponds to 36 lbs. to the yard. This is then the minimum weight permissible, when the rail is supported throughout its whole length; but if supported at intervals it must have much greater weight and strength, their degree depending on the distance between its points of support.

This distance has varied from 3 to 6 feet. It is now generally made less than 3 feet. For Professor Barlow's form of rail, Fig. 122, with a strength of 7 tons, the weight should be 51 lbs. per yard, for a bearing of 3 feet. To attain the same *strength* with a bearing of 6 feet, the weight should be 79 lbs. per yard. But the *deflection* of the rail with 3 tons, which in the former case is only .024 inch, in the latter is .082 inch. Thus the longer bearings, when equally *strong* with the shorter, are much less *stiff*, and therefore much inferior to them. The effect of any depression under a passing load is that the engine, at slow speeds, after sinking into it, has an inclined plane to ascend, and at high speeds it leaps over the hollow, and strikes with great violence upon the other side of it. A rail having been bent half an inch, and then covered with paint, an engine with a train of cars was run over it, and none of the wheels touched the paint for a space of 22 inches.* Strength to resist deflection is therefore as important to a rail as its strength to bear weights. The latter should be double the mean strain or load. The former should not admit of a depression under a passing load of more than $\frac{3}{100}$ of an inch.

The weight of rails has been yearly increasing. The first rails laid on the Liverpool and Manchester railway were only 35 lbs. to the yard; they have been successively replaced by rails weighing 50, 65, and 75 lbs. to the yard. The rail shown in Fig. 123 weighs 75 lbs. to the yard, with bearings 3 feet 9 inches apart. Its whole depth is 5 inches; the top and base are $2\frac{1}{2}$ inches; and the thickness of the middle rib is about $\frac{3}{4}$ of an inch. On the Massachusetts railroads the rails weigh from 56 to 60 lbs. per yard, and rest on cross-sleepers, 2 feet 6 inches apart; the weight on a driving wheel being from 5,000 to 8,000 lbs. English rails weigh from 50 to 80 pounds per yard, and German rails from 50 to 70 pounds. Steel rails can be made much lighter than iron ones, and yet have the same strength. The steel rails on the Hudson River Railroad weigh from 56 to 60 pounds to the yard.

The rails are usually rolled in lengths of from 12 to 20 feet. Their ends have received various shapes. Square or butt ends, Fig. 125, are generally preferred, but cause considerable shock to the wheel. The half-lap joints, Figs. 126 and 127, retain their positions better, but weaken the rail. The form shown in Fig. 128 is

Fig. 125.

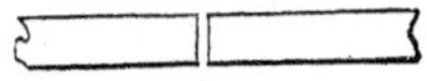

Fig. 126.

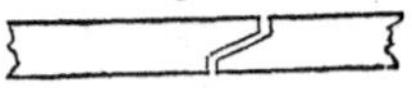

Fig. 127.

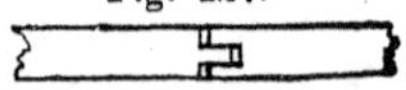

* Lecount, p. 89.

recommended when trains run on each track in only one direction, (as indicated by the arrow) so that they never meet the points of the rails.*

Fig. 128.

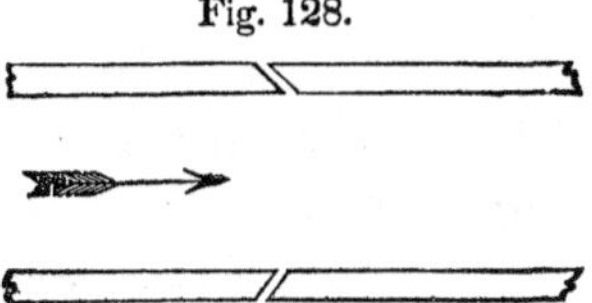

Between the ends of two successive lengths of rails, a space must be left to allow for their expansion by heat. The expansion of a fifteen feet rail may be taken at $\frac{1}{800}$ of an inch for each degree of Fahrenheit, or $\frac{1}{8}$ inch for 100°. If the rails were laid in the coldest weather, a space of one-eighth of an inch should therefore be left between their ends. The force with which iron expands is from 6 to 9 tons per square inch of section, which corresponds to 10 lbs. to the yard; so that the rail of 70 lbs. expands with a force of about fifty tons.

CHAIRS.

The rails may be fastened directly to their supports, or have their ends also held by "chairs," spiked to the blocks or cross-sleepers. The chairs are generally of cast iron, and weigh from 20 to 30 lbs. They are cast in one piece, consisting of a bottom plate, and two side pieces, between which the rail passes, its under surface being about an inch above the block. The opening of the chair must be as wide as the lower part of the rail, in order that it may be removed and replaced without disturbing the chair. Keys of wood, or of iron, must therefore be employed to fill up this opening, and to hold the rail firmly in the chair, but without offering any resistance to its longitudinal motion in expansion and contraction.

On the Liverpool and Manchester Railway the chair,

---

* Lecount, p. 112.

shown in Fig. 129, was employed The rail has on one side of its bottom a projecting rib which enters a notch in the chair, and another notch on the other side receives an iron pin. To prevent its getting loose, that end of the pin which enters first may be split, and opened when driven home.

Fig. 129.

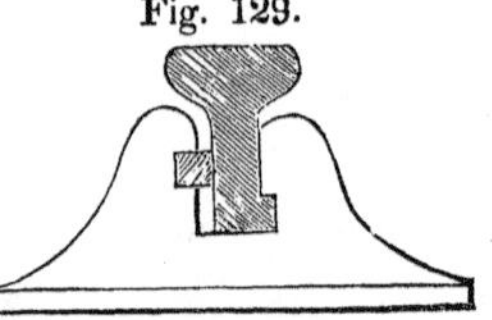

Another good form, shown in Fig. 130, was invented by Mr. Robert Stephenson. In it the rail is confined by two bolts with angular ends, which enter a small score in the rail, and are keyed home by iron keys with split ends.

Fig. 130.

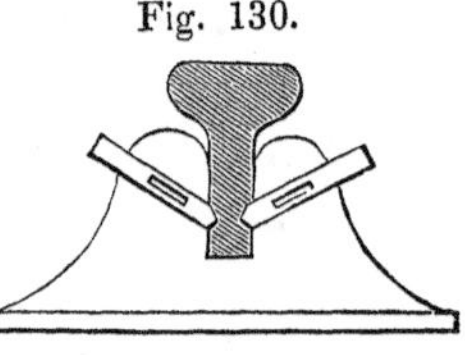

Fig. 131 represents Mr. Barlow's patent hollow iron key applied to fasten a double-headed parallel rail.

Fig. 131

Wooden keys, of similar shape, but solid, have been much used, owing to the great facility which they offer of being tightened and replaced. They should be kiln-dried, cut, and compressed by hydraulic pressure, so that by their swelling, after being driven in, they may hold the rail very tightly.

The chair used for the inverted T rail is shown in Fig. 132. It is $6\frac{1}{4}$ inches wide, $8\frac{1}{4}$ long, $2\frac{1}{2}$ high, and weighs 24 lbs.

Fig. 132.

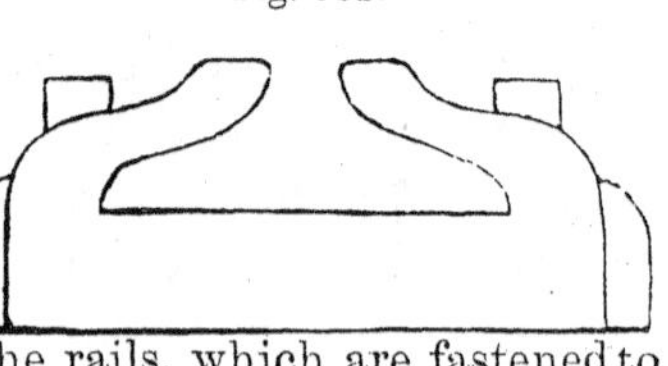

Generally the chairs are placed only at the ends of the rails, which are fastened to the intermediate supports by spikes with bent heads.

The most perfect arrangement for joining the rails, and for keeping them at the same level and in line with each other, is "fish-plates." These are iron plates, bolted on each side of the rails at the joints.

They should be at least 20 inches long, fit closely between the head and base of the rail, and be in contact with the vertical web throughout the entire length of the plates.

They should be secured by four ¾ inch bolts. The holes in the plates and rails should be oblong, to allow for the expansion and contraction caused by the changes of temperature.

STONE BLOCKS.

Stone blocks imbedded in the ballasting, have been till lately the principal supports employed on the English railways. They are usually blocks of granite, or whin stone, two feet square and one foot deep. The customary distances between their centres have been noticed on page 271. They are sometimes placed, as in Fig. 133, with their sides parallel to the line of the road; and sometimes diagonally, as in Fig. 134.

Fig. 133. Fig. 134.

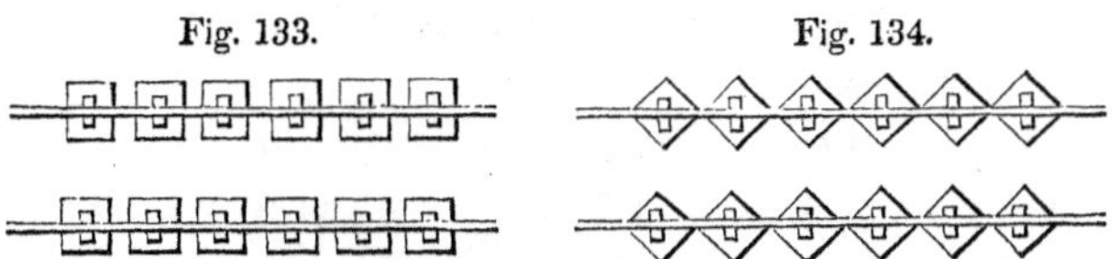

Both plans have their advocates. The former position offers more resistance to motion in the line of the road.* The latter is less stable, but is more convenient for packing the ballast around. Circular blocks have been pro-

* In the proportion of 1728 to 1629 For the investigation, see Lo count, p. 93

posed in order to get equal resistance in all directions, but the gain would not equal the extra expense.

The blocks must be very carefully set precisely level, since even a quarter inch difference in 3 feet, would create an inclined plane of 1 in 144, or 37 feet to the mile.

On curves the blocks on each side of the road must be connected by iron tie-rods, that the exterior ones may not be pressed outward by the centrifugal force of the cars.

Stone blocks have been also laid transversely, with the advantage of preserving the gauge of the road, but with the evils of great rigidity, hardness, and jolting.

WOODEN CROSS-SLEEPERS.

Transverse or cross-sleepers of wood are now considered preferable to stone blocks for many reasons. They tie the rails together and preserve their parallelism, and also make the road less rigid and more elastic than the stone, and therefore much more smooth and pleasant to travellers. Thus the blacksmith puts his anvil on a block of wood to lessen the concussion. The only objection to them is their liability to decay, against which, however, there are many preservatives

They are usually of chesnut, oak, pitch-pine, or red cedar. They may be round sticks, hewn on two sides, so as to leave at least six inches thickness, and more if possible. The longer they are the better, as the extra length on each side of the track lessens the danger of settling. On the Massachusetts roads they are of second growth chesnut, 7 feet long, and 8 inches by 12. They are simply laid on the ballasting, except on new embankments and soft ground in which cases they are laid on longitudinal timbers or sub-sills, which may be of plank 8 inches wide and 3 or 4 thick.

The "ties," or cross-sleepers, can be made to last much longer by preserving them by chemical means. Some of the methods are given on p. 254.

CONTINUOUS SUPPORTS.

When rails are supported at intervals, the less the intervals and the nearer the supports, the less will be the yielding and deflection of the rails. Carrying out this principle, and continually lessening the intervals, we at last arrive at *continuous* supports. The advantages of such solid bearings for the rails would seem to admit of no dispute. It is evident that an iron bar, laid on a series of points, will be much more easily bent, either laterally or vertically, by the heavy blows or jolts of a carriage, than when the same bar is made to form a part of the solid roadway. The system of continuous supports of longitudinal timbers is therefore superior to any other in strength, solidity, and ease of motion. It has been of late increasing in popularity in England, in spite of the cost of timber in that country, while with us it has been abandoned on our best roads for the system of supports at intervals. This has probably arisen from the circumstance that most American roads with longitudinal timbers have been laid with plate rails, so thin that their ends sometimes spring up so as to form " snake-heads," and have thus received the scarcely caricatured description of " A hoop tacked to a lath." Such roads have the defects of instability, insecurity, inequality of surface, waste of power, resistance to speed, and great expense of maintenance. But these faults do not belong to the system itself, but to its imperfect execution. The rails should be heavy edge rails, of suitable form, and in contact with the timbers for their whole length ; and the longitudinal timbers should be tied together by cross-sleepers. The best rail road in the world, the "Great Western," has such con tinuous bearings. The wood may be preserved from decay by any of the methods noticed on page 234.

Fig. 135.

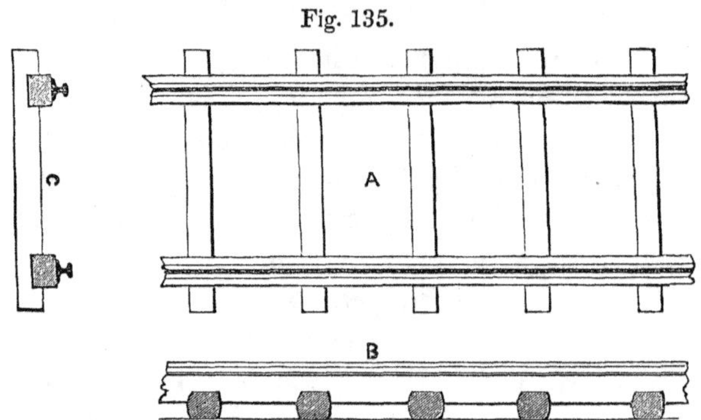

In the above figure, A is the ground plan, B the side view, and C the end view, of such a system of railroad.

For these longitudinal bearings, chairs are unnecessary, and peculiarly shaped rails are preferable. A favorite form is that shown in Fig. 136, which has been made to weigh from 35 to 60 lbs. per yard.

Fig. 136.

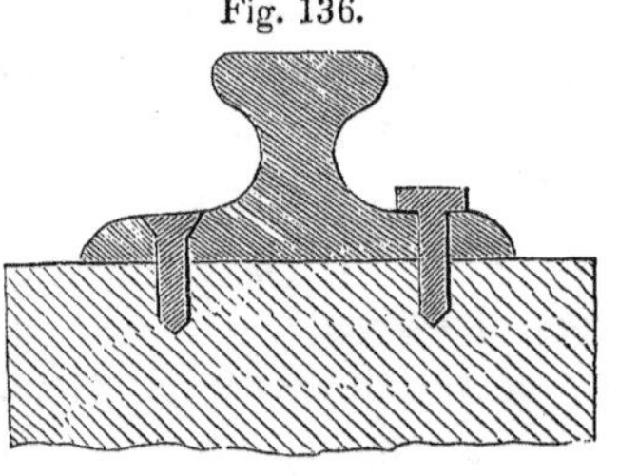

It is fastened by screws, 4 inches long, the heads of which are countersunk on the inner side, so as to be out of the way of the flange of the wheel. At the joints, four screws are employed.

Sometimes the rails are fastened by spikes with bent heads, driven just outside of them, and clasping them firmly.

The greater difficulty of packing the gravel around

such longitudinal sleepers, and of removing and replacing them, is the chief cause of the general preference of cross-ties, or transverse sleepers.

Fig. 137.

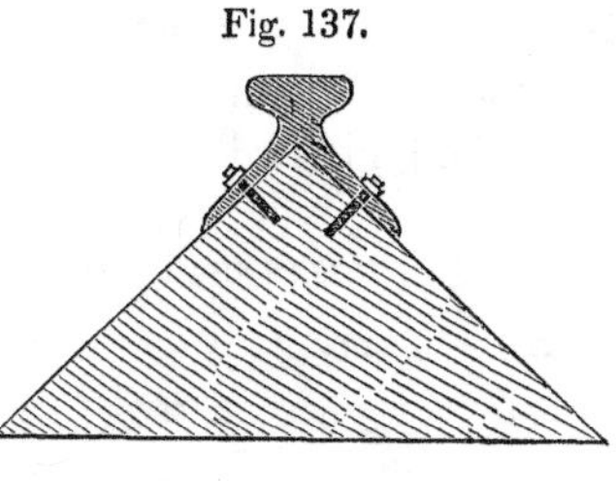

Triangular sleepers have been employed, with a rail forked at bottom, as in the figure. The rail can thus be very firmly attached to the sleeper, the shape of which gives it much stability.

Evans' method of fastening is warmly recommended by Professor Vignoles. The rails are rolled with a slit, or groove, of a dove-tailed shape, (in its cross-section) running on their under side for their whole length. The bolts have heads of corresponding shape, and are slipped into the end of the groove, passed along it, and dropped through holes made at proper intervals in the longitudinal timbers. The lower ends of the bolts are cut into screws, and washers and nuts draw the rails close down to the timbers. They are easily tightened, and not exposed to injury, while spikes and screws get loose, and their heads are in the way.

Upon the Great Western Railroad, between Bristol and London, (on which Mr. I. K. Brunel first introduced into England the system of longitudinal bearings) the hollow rail, shown in Fig. 138, was adopted. The original rails weighed only 44 lbs. to the yard, and were $1\frac{1}{4}$ inch high, the head of the inner screw being countersunk. The later ones weigh 70 lbs. to the yard, and are $2\frac{1}{2}$ inches high; the increase of height be-

Fig. 138.

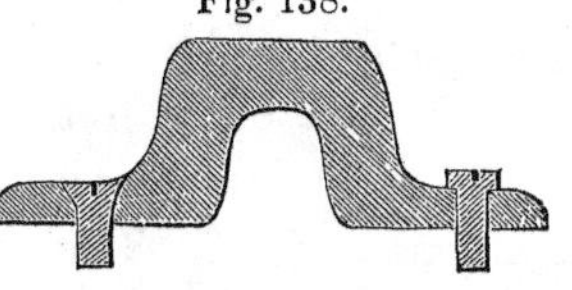

ing intended to compensate for not countersinking the nut of the inner screw. The longitudinal timbers are 15 by 9 inches, and the cross-ties bolted to them at intervals of 9 or 10 feet, are 5 by 8 inches. With such rails, and the broad gauge, this railroad combines speed and ease of motion in the highest degree yet attained.

INCLINATION OF THE RAILS.

The wheels having a conical shape, they would touch a level rail only on a narrow line, and both would soon be worn into grooves. To prevent this, the rails are sometimes inclined inward, so as to meet the cone of the wheel more directly, and to present a broader bearing surface. The usual inclination is from 1 in 29 to 1 in 20. It may be given by sloping the blocks, or by cutting the sleepers which support the rails, or may be formed in the original rolling of the rail. An objection to this breadth of contact is that a rubbing and grinding action is constantly caused by the unequal velocities with which the outer and inner parts of the coned wheels revolve, and produces the same effect as if the train was dragged a certain distance with its wheels locked.

ELEVATION OF OUTER RAIL.

When a railroad car enters upon a curve, the centrifugal force tends to force the flanges of its wheels against the outer rail. To resist this tendency, the outer rail is made higher than the inner one, so that an inclined plane may be formed beneath the cars, down which they will tend to slide in an inward direction, in oppos'tion to their centrifugal impulse. The inclination should be such that the two antagonist forces may just balance each other. It will vary with the radius of the curve, the velocity upon it, the gauge of the road, and the "cone" of the wheels With these elements it may be readily calculated.* Some results (with the usual data) are given in the following table:

RADIUS OF THE CURVE.	ELEVATION OF THE OUTER RAIL.		
	At 10 miles per hour.	At 20 miles per hour.	At 30 miles per hour.
Feet.	Inches.	Inches.	Inches.
250	1.14	5.60	12.99
500	0.57	2.83	6.56
1000	0.29	1.43	3.30
2000	0.15	0.71	1.65
3000	0.10	0.47	1.10
4000	0.07	0.36	0.83
5000	0.06	0.28	0.66

An approximate rule for finding the elevation is this: "Multiply the square of the velocity, in feet per second, by the gauge of the railroad in inches; and divide the product by the accelerating force of gravity, multiplied by the radius of curvature in feet, and the quotient will be the elevation in inches."

For a velocity of 30 miles per hour on a curve of 1000 feet, this rule gives $\frac{44^2 \times 56\frac{1}{2}}{32 \times 1000} = 3.4$ inches.

The old practice made the greatest elevation 1 inch; which is that due to a velocity of 30 miles per hour on a curve of two thirds of a mile radius. When the cars go faster than the velocity assumed in the calculation, which has determined the elevation, their flanges press the outer rail; when slower, they press the inner one.

SIDINGS, CROSSINGS, ETC.

On railroads which have only a single track, a second one, called a *siding*, is occasionally laid for a short dis

* See Pambour. pp. 277–290; and Lecount, pp. 135–140.

tance, to form a passing-place for meeting trains. *Crossings* are the arrangements by which cars pass from one track to the other. The angle of their divergence should not exceed 1½° or 2° for speeds of 20 or 30 miles per hour, but when the speed, as in mines, is not more than 8 miles per hour, the angle may be as many degrees.* They are always dangerous, and therefore the fewer of them that are employed the better. The misplacing of them, carelessly or malevolently, causes a large portion of the accidents on railways. Their simplest form is that of two "points" or "switches," which are attached at one end to the main track, and are moveable at the other, so as to continue the principal line, or to connect it at pleas-

Fig. 139.

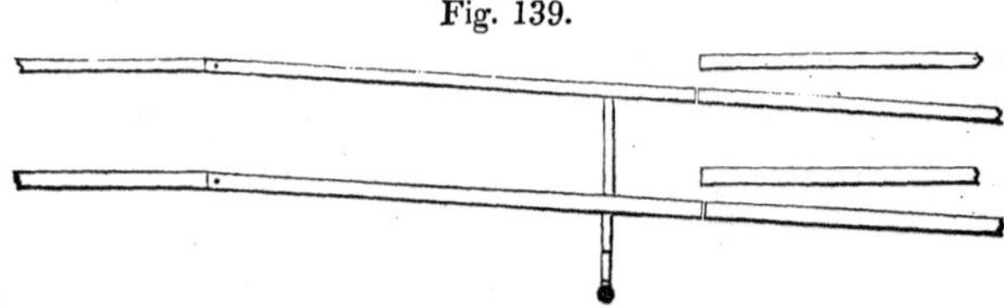

ure with the side-track. The switches are usually moved by hand, with either a lever or an eccentric. A signal plate at the top of the lever, with which it moves, by its position shows to the engine-driver, as he approaches, to which track it is prepared to turn the train. Self-acting switches, kept in place by powerful spiral springs, and moved by the flanges of the engine wheels, have been tried; but the system of manual operation is preferred, with all its uncertainties, owing to the self-acting arrangement rendering it impossible for the conductor to know whether the switches are in place or not until he is upon

* Cresy, Encyclopedia of Civil Engineering, p. 1576.

them, when any precaution which might be required would be too late.*

Turn-tables, or Turn-plates, are platforms, turning on rollers upon an underground circular railroad, and forming a very convenient substitute for switches, in transferring carriages from one set of rails to another.

A *Hydraulic Traversing Frame* has been used instead of Turn-tables. It consists of a wrought-iron frame, under each corner of which is a cast-iron hydraulic press, operated by force pumps. The frame is pushed under the carriage to be moved, the pumps are worked, and raise the flanges clear of the rails. The carriage is then moved to the desired spot and there let down.†

SINGLE RAIL RAILROAD.

In this arrangement a single rail is supported on posts at a suitable height above the ground, and passes through the middle of the cars, which hang from it on each side, like two saddle-bags on a horse. The advantages of thus lowering the centre of gravity are considerable; the cars can never leave the track; and the expenses of construction are much reduced. In some situations this system might be very conveniently employed.

The engines devised by Mr. Fell require three rails, the centre one being gripped by horizontal driving wheels. This system not only gives great increase of tractive power, but is much safer, the central rail making it almost impossible for the engine to leave the track. It was used for the railroad *over* Mount Cenis.

---

* Ritchie, p. 115.

† Cresy, Encyclopedia of Civil Engineering, p. 1282.

IV. MOTIVE POWERS.

The principal powers which have been employed to move carriages on railroads are Horses, Stationary Engines, Locomotives, and Atmospheric Pressure.

1. HORSE POWER.

The power of a horse in moving heavy loads at a slow rate, has been given on page 67; the usual conventional assumption being 150 lbs. moved at the rate of $2\frac{1}{2}$ miles per hour for 8 hours a day. At greater speeds his power very rapidly diminishes, a large portion of it being expended in moving his own weight. The following table shows the results obtained by different authors; those of Tredgold being for 6 hours daily labor, and those of Wood for 10 hours.

VELOCITY.	FORCE OF DRAUGHT; ACCORDING TO		
Miles per hour.	Leslie.	Tredgold.	Wood
2	100	166	125
3	81	125	83
4	64	83	62
5	49	42	50
6	36		42
7	25		36
8	16		31
9	9		28
10	4		25
11	1		23
12	0		21

From the above table it appears that, according to Wood, at 4 miles per hour a horse can draw only half his load at 2 miles; at 8 miles only a quarter; and so on.

At 10 miles per hour Tredgold considers the power of

a horse to be 37 lbs. moved 10 miles per day. At the same velocity Sir John Macneill estimates it at 60 lbs., moved 8 miles per day.

The power of a horse is also very rapidly diminished upon an ascent. On a slope of 1 in 7 ($8\frac{1}{4}°$) he can carry up only his own weight, without any load.

It is consequently very desirable to find a motive power on railroads, so much of which would not be uselessly lost at the high speeds which their diminution of friction renders possible. Steam has been therefore employed, through the medium of Stationary Engines, and of Locomotives.

2. STATIONARY ENGINES.

Stationary steam engines were once the rivals of locomotives, as motive powers for railroads, and were recommended by two distinguished engineers, less than twenty years ago, for adoption on the Liverpool and Manchester railway. It was proposed to place fixed engines along the line, at stations $1\frac{1}{2}$ miles apart. These engines were to turn large drums, or cylinders, around which were wound ropes, 4 or 5 inches in diameter, stretched along the road between the rails, and supported on rollers. The wagons were to be hooked to the ropes, and would be drawn onwards with them, as they were wound up on the revolving cylinders. An endless rope might also be employed, and two trains of cars be drawn at the same time in opposite

Fig. 140.

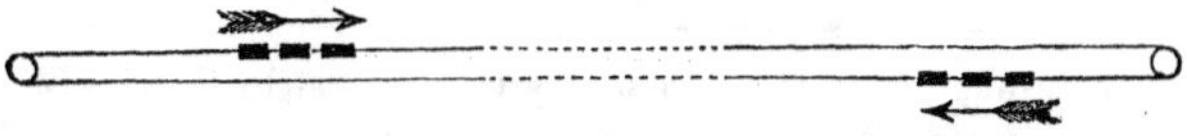

directions, as indicated by the arrows in the figure. When the cars had passed over the mile and a half, and reached the end of one rope, they could be detached from it, and attached to the succeeding one, without any stoppage.

The system has some advantages for short lines over which the travel must pass at brief intervals, owing to the economy of working stationary engines ; but it is utterly unsuited for general use. Its radical defect is, that the disarrangement of any single length of it, by any accident, must stop the travel on the whole line. It is a chain, the failure of any link of which will render the whole useless. It is therefore now seldom employed, except on inclined planes.

A very convenient application of the system is, however, seen on the London and Blackwall railway, 3½ miles long, with a fixed engine at each end. In this short distance there are five intermediate stations ; but no detention is caused by them, for a car is appropriated to each, in proper order—the last of the train being the one belonging to the first station, and so on. On reaching it, the sort of pincers by which the car is attached to the rope, is opened, and the car there stops, while the others of the train move on.

A railroad worked by a stationary engine, would be the most convenient method of relieving the rush of travel through Broadway. The railroad track should be supported on iron columns, out of the way of carriages, as in the figure. These columns might be placed on the edges of the sidewalks, where now are the lamp and awning posts, and by extending over the gutter they would have a base of 3 feet.* Their lower extremities should be set

---

* This arrangement of the columns was suggested by Charles Ellett, Jun., C. E., in 1844, for an " Atmospheric Railroad." In 1834, Mr. J. H. Patten proposed to use a secondary street, and to connect the columns by arches across the street, forming a flooring on which horses should travel.

Fig. 141

BROADWAY RAILROAD.

in heavy masses of masonry. At top they should spread outward, a foot on each side, which would give sufficient width for the railroad track. The columns should be set at distances of 15 or 20 feet, and connected by flat arches. There would be no flooring over the street, and the rails would intercept no more light than do the boards which now connect the awning posts. No locomotives, or even horses, would pass over the road ; but an endless rope would continually run over pulleys, and light cars would be under the most perfect control, and could be attached to it, or disengaged, at will, and stopped more easily than an ordinary omnibus. At the upper end of Broadway, a stationary engine, or the water-power of the Croton, would easily and cheaply keep up the circulation, which would pass up one side of the street and down the other. At each corner might be a platform, to which there would be a short flight of steps from the sidewalk, the ascent of which would be very easy ; or a certain number of corner houses might be used as dépôts, so that passengers might step into the cars from their second-story windows. As these cars would replace the omnibuses, the entire street would be left for miscellaneous travel.

A railroad on the surface of the ground, with its continual stream of cars stopping up the cross-streets every minute, would create a worse evil than that which it was intended to remedy ; and the endless rope could not be applied to it. If a railroad were made through a secondary street, passengers would not generally leave Broadway to avail themselves of it. A *surface* railroad being thus out of the question, two alternatives remain. The underground one will find few advocates ; and the only feasible arrangement seems to be the column and endless rope system. With its cheap construction, economical working, and thronged travel, it could scarcely fail to be the most profitable railroad ever built, and might be made to add largely to the city revenue.

3. LOCOMOTIVE ENGINES.

When a steam engine is required to move from its place, and to travel with its load, as do horses of flesh and blood, its usual weighty appendages of cold-water cistern, walking-beam, fly-wheel, &c., must be dispensed with. High-pressure steam must therefore be employed in order to enable the engine to combine the necessary compactness, lightness, and powei

HISTORY.

The first locomotive engine was constructed in 1802, by Richard Trevithick, who took out a patent in conjunc tion with Andrew Vivian.* Both were Cornwall engi-

* In 1759, however, Dr. Robison, then a student in the University of Glasgow, suggested to Wa · the application of the steam engine to moving

neers. This engine was tried on common roads, but in 1804, Trevithick applied a second one to a tram-road in South Wales, on which it drew ten tons of iron at the rate of 5 miles an hour.

Many years elapsed before any considerable improvements were made, owing in a great degree to useless efforts to overcome a difficulty which never had any real existence. When steam is applied to propel a wheel carriage, each piston-rod, to which the steam gives a backward and forward motion, is attached to a pin on one of the wheels, called the driving-wheel, and turns it by a crank, as a man turns a grindstone. If there was no friction between the wheel and the road, the wheel would turn around, while the carriage would remain stationary But the friction, which does exist, prevents the wheel from slipping, and it is enabled to turn only by propelling the carriage forward over a distance equal to the circumference of the wheel for each complete revolution of it. The imaginary difficulty referred to, was the notion that the adhesion or "bite" between the wheel and the rail, was so slight that with a load, and particularly on an ascent, the wheels would slip, slide, or "skid," either completely or partially, and thus fail to propel the engine.

Great ingenuity was expended in devising remedies for this non-existent evil. Wheels were at first made with knobs and claws to take hold of the ground; in 1811 a toothed rack was laid along the road, and a wheel with teeth was attached to the engine and fitted into the rack;

---

wheel carriages. In 1782, Murdoch, to whom Trevithick was a pupil, made a model of a steam-carriage; and in 1784 Watt described such an application in his patent. In 1801, Oliver Evans in Philadelphia moved a steam-dredging machine a mile and a half on wheels turned by its own engine.

and in 1812 a chain was stretched between the extreme ends of the road, and passed around a grooved wheel fixed to the engine and turned by it. But the most singular and ingenious contrivance was patented in 1813 by Mr. William Brunton. He attached to the back of his engine two legs, or propellers, which, being alternately moved by the engine, pushed it before them. The propellers imitated the legs of a man, or the fore legs of a horse, as shown in the figure.

Fig. 142.

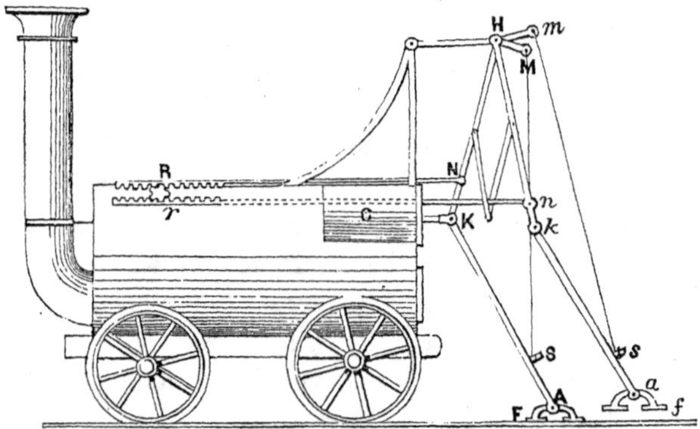

The legs are indicated by HKF, and H*kf*. H represents the Hip-joint, K and *k* the Knee-joints, A and *a* the Ankle-joints, and F and *f* the Feet.

We will first examine the action of the front leg. The knee, K, is attached to the end of the piston-rod, which the steam drives backward and forward in the horizontal cylinder C. When the piston is driven outward, it presses the leg, KF, against the ground, and thus propels the engine forward as a man shoves a boat ahead by pressing with a pole against the bottom of a river. As the engine advances, the leg straightens the point H is carried forward and the extremity, M, of the bent lever, HM, is raised. A cord, MS, being attached to S, the shin of the leg, the motion of the lever tightens the cord, and finally raises the foot from the ground and prepares it to take a fresh step when the reversed action of the piston has lowered it again.

The action of the other leg is precisely similar, but motion is communicated to it from the first one. Just above the knee of the front leg, at N, is attached a rod, in which is a toothed rack, R. Working in it is a cog-wheel, which enters also a second rack, *r*, below it, which is connected by a second rod, with a point, *n*, of the other leg. When the piston is driven out and pushes the engine from the knee, the rack, R, is drawn backwards and turns the cog-wheel, which then draws the lower rack, *r*, forwards, and operates on the hind leg, precisely as the piston-rod does on the front one, and thus the two legs take alternate steps, and walk on with the engine.

This locomotive, or "mechanical traveller," as it was termed by its inventor, moved on a railway at the rate of 2½ miles per hour, with the tractive force of 4 horses.

All these contrivances were, however, rendered useless by the discovery in 1814, by actual experiment, that the adhesion, or friction, of the wheels was amply sufficient for propelling the engine, even with a heavy load attached to it, and up a considerable ascent. Even if the adhesion were less than it is, it could be increased to an almost unlimited extent, by inducing a galvanic action between the engine and the rails.*

The first really successful locomotive was constructed

---

* Lecount, p. 352.

by Mr George Stephenson in 1814. By applying the "steam blast," he doubled its power and enabled it to run 6 miles per hour, and to draw 30 tons

Still no great progress was made in the application of steam to locomotion, until, in 1829, the directors of the Liverpool and Manchester railway resolved to employ locomotives in preference to stationary engines, and offered a premium for the best engine, not heavier than six tons, which should be able to draw twenty tons at the rate of ten miles per hour, and should fulfil certain other conditions. Four engines appeared, but the "Rocket" engine made by Mr. Robert Stephenson, won the prize, having run at an average speed of 15 miles per hour, and having performed one mile at the rate of 29½ miles per hour.

Since that time the progress of improvement has been onward, and one engine has travelled 75 miles per hour; another,* weighing 15¾ tons, has drawn 1268 tons (in a train of 158 coal-cars, 2020 feet long) 84 miles in 8 hours, over a line of which 40 miles were level, and which had curves of only 700 feet radius; and a third,* weighing only 8 tons, has drawn 309 tons on a level, and 16 tons up an inclined plane which rose 369 feet to the mile. The Rocket, however, contained the germs of all the principles which have been so wonderfully developed in its successors, and which will now be briefly noticed.

---

* Made by William Norris: Philadelphia.

PRINCIPLES.

The power of an engine is proportional to the quantity of steam which it can generate in a given time; for each revolution of the wheels corresponds to a double stroke of each piston, and consequently to four cylinder-fulls of steam. It is therefore necessary to expose the largest possible surface of water to the action of heat. This is most effectually attained by a *tubular boiler*, patented by Mr. Seguin in 1828, but perfected by Mr. Stephenson in 1829. Through the boiler, which occupies the principal mass of the engine, run a great number of small brass tubes, and through them the flame and heated air pass from the fire-box to the chimney. The tubes are about 6 feet long, 2 inches in diameter, and from 90 to 120 in number. They have been made 300 in number, and 1½ inches in diameter.* By this contrivance, and by surrounding the fire-box with a double casing, containing water, all the heat is absorbed by the water before it reaches the chimney.

The introduction of such tubes tripled the evaporating power of the engine, and caused a saving of 40 per cent of the fuel. But the abstraction of all the heat from the air, destroyed the draught of the chimney, and therefore the activity of the fire. This evil seemed insurmountable, in spite of the use of fanners, till George Stephenson used the waste steam, which passed from the cylinder after working the engine, to create an artificial draught, by discharging it into the chimney. This *steam blast* has been termed the life-blood of the locomotive machine.

To economize heat still farther, the cylinders are some-

* The tubes being very perishable, the Earl of Dundonald and others have proposed to construct boilers with the water in the tubes to be heated, instead of the fire in the tubes.

times placed within the smoke-box, or bottom of the chimney, so that none of their steam is condensed by the cold atmosphere. In this position, besides being nearer the centre of resistance, they act with a less injurious strain; although, two pistons being necessary to pass the "dead-points" of the crank, their action is unavoidably unequal on each side in turn. But this arrangement gives less room for the machinery, and renders necessary a double-cranked axle, which is consequently much weakened, though cut from a solid mass of iron. Both the outside and inside arrangements have their advocates.

Six wheels are generally employed, the two largest being the driving-wheels to which the power is applied. These are from 5 to 7 feet in diameter, the others being from 3 to 4. Sometimes all are made of the same size. Eight wheels are also used, four large and four small, the latter being under a truck, which supports one end of the engine, and is attached to it by a pivot in its centre, around which it can readily turn when on a curve. The springs are so adjusted that the principal part of the weight of the engine is thrown upon the driving-wheels. Sometimes two pairs of wheels are coupled together to obtain greater adhesion in ascending inclined planes, but this arrangement produces an unequal strain.

The eccentrics which open and shut the slide-valves to admit the steam to each end of the cylinders in turn, are so adjusted as to shut off the steam from one end of the cylinder and admit it to the other a little while before the piston has finished its stroke, so as to permit the expansive action of steam, and to form a sort of steam spring, to deaden the jerks of the engine. The degree of opening of the valve in advance is termed the "lead," and is usually from one-eighth to one-fourth of an inch.

Fig. 143

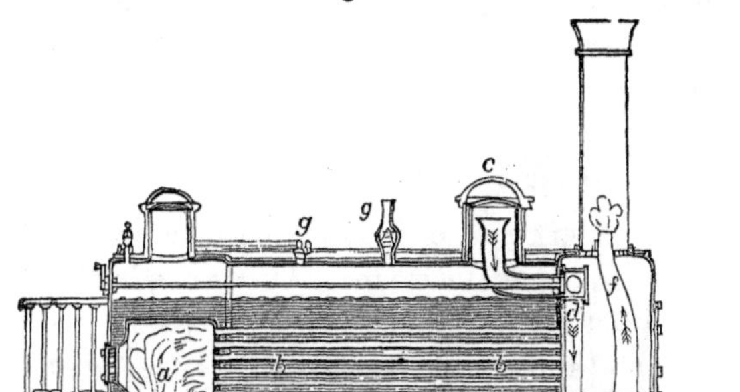

The above figure is a longitudinal section through a modern locomotive engine, in one of its very varied forms. *a* represents the fire-box; from which the flames and heated air pass, through the tubes *b b*, into the smoke-box at the other end of the engine. The water of the boiler, (which is cased with wood to prevent loss of heat by radiation) surrounds the fire-box and the tubes, and the steam generated by the heat thus absorbed, is collected in the steam chamber *c*. Thence it passes, through *d*, to the cylinder *e*, and being admitted, by the slide-valve, alternately before and behind the piston, it gives to it the reciprocating motion, which the crank on the axle of the driving-wheel converts into the revolution which propels the engine. The blast-pipe, *f*, conveys the waste steam from the cylinders into the chimney, to increase its draught. *g g* are safety-valves, one of which should be locked up, so as to be out of the control of the engine-driver. *h* is one of the feed-pipes which conduct the water from the tender to the boiler, into which it is pumped by small force-pumps, which are worked by the engine, and the derangement of which has produced serious accidents.

SPEED AND POWER.

The *speed* of an engine depends on the rapidity with which its boiler can generate steam. One cylinder full of steam is required for each stroke of each of the pistons. Each double stroke corresponds to one revolution of the driving-wheels, and to the propulsion of the engine through a space equal to their circumference. Wheels seven feet in diameter pass over twenty-two feet in each complete revolution. To produce a speed of seventy-five miles per hour, they must revolve exactly five times in a second; and to effect this number of revolutions each piston must make double that number of strokes in the same time. In this way does this ponderous machine divide time into tenths of seconds, almost as precisely as the delicate chronometer of the astronomer.

This rapid reciprocating motion of the pistons is very destructive to the machinery, and is too great to attain the maximum effect of the power expended. It would therefore be very desirable to lessen this rapidity, and to provide some means of multiplying the motion of the pistons, as by chains on pulleys, &c.

High velocities are also very expensive, in consequence of the rapidity with which the steam must be generated, and rammed, as it were, into the cylinders. The same effect might be produced by one quarter of the quantity of steam, if time were given it to act expansively.

The *power* of an engine in drawing loads, depends on the pressure of the steam, which is usually between 80 and 120 lbs. to the square inch It is also limited by the adhesion between the road and the driving-wheels, which is proportional to the weight pressing upon the latter; so that instead of the weight of the engine being an obstacle, it is one of the principal elements of power. The average adhesion may be considered to be $\frac{1}{8}$th the weight. The tractive power of an engine of 20 gross tons weight, with 16 tons resting on the driving-wheels, would, on this assumption, be $16 \times 2240 \div 8 = 4480$ lbs. If the friction be 10 lbs. to the ton, its gross load, exclusive of its own weight, would be $4480 \div 10 = 448$ tons. If the ratio of the weight of the freight to the joint weight of the car and freight be as 6 to 10, the quantity of freight which such an engine could convey on a level would be $\frac{6}{10} \times (448-10) = 263$ tons; the weight of the tender, 10 tons, being deducted from the gross load.

The diminution of this power on inclinations has been noticed on page 276, but is more fully shown in the following *Table,* which is calculated for an engine of 20 tons, all resting on the driving-wheels, and for a friction of $8\frac{1}{2}$ lbs. to the ton.

Ascent, in feet per mile.	Tons of freight transported.	Fractional part of the full load on a level.	Number of engines necessary to transport the full load.
Level.	389	1.000	1
10	254	.653	$1\frac{1}{2}$
20	185	.476	$2\frac{1}{4}$
30	145	.372	$2\frac{3}{4}$
40	118	.304	$3\frac{1}{4}$
50	98	.252	4
60	84	.215	$4\frac{3}{4}$
70	71	.180	$5\frac{1}{2}$
80	63	.160	$6\frac{1}{4}$

The friction of 8½ lbs. to the ton, with which the preceding table was calculated, was found (p. 265) to be too small. The following table has been calculated by taking the friction at 10 lbs. per ton, (the average amount for a slow freight speed,) the other data remaining unchanged. The principles of the calculation are found on pages 325 and 276–7. The *adhesion* is taken at one-eighth of the weight resting on the driving wheels.*

Ascent, in feet per mile.	Tons of freight transported.	Fractional part of the full load on a level.	Number of engines necessary to transport the full load.
Level.	330	1.000	1
10	227	.688	1½
20	170	.515	2
30	135	.401	2½
40	111	.333	3
50	94	.284	3½
60	80	.242	4
70	70	.211	4¾
80	61	.185	5½

The above table shows that, with its data, on an ascent of 20 feet per mile, two engines will be required to transport the load which one could draw on a level; that three engines would be required on an ascent of 40 feet per mile, and so on.

A comparison of the two tables also shows that by assuming a small amount of friction, ascents are made to appear much more objectionable, relatively, than if a larger amount of friction had been employed. Thus, on an ascent of 50 feet per mile, according to the former table, (calculated with the insufficient, though commonly assumed friction of 8½ lbs. to the ton,) four engines are required to do the work of one upon a level; but the latter and more correct table shows that only three and a half are needed. For higher speeds, and consequently greater resistances, the same ascents would be found to be relatively much less injurious, as has been shown, on page 34, with reference to common roads.

* The greatest adhesion of iron upon iron is about one-sixth of the insistent weight; but in wet and freezing weather becomes almost nothing. It lessens with the increase of the slope of the road, nearly as the sine of the angle of inclination. It would evidently be nothing, if the road were vertical.

WORKING EXPENSES.

All the expenses of working the road for any given time are usually added together, and divided by the total number of miles run in that time by engines drawing trains. In this way is obtained the common average of working expenses, which are thus measured by *the cost of running trains per mile.* But this principle of comparison is evidently faulty, since a train may be run for a very small cost per mile, but carry few passengers and little freight; and thus its expenses, though small absolutely, may be ruinously great relatively. On the other hand, a heavy train may cost much more per mile, but carry so great an amount of freight or passengers, as to be run very cheaply, relatively to them. Fifty tons, carried for 75 cents per mile, would cost $1\frac{1}{2}$ cents per ton, while a hundred tons carried for even $1 per mile, would cost but 1 cent per ton.

The cost of transport per mile for each *passenger* or *ton* of freight carried, is therefore a preferable standard, with certain restrictions, as affording a means of direct comparison between the expenses and the receipts, which are the final objects of all the operations. But this, again, does not of itself show the comparative economy of the working of different roads, for a road may be worked very cheaply per mile run, but, having little business, at a great cost per passenger or per ton, since a large part of the expenses are the same for one passenger or for a hundred. The converse of this takes place on a heavy road, worked expensively per mile, but cheaply per passenger or ton of freight.

Both these methods of comparison ought, therefore, to be employed in conjunction.

The complete average expense per train per mile of running eleven New York railroads during 1850, was 67 cents for passenger trains, (ranging from 34 to 94 cents,) and 87 cents for freight trains, (ranging from 37 to 159 cents;) including in this the expenses of maintaining the road, of repairing machinery, and of operating the road.

Upon the same roads, the average cost per passenger per mile was $1\frac{29}{100}$ cents; the lowest being $\frac{78}{100}$, and the highest $2\frac{40}{100}$. The average cost of freight per ton was $3\frac{16}{100}$ cents; the lowest being $1\frac{64}{100}$, and the highest $4\frac{93}{100}$ cents.*

Upon five leading Massachusetts Railroads, the average expenses, for *Passenger* trains, per mile run, was 74 cents, (from 63 to 93 cents,) and per passenger per mile 1 cent, (from $\frac{62}{100}$ to $1\frac{23}{100}$.)

Upon the same five roads, the *Freight* expenses, per mile run, averaged 89 cents, (from 81 to 96 cents,) and per ton per mile $1\frac{45}{100}$ cents, (from $\frac{90}{100}$ to $1\frac{77}{100}$.)

The expenses on the Utica and Schenectady road are classified thus:*

Utica and Schenectady Railroad.	Passengers per mile run.	Freight per mile run.
1. *Maintaining road* .... (Repairs and depreciation, taxes, &c.)	22	22
2. *Repairs of machinery* .... (Engines, cars, tools, &c.)	21	26
3. *Operating the road* .... (Office expenses, Laborers, Conductors, Enginemen, &c., Fuel, Oil, and Waste, Damages, Superintendence, Contingencies.)	33	94
	76	142

* N. Y. State Engineer's Report on Railroad Statistics, Jan. 7, 1851.

Upon the Eastern Railroad (Boston to Portsmouth, 54 miles) the expenses were thus classified for the year ending June 30, 1850: 1,037,000 passengers, and 71,000 tons of freight having been carried:*

Eastern Railroad.	Per mile run.	Per cent.
Machine shop	0.3	
Maintenance of way	12.5	20
Locomotive power	22.9	37
Train expenses	9.9	16
Office expenses	6.0	10
Station do.	8.3	13
Mail	0.3	
Ferry	2.3	4
	62.5	100

With careful management in every department, trains carrying average loads of from 100 to 150 tons can be moved, on ordinary grades, at a cost of 80 cents per mile. Such economy of transport depends mainly, however, upon the certainty of always carrying full loads. For this reason the Baltimore and Ohio Railroad carried coal, by contract, for $1\frac{1}{3}$ cents per ton, while their ordinary traffic, giving the engines only half a load, cost them over $2\frac{1}{2}$ cents. The Reading Railroad is said to be able to carry coal for 6 mills per ton per mile, because fully loaded on the down trips.

The present cost of transport on the Erie Canal is $1\frac{56}{100}$ cents per ton per mile, of which the State receives $\frac{7}{10}$ cent, or nearly one-half. On the Enlarged Canal, the cost is estimated at 7 mills, 3 of these being tolls.†

* Report of the President, D. A. Neal.

† N. Y. State Engineer's Report on Canals, Feb. 7, 1851.

SAFETY OF TRAVELLING.

The comparative safety of railroads is one of their most valuable attributes, though the one least appreciated and most imperfectly realized. The popular impression is generally the reverse of the truth, for an accident to a stage-coach is seldom heard of beyond the immediate scene of its occurrence, while any railroad disaster is passed from paper to paper over the whole land.

There are many reasons why travelling on railroads should be safer than on common roads. The former are level instead of hilly, and smooth instead of uneven; and all miscellaneous travel is excluded.

The cars are safer than coaches, because their centres of gravity are lower; their axles are less exposed to violent shocks, and therefore are less subject to break; and they are altogether less exposed to be overturned.

Locomotive engines are safer than horses, because they are not liable to take fright, shy, or run away; and can be stopped at once by a brake, tamed down by opening a valve, and backed by simply moving a lever.

The statistics of railroads fully confirm the conclusions of theory. On the English railroads, according to the parliamentary returns, between 1840 and 1845, both inclusive, more than 120,000,000 of passengers were carried, and of these only 66 were killed, or one in nearly two millions; and only 324 others were in any way injured, or one in nearly four hundred thousand.

On the Belgian railroads, 6,609,215 persons travelled between 1835 and 1839, and of these 15 were killed and 16 wounded. But of these, 26 were persons employed on the railroads, and only 3 *passengers* were killed and 2 wounded. In 1842, of 2,716,755 passengers, only three were killed, and of these one was a suicide, and the other two met their deaths by crossing the line.

On French railroads, 212 miles in length, of 1,889,718 passengers who travelled over 316,945 miles, in the first half of 1843, not one was either killed or wounded, and only three servants of the railroad suffered.

Comparing with this the travelling by horse-coaches in the same region, we find that in seven years, from 1834 to 1840, 74 persons were killed, and 2073 wounded!

But few as are the accidents on railways they are still much more numerous than they need be. They may be divided into those which arise from mismanagement and negligence, and those which are caused by inherent faults in the construction and working of the railroad.

To the former class belong accidents from collision. When two engines and trains meet each other, or when one overtakes another, the destructive consequences, which so often ensue, are generally due to the carelessness or ignorance of the conductors, or engine-drivers of the train; and are finally attributable to the false economy of employing at a low salary incompetent persons. The danger of collision would also be much lessened, if trains running in different directions were confined invariably to one line of rails.

Many accidents have arisen from a slow train being overtaken by a faster one. There is extreme danger in permitting one engine to follow another, except at very considerable distances; and a mile is a very short distance when measured by the brief time in which a locomotive can pass over it.

The practice of attaching an engine behind a train to assist the front one in the ascent of a steep grade, is also fraught with danger; for any derangement of either engine makes it the anvil on which the other one falls like a trip-hammer, crushing every thing between them.

The excessive speed demanded by the impatience of the travelling public diminishes the controlling power, and makes the consequences of any negligence or malicious obstruction proportionally destructive.

Wilful disobedience of orders on the part of engine-drivers and conductors, (as to time, turning-out places, waiting for other trains, &c.) reckless exposure to possibilities of collision, and insane confidence in good-luck, are causes of the majority of accidents ; and though no faithful superintendent would permit such men to have a second opportunity for similar misconduct, yet the disastrous effects of even the first faults might be generally avoided by employing only men of undoubted intelligence, experience, sobriety, and self-control, and securing the services of the very best of their class by liberal compensations.

The second division of accidents included those caused by inherent faults in the construction and working of the railroad. These may be in a great degree guarded against, by careful and continual inspection of the line of the road, and examination of all parts of the engines and cars.

The explosion of the locomotive boiler is often injurious, if not fatal, to the engine-men, and by its stoppage of the train may cause a collision with a following one.

The settling of an embankment may cause a depression of one side of the road, which will compel the engine to run off. The looseness of a rail, its breakage, (when supported only at intervals) the misplacing of a switch, &c., may produce a similar result. The destructive consequences would be much lessened, if means were provided for instantly detaching the train from the engine, or if they were so coupled that they would be separated by any lateral strain.

The breakage of the axles of the engine or carriages has caused many accidents; but this danger is greatly lessened by the eight wheels of the American cars,* and by the appendages of "Safety-beams."

The sparks from the locomotive chimney frequently communicate fire to the train, and have thus, in one instance, caused great loss of life, increased by the impossibility of communicating the intelligence to the engine-driver in time to arrest the disaster.

SIGNALS.

Many of the accidents which occur with the locomotive system might be prevented by a uniform, simple, and complete plan of signals. *Red* flags and lights for imminent danger; *green* for caution; and *white* for safety, are leading features in all the systems. The signals are made by the policemen, who are, or should be, stationed along the line, to see that the rails are clear, to communicate intelligence, to work the signals, &c.

The Danger signal is a *red* flag by day, or red glass lamp by night, waved backwards and forwards. The engine should be stopped the moment this signal is seen. *Any* signal, *violently* waved, should also cause an immediate stoppage.

The Caution signal is a *green* flag, or light, and should be obeyed by slackening the speed of the engine. When

---

* In this respect we are far in advance of European Railroads, and a writer in the Westminster Review lately suggested, as an improvement of the highest importance, a peculiar style of car, which was almost precisely identical with those which have been for many years in general use on American Railroads.

the green flag is held so as to point upwards, it indicates that another engine is less than five minutes in advance of the one to which the signal is made. When held pointing downwards, it enjoins a slow rate of speed as a precaution against defects in the rails at that place.

The Safety, or "All-right" signal, is a white lamp at night, and by day the upright position of the policeman with his flags furled.

These signals are made by the policemen, either with hand flags and lamps, or by arms which are moveable on signal posts, and worked by cords.

In the absence of these conveniences the policeman makes the signal "All right," by extending his arm horizontally; the Caution signal by holding one arm straight up; and the Danger signal by holding both arms straight up, or by waving violently a hat, or any other object.

The Danger signal is always to be made immediately after any engine or carriage has passed along the line, and is to be continued for five minutes; it is also to be made whenever there is any obstruction on the line, or any danger of it.

The Caution signal is always to follow the Danger signal, and to be continued for five minutes; it is also to be made wherever there is any reason for slackening the speed.

The All-right signal is to be made only when the signal-man has satisfied himself that the line is clear, unobstructed, and free from any suspicion of danger. Every signal-man should immediately report to his nearest superior officer any instance of disobedience to the signals which he had made.

In foggy weather both day and night signals are given;

and in addition, when any emergency requires the *immediate* and certain stoppage of any train, a detonating compound, packed in a small box, is fastened to the rail with slips of lead, and explodes with a tremendous noise when a wheel passes over it, giving an unmistakeable signal for instant stoppage.

White and red lights on the front and back of a train at night should be so arranged and combined as to indicate the direction, speed, &c. of the train. But all these precautions are finally dependent for their complete success upon the character of the persons in the employ of the company.

4. ATMOSPHERIC PRESSURE.

The pressure of the atmosphere is usually assumed to be 15 lbs. on every square inch of surface, and though the equality of this pressure in all directions renders it generally insensible, it becomes very apparent to the senses when the hand is held on one end of a cylinder from the interior of which the air is drawn out by an air-pump. It is this pressure which is the motive power of the ATMOSPHERIC RAILWAY.

The first idea of such a construction seems to have originated in 1805, in which year an Englishman, named Taylor, proposed to employ atmospheric pressure for sending letters and parcels from town to town. His plan was to lay a long tube, like a gas or water pipe, between the places, and to fit into it an air-tight piston. If the air was pumped out from one end of such a tube, the pressure of the atmosphere would force forward the piston, and any thing attached to it.

In 1810, Medhurst proposed to make a tube, archway, or tunnel, large enough to contain carriages with passengers, to be propelled in a similar manner. But this scheme was never put into practice, for travellers did not relish the idea of being shot through a tube, like pellets in a popgun.

The problem was now to devise some means of communicating the motion of a piston, blown through an air-tight tube, to a carriage on the outside of this tube.

Medhurst, in 1827, proposed to make the desired communication and application of power, through a channel, or groove, on the top of the tube, filled with water to make it air-tight. He also suggested the use of a square iron tube, with half its top rising and falling on hinges, and an arm coming through the opening to connect the piston to the carriage.

Vallance, in 1824, patented a variation of the tunnel of Medhurst.

Pinkus, an American residing in London, in 1834 proposed the use of a tube with a slit in its top and a sort of rope for the covering valve.

But no substantial success was attained till Clegg, in 1839, invented his flap valve, and, in conjunction with Samuda, developed the present system. Fig. 144 is a

Fig. 144.

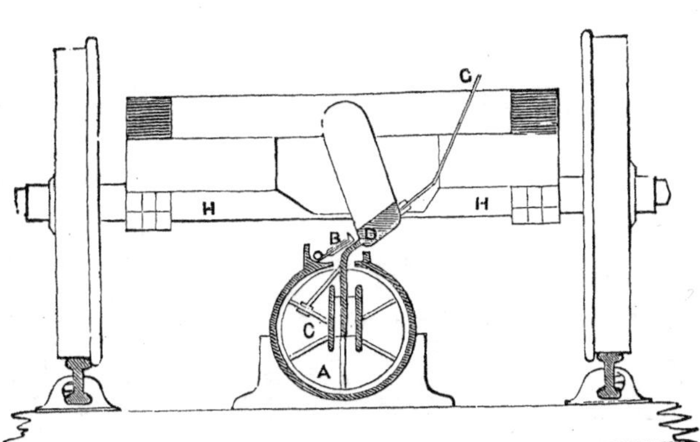

cross-section of the pipe, valve, &c. The pipe, A, is of cast iron, and about eighteen inches in diameter. It is laid between the rails on which the carriages are to run. Along its top is a continuous slit, or longitudinal opening, through which is to pass obliquely the iron bar, or arm, D, which connects the piston C, with the carriage, of which HH is an axle. The valve which covers this slit, and which is shown in cross-section at B is essentially a strip of leather, one edge of which is fastened to one side of the slit, so that the rest of it can rise and fall, and thus alternately open and close the slit. In the figure it is represented as open. To strengthen it, plates of iron, each eight inches long, are attached to its upper and under sides. The under ones are just wide enough to fit into the slit; the upper ones are a little wider, to prevent the valve from being pressed into the pipe. On each side of the slit is a rib, or projection, cast with the pipe, and forming a sort of trough, at the bottom of which the valve lies when shut. This trough is filled with a mixture of tallow and bees-wax, which, after being melted and cooled, adheres to the edge of the valve and makes it perfectly air-tight.

Fig. 145.

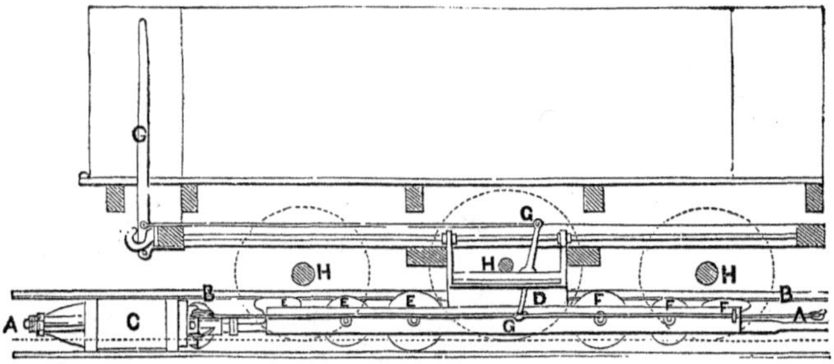

Fig. 145 is a longitudinal section of the pipe, piston, and leading carriage. The same letters of reference are employed as in Fig. 144.

A steam engine, at the end of a length of 3 miles of the pipe, works an air-pump, which draws out a portion of the air from the pipe, AA. The air behind the piston, (shown at C) being no longer balanced by the air before the piston, presses it forward. The small wheels, EEE, behind the piston, raise the edge of the valve in order to make way for the connecting arm, D, which draws the carriage (of which HHH are the axles) onward with the piston. The small wheels, FFF, behind the arm, lift up the valve to admit the air more freely to press on the back of the piston. The piston and carriage thus proceed as long as there is a greater pressure of air behind than before them.

To re-seal the valve, after the piston has passed, in readiness for being again exhausted, the second carriage of the train carries under it a small steel wheel which presses down the valve, and which is followed by a heater, or copper tube, five feet long, and filled with burning charcoal, which melts the composition in the trough and solders down the edge of the valve.

To stop the train the brake may be applied; or the lever, shown at G in Figs. 144 and 145, may open a valve in the piston, and admit air in front of it to destroy the vacuum, and consequently the propelling power.

When the carriage has reached the end of one length of 3 miles, it passes into the next length of pipe by an entrance, or equilibrium valve, ingeniously contrived to permit the change without affecting the vacuum.

The power of this system depends upon the size of the pipe, and the perfection of the vacuum in front of the piston. If the pipe be 18 inches in diameter, the area of the piston will be 254 square inches, and if a perfect vacuum could be attained, the pressure of the atmosphere upon this surface would be $254 \times 15 = 3810$ lbs. Calling the friction 10 lbs. to the ton, this power would be sufficient to move 381 tons. In practice, however, the vacuum is seldom reduced below 8 lbs. to the square inch, or half an atmosphere, there being an unavoidable leakage.

The speed is proportioned to the rapidity with which the air-pump exhausts the pipe, and therefore to the velocity with which the air-pump piston moves, and to the ratio between its area and that of the travelling piston. Air rushes into a vacuum with a velocity of 800 miles per hour, and this is therefore the maximum limit of speed. It is probable, however, that a railroad which approximated to this speed would find but few passengers, and a mile in 62 seconds, or 58 miles per hour, is the nearest approach to it yet made.

The vacuum may be made not only by working an air-pump by a steam engine or by water-wheels, but by filling an air-tight vessel with water, subsequently allowed to run out at a depth greater than that at which the atmosphere will support a column of it.

The time required to exhaust a 3 mile length of pipe, by the usual air-pump, is 4 minutes. Allow 5 minutes for the train to pass, and the 4 minutes needed to exhaust the pipe again, would give 9 minutes as the least possible interval between the starting of trains, since only one train at a time can be on any one length of pipe. The application of this system to a Broadway railway, as has been suggested by some projectors, would, for this reason, be wholly impracticable.

The principal *advantages* claimed for the Atmospheric Railroad by its advocates are the following:

Its cars can ascend any inclination however steep; since the force capable of being applied does not depend at all upon the adhesion of the wheels to the rails, as in the case of locomotives. At a certain degree of steepness locomotive engines could not carry up themselves, much less a load; while the piston of an Atmospheric Railroad would exert equal force if its pipe were even vertical, though of course with much less profitable effect.

The engine and tender being dispensed with, the force which would have been expended in moving their weight of 20 or 30 tons, is so much clear saving.

The rails may be made much lighter and will last much longer, where they have not to sustain the shocks of the locomotive, which is the most powerful agent in their de struction.

High speed with locomotives involves great waste of power, in consequence of the disadvantageous velocity with which the pistons must move. It is not so with the atmospheric system.

But greater safety is one of the most important recommendations of this system; for the cars cannot run off the track, being securely attached to the pipe; nor can they ever come into collision with each other, for no two trains can be on the same length of pipe at once.

On the other hand, if any obstacle be on the track, there is less power of stopping them, and none at all of reversing their motions; and the great objection to the stationary engine system—that the failure of one link deranges the whole chain—applies to this plan also.

But the comparative economy of the Atmospheric and Locomotive systems is the principal element in determining their relative merits. Much greater cheapness of working is claimed, by its partisans, for the atmospheric system, but this is strenuously denied by other engineers, and the testimony is so conflicting and varying, in consequence of the insufficiency of the data, that no satisfactory conclusion can be arrived at. The balance of argument seems, however, to be against the profitable employment of the system in ordinary cases. Under some peculiar circumstances, however, such as the case of a line with steep grades, on which light trains must be run at short intervals, it may probably be advantageously applied.

The longitudinal valve being the weak point of the system, several attempts have been made to dispense with it. The most successful inventions have been those of Pilbrow; and of Julien and Vallirio.

Compressed air, Carbonic acid gas, Electro-magnetism &c., have been also proposed as motive powers for railroads, but none of them seem likely to rival, in power, speed, or economy, that most magnificent and life-like of all human creations, the Locomotive Engine.

CHAPTER VI

THE MANAGEMENT OF TOWN ROADS.

"The money levied is more than double of what is necessary for executing in the completest manner the work, which is often executed in a very slovenly manner, and sometimes not executed at all."

ADAM SMITH.

A WISE and well-regulated system of managing the repairs of roads, and of obtaining the greatest degree of improvement with the least amount of labor, is as important as their judicious construction. The "*Road-tax*" system, of personal service and commutation, though nearly universal among us, is unsound in its principle, unjust in its operation, wasteful in its practice, and unsatisfactory in its results. Borrowed from the "statute-labor" of England, and the "*Corvée*" or "*Prestation en nature*" of France, like them it is a remnant of the times of feudal vassalage, when one of the tenures by which land was held was the obligation to make the roads passable for the troops of the lord of the manor. The evil consequences of the system will be examined, when we have briefly explained its organization in the state of New York, where it has been rendered as perfect as its nature permits.*

* A convenient edition of the revised road act, with commentaries, &c., was published at Rochester in 1845.

The directing power is vested in "*Commissioners* of Highways," who are chosen in each town at the annual town meeting, and have "the care and superintendence of the highways and bridges therein." Subordinate to them are "*Overseers*," of whom are chosen, at the annual town-meeting, as many as there are road districts in the town. The commissioners have the authority to direct the overseers as to the grade of the road, how it should be shaped and drained, and the like. They may also lay out new roads. The principal duties of the overseers are to summon the persons subject to perform labor on the roads, to see that they actually work, and to collect fines and commutation money. The commissioners are to estimate the cost of improvements necessary on the roads and bridges of the town, and the board of supervisors are to cause the amount to be levied, but within the limit, for any one year, of two hundred and fifty dollars. But, if a legal town meeting so vote, the supervisors may levy "a sum of money, in addition to the sum now allowed by law, not exceeding five hundred dollars in any one year."

"Every person owning or occupying land in the town in which he or she resides, and every male inhabitant above the age of twenty-one years, residing in the town where the assessment is made, shall be assessed to work on the public highways in such town." The lands of non-residents are also to be assessed. The whole number of days' work to be assessed shall be at least three times the number of taxable inhabitants in such town; and may be as many as the commissioners shall think proper.

Persons assessed to work on the highways, upon receiving twenty-four hours' notice from the overseers, must appear either in person, or by able-bodied substitutes; or pay a sum of one dollar for each day's neglect, unless they shall have previously *commuted* at the rate of sixty-two and a half cents per day. A team, cart, wagon, or plough, with a pair of horses or oxen, and a man to manage them, satisfies an assessment of three days.

Such are the principal features of the present system. They are all defective in a greater or less degree.

In the first place, the condition of the roads, which is so important an element of the wealth and comfort of the whole community, should not be allowed to remain at the mercy of the indolence, or false economy, of the various small townships through which the roads pass. In one town, its public spirit, wealth, and pride, may induce it to make a good road ; in the adjoining town, a short-sighted policy, looking only to private interest in its narrowest sense, may have led the inhabitants to work upon the roads barely enough to put them into such a condition as will allow a wagon to be slowly drawn over them.

In the next place, the "commissioners" who have the primitive direction of the improvements and repairs, should be liberally compensated for the time and attention which they give to the work. Gratuitous services are seldom efficient ; at best they are temporary and local, and dependent on the whims, continued residence, and life of the party ; and if the compensation be insufficient, the same evils exist though in a less degree. Skill, labor, and time cannot be obtained and secured without being adequately paid for.

The third defect in the system is the annual election of the commissioners and overseers. When men of suitable ability, knowledge, and experience have been once obtained, they should be permanently continued in office. On the present system of annual rotation, as soon as the overseer has learned something in his year's apprenticeship, his experience is lost, and another takes his place, and begins in his turn to take lessons in repairing roads at the expense of their condition. In other occupations, an apprenticeship of some years is thought necessary before a person is considered as qualified to practise with his own capital; while a road overseer, the moment that he is chosen, is thought fit to direct a work requiring much science, at the expense of the town's capital of time, labor, and money.

In the fourth place, the *fundamental principle* of the Road-tax is a false one. Its contemporary custom of requiring rents to be paid in kind, has long since been found to be less easy and equitable than money rents. Just so is work paid for by the piece preferable in every respect to compulsory labor by the day. Men are now taken from their peculiar occupations in which they are skilful, and transferred to one of which they know nothing. A good ploughman does not think himself necessarily competent to forge the coulter of his plough, or to put together its woodwork. He knows that it is truer economy for him to pay a mechanic for his services. But the laws assume him to be a skilful road-maker—a more difficult art than plough-making—and compel him to act as one; though his clumsiness in repairing his plough would injure only himself, while his road-blunders are injurious to the whole community. Skill in any art is only to be acquired by practical and successful experience, aided by the instructions of those who already possess it. An artisan cannot be extemporized.

Fifthly, labor by the day is always less profitable than that done by the piece, in which each man's skill and industry receive proportionate rewards. Working on the roads is generally made a half holiday by those who assemble at the summons of the overseer. Few of the men or horses do half a day's work, the remainder of their time being lost in idleness, and perhaps half of even the actual working time being wasted by its misdirection.

Lastly, it follows from the preceding, that the commutation system operates very unfairly and severely upon those who commute; for they pay the price of a full day's work, and their tax is therefore doubled.

Such are the principal defects of the present system of managing the labor expended on town roads. But it is much easier to discover and to expose, than to remove them. In the following plan the writer has endeavored to combine the most valuable features of the various European systems, and to adapt them to our peculiar institutions.

In each State, a general legislative act should establish all the details of construction, and determine definitely "What a road ought to be," in accordance with the theory and practice of the best engineers. Surveys should be made of all the leading roads, and plans and profiles of them prepared, so that it might be at once seen in what way their lines could be most efficiently and cheaply improved.

The personal labor and commutation system should be entirely abolished. If the town-meeting would vote a tax in money of *half* the amount now levied in days' work, its expenditure under the supervision to be presently described, would produce a result superior to the present one. When the road is a great thoroughfare, extending far beyond the town, it would be unjust to levy upon it all the expense; and a county tax, or, in extreme cases, a state appropriation, should supply what might be necessary.

In regulating the expenditure of the money raised, the fundamental principle, dictated by the truest and most far-sighted economy, should be *to sacrifice a portion of the resources of the road to ensure the good employmen. of the remainder.* The justice of this principle needs no argument; its best mode of application is the only diffi culty. The first step should be to place the repairs of the roads under the charge of a professional Road-maker of science and experience On his skill will depend the condition of the roads, more than on local circumstances or expenditures. His qualifications should be tested by a competent board of examiners, if he should not have received special instructions in the requisite knowledge, such as might well form a peculiar department of education in our Colleges and Normal schools. As each town by itself could not afford to employ a competent person, a number of them (more or less according to their wealth and the importance of the roads within their bounds) should unite in an association for that purpose.

The engineer thus appointed should choose, in each township, an active, industrious man, of ordinary education, to act as his deputy in making the expenditures in that town, and as foreman of the laborers employed during the season of active labor on the roads. This deputy might be busily and profitably employed during the entire remainder of the year, in constantly passing over in due rotation the whole line of road under his care, and making, himself, the slight repairs which the continual wear and tear of the traffic would render necessary. If taken in time, he himself could perform them; but if left unattended to, *as is usual,* till the season of general repairs, the deterioration would increase in a geometrical ratio, and perhaps cause an accident to a traveller, which would subject the town to damages tenfold the cost of repairs.

The laborers hired by the deputy in each town should be employed by piece-work as far as is possible. This can be carried out to a great extent, when the superin tendent is competent to measure accurately the various descriptions of work, and to estimate their comparative difficulty. When the work cannot be properly executed by portions allotted to one man, it may be taken by gangs of four or five, who should form their own associations, make a common bargain, and divide the pay. In work not susceptible of definite calculation as to quantity or quality, and in such only, day-labor may be resorted to under a continual and vigilant superintendence.

In such a system as has been here sketched, the money-tax would be found to be not only more equitable than the personal-labor system, but even less burdensome. None of it would be wasted; and those who had skill and strength for road-work would receive back, in wages, more than their share of it; those who were skilful in other work might remain at that which was most profitable to them, and pay only their simple share of the road-tax, not double, as when they now commute; and the only losers by the change would be the indolent, who were useless under the old system, but under this, would be obliged to contribute their share; while great gain in every way would ensue to the community at large. The subject urgently demands legislative attention.

APPENDIX A.

EXCAVATION AND EMBANKMENT.

IT is required to find the content of a mass of earth, filled into a hollow or dug out of a hill.

Since we cannot really measure the dimensions of the mass after the work is done, it is necessary to determine the height of the original surface of the ground, above (or below) some datum, before the cutting or filling is made. This is done by taking "levels" (or cross-sections) at all points where the ground changes slope.

After the cutting or filling has been made, "levels" are again taken over the new surface, generally exactly above, or below, the original ones. The difference of the corresponding levels gives the desired heights, or depths. Then the proper rules of mensuration can be applied. There are several cases to be considered.

CASE I. When the ground is level transversely, "One level."

CASE II. When it is sidelong, that is, has a transverse slope, "Two level."

CASE III. "Three level" ground.

CASE IV. Irregular ground.

CASE V. On curves.

CASE I.—*When the ground is level transversely, "One level."*

It is then sufficient to take a single "level" at each point of the line at which the ground changes slope longitudinally (as on p. 116). The content is then calculated by one of the following methods:

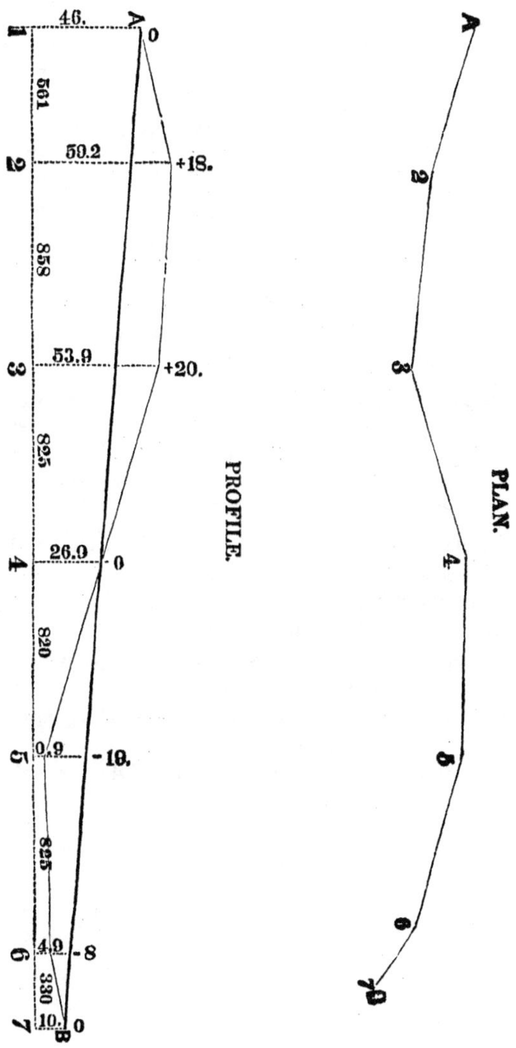
PROFILE.
PLAN.
A
B
1
2
3
4
5
6
7
46.
561
59.2
+18.
858
53.9
+20.
825
26.9
0
820
0.9
-10.
825
4.9
-8
330
10.
0

CALCULATION OF EXCAVATION AND EMBANKMENT.

1	2	3	4	5	6	7	8	9	10
Station.	Distance.	Height of ground above datum line.	Rise or fall for each distance.	Height of grade above datum.	Cut. +	Fill. —	End Areas.	Excavation. Cubic feet.	Embankment Cubic feet.
1		46.0		46.0	0		0		
2	561	59.2	— 4.8	41.2	18		1386	388,773	
3	858	53.9	— 7.3	33.9	20		1600	1,280,994	
4	825	26.9	— 7.0	26.9	0	0	0	660,000	
5	820	0.9	— 7.0	19.9		19	1672		685,520
6	825	4.9	— 7.0	12.9		8	528		907,500
7	330	10.0	— 2.9	10.		0	0		87,120
	4219		36.0					2,329,767	1,680,140

In the above tabular view, the first seven columns are transferred from page 116. The remaining columns of areas and cubical contents are filled up by the following calculations, assuming at 50 feet the width of road-bed ; which will be the bottom of a cutting, or the top of an embankment, at a height just sufficient to equalize the elevations and depressions of the final transverse profile of the surface of the road. The side-slopes of the excavations are supposed to be $1\frac{1}{2}$ to 1, and those of the embankments 2 to 1. We are now prepared to take up, in turn, each of the four usual methods of calculation.

1. CALCULATION BY AVERAGING END-AREAS.

At station 1 there is neither cutting nor filling. The end-area in column 8, opposite that station, is therefore 0.

Fig. 146.

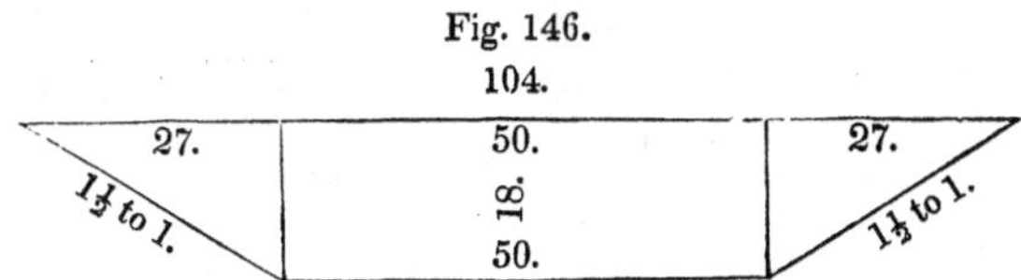

At station 2, the cross-section of the excavation is shown in the figure. The "Distances out" of the side-slopes are $1\frac{1}{2} \times 18 = 27$ feet. The top width is therefore $27 + 50 + 27 = 104$ feet. The area equals $\frac{104 + 50}{2} \times 18 = 1386$; or otherwise, since the two triangular portions equal a rectangle of the same base and height as one of them, the area $= (50 + 1\frac{1}{2} \times 18) \times 18 = 1386$.

At station 3, the area equals $(50 + 1\frac{1}{2} \times 20) \times 20 = 1600$.

At station 4, the Excavation ends, or "runs out," and the area $= 0$

Fig. 147.

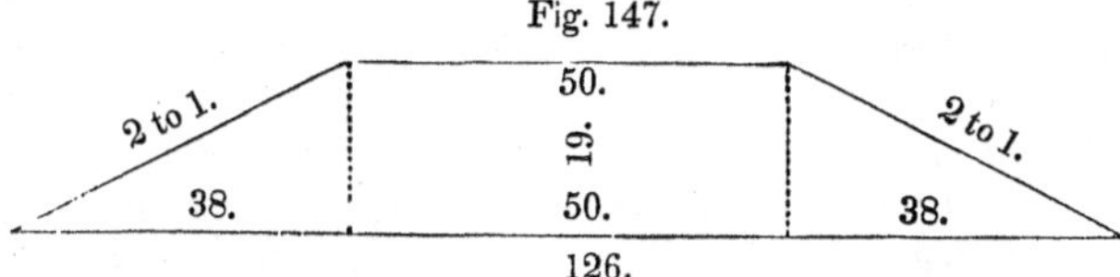

At station 5, the section of the embankment is shaped as in the figure, and has an area $= (50 + 2 \times 19) \times 19 = 1672$.

At station 6, the area $= (50 + 2 \times 8) \times 8 = 528$.

At station 7, the area $= 0$.

The column of *End-Areas* is thus filled.

The *Cubical Contents* are next to be calculated.

The mass between stations 1 and 2, has an area of 0 at one end, and of 1386 at the other, and is 561 feet long. Its contents, by the method which we now employ, will equal the average of the two areas, multiplied by the length; i. e., $\frac{0 + 1386}{2} \times 561 = 388{,}773$ cubic feet

The contents of the second mass, that between 2 and 3, equals $\frac{1386 + 1600}{2} \times 858 = 1,280,994$ cubic feet.

The third mass $= \frac{1600 + 0}{2} \times 825 = 660,000$ cubic feet.

Here the excavation ends, and the embankment begins.

The fourth mass $= \frac{0 + 1672}{2} \times 820 = 685,520$ cubic feet.

The fifth mass $= \frac{1672 + 528}{2} \times 825 = 907,500$ cubic feet

The sixth mass $= \frac{528 + 0}{2} \times 330 = 87,120$ cubic feet.

These results, being in cubic feet, should be divided by 27, to reduce them to *cubic yards*, the denomination in which estimates are made and contractors paid. This reduction would be facilitated, if the measuring tapes and rods were divided into yards and their decimal parts; or if the distances of the stations were always some multiple of 54 feet.

The results thus obtained, by averaging the end-areas, exceed the correct amount, as will appear from an inspection of the figure on the following page, from which may also be deduced the correction to be applied.

This figure presents a perspective view of a tapering prismoidal mass, such as is an excavation of unequal size at its two extremities; ABCD being the area of its largest end, and EFGH of its smallest. Conceive a plane, parallel to the base of the cutting CDHG, to be passed through EF. It would cut the larger end in the line IJ, leaving below it a quadrangular prism, with equal bases EFGH and CDIJ. Subdivide the remaining figure, by raising the vertical lines IL and JK, and passing a plane through IL and E, and another through JK and F. The interior body thus formed appears wedge-shaped, but is a triangular prism, equal to half the quadrangular prism, which has IJKL for base, and IE or JF for height. There remain two triangular pyramids,—one with base ALI and vertex E, and the other with base BJK and vertex F.

The prismoid being thus dissected, the contents of the quadrangular and of the triangular prisms would be correctly obtained by multiplying the sum of the bases or end-areas by *one-half* the

Fig. 148.

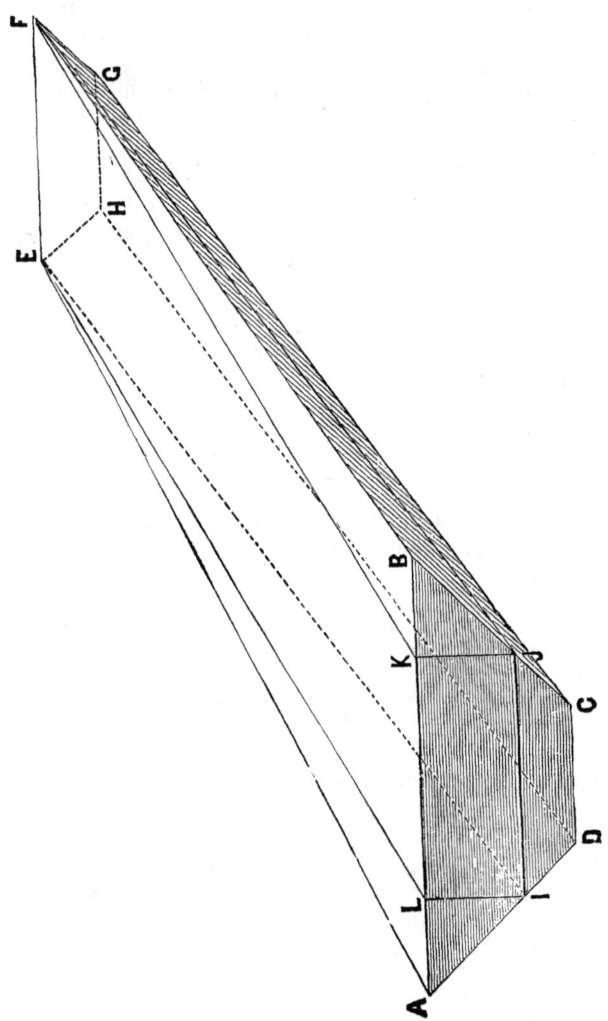

length; but to find the contents of the pyramids, their bases should be multiplied by *one-third* of their length. The method of calculation which we have employed multiplies the sum of the end-areas of the original figure, (which is composed of the prisms and pyramids which we are discussing) by one-half the length; and therefore gives a result too large by the difference between a half and a third—*i. e.*, by a sixth—of the product of the bases of the pyramids by their length: *i. e.*, $\frac{JK \times KB + IL \times LA}{2} \times \frac{JF}{6}$.

Representing by d the difference of the depths of the end cuttings, the ratio of the side-slopes by s to 1, and the length of the cutting or filling by l, the error in excess will be

$$\frac{d \times sd + d \times sd}{2} \times \frac{l}{6} = \frac{sd^2 l}{6}.$$

If this be calculated for each mass, and subtracted from the results previously obtained by averaging end-areas, the remainder will equal the result obtained by the correct prismoidal formula, to be hereafter examined. Thus, for the mass between stations 1 and 2 the correction is $\frac{1\frac{1}{2} \times 18^2 \times 561}{6} = 45,441$,—giving a remainder $= 388,773 - 45,441 = 343,332$, which is the correct amount. The original and corrected amounts are presented below in a tabular form:

ORIGINAL AMOUNTS.		CORRECTIONS.			CORRECTED AMOUNTS.	
Excavation.	Embankment.	Formulæ.	Amounts.		Excavation.	Embankment.
388,773		$\frac{1\frac{1}{2} \times 18^2 \times 561}{6}$	45,441		343,332	
1,280,994		$\frac{1\frac{1}{2} \times 2^2 \times 858}{6}$	858		1,280,136	
660,000		$\frac{1\frac{1}{2} \times 20^2 \times 825}{6}$	82,500		577,500	
	685,520	$\frac{2 \times 19^2 \times 820}{6}$		98,673		586,847
	907,500	$\frac{2 \times 11^2 \times 825}{6}$		33,275		874,225
	87,120	$\frac{2 \times 8^2 \times 330}{6}$		7,040		80,080
2,329,767	1,680,140		128,799	138,988	2,200,968	1,541,152

We thus see that the method of calculating excavation and embankment by averaging the end-areas, though very generally used, is so incorrect that in the present example its excess over the truth is nearly 130,000 cubic feet in the excavation, and 140,000 in the embankment, or 270,000 in the whole, equal to 10,000 cubic yards. If this method had been used in estimating the payment due to a contractor at 10 cents per yard, he would have been consequently overpaid $1000.

2. CALCULATION BY THE MIDDLE AREAS.

The second method of calculation is to deduce the *middle area* of each prismoidal mass from the middle height, or arithmetical mean of the extreme heights, and multiply it by the length.

Applying this method to the preceding example, and adopting the columns 1, 2, 6, and 7 of the table on page 116, we obtain the results exhibited in the last three columns of the following table.

Station.	Distance.	Cut.	Fill.	Middle Heights.	Middle Areas.	Excavation.	Embankment.
1		0		9	571.5	320,611	
2	561	18.		19	1491.5	1,279,707	
3	858	20.		10	650.	536,250	
4	825	0	0	9.5	655.5		537,510
5	820		19.	13.5	1039.5		857,587
6	825		8.	4	232.		76,560
7	330		0.				
						2,136,568	1,471,657

The following formulæ show the method of obtaining the "middle areas" in the sixth column of the above table.

Middle height $= 9$. Middle area $= (50+1\frac{1}{2}\times 9)\times 9 = 571.5$

" " $= 19$. " " $= (50+1\frac{1}{2}\times 19)\times 19 = 1491.5$

" " $= 10$. " " $= (50+1\frac{1}{2}\times 10)\times 10 = 650.$

" " $= 9.5$ " " $= (50+2\times 9.5)\times 9.5 = 655.5$

" " $= 13.5$ " " $= (50+2\times 13.5)\times 13\ 5 = 1039.5$

" " $= 4.$ " " $= (50+2\times 4)\times 4 = 232.$

The cubical contents are then calculated as follows :

$571.5\times 561 =$ 320,611.5 cubic feet.

$1491.5\times 858 =$ 1,279,707. " "

$650.\times 825 =$ 536,250. " "

$655.5\times 820 =$ 537,510. " "

$1039.5\times 825 =$ 857.587.5 " "

$232.\times 330 =$ 76,560. " "

The results thus obtained are too small their deficiency being equal to just half the excess of the first method. This will appear by again referring to the figure on page 352. It will be seen that the contents of the prisms in that figure will be correctly given by this method, but that the deficiency is in the pyramids. Calling their middle heights $\frac{d}{2}$; their middle widths will be $s\frac{d}{2}$; their middle areas $s\frac{d^2}{8}$; the contents of one of them $sl\frac{d^2}{8}$; and of the two $sl\frac{d^2}{4}$. But the true contents of the pyramids is $2\left(\frac{d \times sd}{2} \times \frac{l}{3}\right)$ $= sl\frac{d^2}{3}$; and the deficiency of the method of middle areas is therefore the difference between a third and a fourth—*i. e.* a twelfth—of the product of the bases of the pyramids by their length, or $\frac{sd^2l}{12}$. Corrections thus calculated, and added to the above results, would make them coincide with the true ones given by the prismoidal formula, which we will next consider.

3. CALCULATION BY THE PRISMOIDAL FORMULA.

The mass, of which the volume is demanded, is a true *Prismoid*, and its correct contents will therefore be given by the well-known prismoidal formula, which is as follows:

Find the area of each end of the mass, and also the middle area corresponding to the arithmetical mean of the heights of the two ends. Add together the area of each end, and four times the middle area. Multiply the sum by the length, and divide the product by 6. The quotient will be the true cubic contents required.

Applying this method to the original example, and adopting columns 1, 2, 6, 7, 8, from page 349, and the middle areas from page 354, we may prepare the following table:

Station.	Distance.	Cut.	Fill.	End Areas.	Middle Areas.	Excavation.	Embankment.
1		0		0	571.5	343,332	
2	561	18		1386	1491.5	1,280,136	
3	858	20		1600	650.	577,500	
4	825	0	0	0	655.5		586,847
5	820		19	1672	1039.5		874,225
6	825		8	528	232.		80,080
7	330		0	0			
						2,200,968	1,541,152
						1,541,152	
						659,816	

The manner of obtaining the amounts in the last two columns is as follows:

$$(0 + 1386 + 571.5 \times 4) \times \frac{561}{6} = 343,332.$$

$$(1386 + 1600 + 1491.5 \times 4) \times \frac{858}{6} = 1,280,136.$$

$$(1600 + 0 + 650 \times 4) \times \frac{825}{6} = 577,500.$$

$$(0 + 1672 + 655.5 \times 4) \times \frac{820}{6} = 586,847.$$

$$(1672 + 528 + 1039.5 \times 4) \times \frac{825}{6} = 874,225.$$

$$(528 + 0 + 232 \times 4) \times \frac{330}{6} = 80,080.$$

Whatever the shape of the mass of earth intercepted between two parallel cross-sections, it may be divided into prisms, pyramids, wedges, or frustra of pyramids, to all which, and therefore to the entire mass, the prismoidal formula may be correctly applied.*

The labor of the calculation may be much lessened by the use of tables, such as those of Macneill, Bidder, Fourier, Johnson, &c. A specimen of Macneill's is given at the end of the volume.

The prismoidal formula may be readily deduced from the dissected figure on page 354. Call the height of the lesser end h; of the greater end g; the breadth of base b; the ratio of the side-slopes to unity s; and the length l. Then we may proceed thus:

* Journal of the Franklin Institute, January and June, 1840

Area of the smaller end EFGH $= h\,(b + sh) = bh + sh^2$.

$\therefore$ Content of the lower prism $= (bh + sh^2) \times l$, . . . [A]

Area of rectangle IJKL $= (b + 2sh)\,(g - h) = bg + 2sgh - bh - 2sh^2$.

$\therefore$ Content of the upper prism $= (bg + 2sgh - bh - 2sh^2) \times \frac{l}{2}$, [B]

Bases of the two pyramids $= (g - h) \times s\,(g - h) = sg^2 - 2sgh + sh^2$.

$\therefore$ Contents of the pyramids $(sg^2 - 2sgh + sh^2) \times \frac{l}{3}$, . . . [C]

Uniting the expressions for the partial contents [A], [B], and [C], and reducing them to a common denominator, we get for the contents of the prismoid,

$$(6bh + 6sh^2 + 3bg + 6sgh - 3bh - 6sh^2 + 2sg^2 - 4sgh + 2sh^2) \times \frac{l}{6}$$

$$= (3bh + 3bg + 2sgh + 2sg^2 + 2sh^2) \times \frac{l}{6} \quad . \; . \; . \; . \; . \; . \; . \quad [D].$$

This expression may be decomposed into the following:

$$(bh + sh^2 + bg + sg^2 + 2bg + 2bh + 2sgh + sg^2 + sh^2) \times \frac{l}{6}.$$

The first two terms express the area of the smaller end of the prismoid, and the next two the area of the larger end. The remaining five terms may be transformed into

$$4\left(\frac{b\,(g+h)}{2} + s\,\frac{(g+h)^2}{4}\right) = 4\left[\frac{g+h}{2} \times \left(b + s\,.\,\frac{g+h}{2}\right)\right]$$

which is the expression for 4 times the middle area; thus giving the *prismoidal formula.*

The formula [D], giving the contents of the prismoid, may be transformed into another, more convenient for calculation than the usual prismoidal one. By separation into factors, it becomes,

$$[2s\,(gh + g^2 + h^2) + 3b\,(h + g)] \times \frac{l}{6} \quad . \; . \; . \; . \; . \; . \; . \quad [E]$$

which gives the following

RULE.

Add together the squares of the heights at each end, and their product. Multiply the sum by twice the ratio of the side-slopes to unity Reserve the product. Multiply the sum of the heights by three times the breadth of base, and add the product to the reserved product. Multiply their sum by the length or distance between the two cross-sections, and divide by six.

Applying the rule to the mass between stations 2 and 3, we find $g = 20$, $h = 18$, $b = 50$, $s = 1\frac{1}{2}$, $l = 858$, and the calculation is made thus:

$$
\begin{array}{rl}
18^2 = & 324 \\
20^2 = & 400 \\
18 \times 20 = & 360 \\
\hline
& 1084 \times 2 \times 1\tfrac{1}{2} = 3252 \\
\hline
& 18 \\
& 20 \\
\hline
& 38 \times 3 \times 50 = 5700 \\
\hline
& 8952 \\
& 858 \\
\hline
& 6)\ 7680816 \\
\text{Cubical contents} = & 1280136
\end{array}
$$

Formula [E] may be also transformed into the following formulæ, either of which is more convenient for calculation than the usual prismoidal formula.

$$[2s\,(g - h)^2 + 3b\,(g + h) + 6sgh] \times \frac{l}{6} \quad . \quad . \quad . \quad [F]$$

or $$[2s\,(g + h)^2 + 3b\,(g + h) - 2sgh] \times \frac{l}{6} \quad . \quad . \quad . \quad [G]$$

When the side-slopes are $1\frac{1}{2}$ *to* 1, the preceding formulæ are much simplified, for $2s = 3$, and the factor *three* may therefore be eliminated from each term, and one-half, instead of one-sixth of the length be used as a multiplier.

Formula [G] then becomes

$$[(g + h)^2 + b\,(g + h) - gh] \times \frac{}{2} =$$

$$= [(b + g + h)\,(g + h) - gh] \times \frac{l}{2}. \quad . \quad . \quad [H]$$

This formula gives the following

RULE.

When the side-slopes are $1\frac{1}{2}$ to 1, add together the breadth of base and the heights at each end of the mass. Multiply this sum by the sum of the two heights. From the product subtract the product of the two heights. Multiply the remainder by half the length.

The calculation of the preceding example will then be made thus:

$$\begin{array}{rcrcr} 50 & & & & \\ 18 & & 18 & & \\ 20 & & 20 & & \\ \hline 88 & \times & 38 & = & 3344 \\ 18 & \times & 20 & = & 360 \\ & & & & \overline{2984} \\ 858 & \div & 2 & = & 429 \\ \end{array}$$

Cubical contents $= 1{,}280{,}136$

When the height and therefore area at one end $= 0$, h vanishes from the formula [E], which thus becomes

$$(2sg^2 + 3bg) \times \frac{l}{6} = (2sg + 3b)\,\frac{gl}{6} \quad \ldots\ldots \quad [I]$$

giving the following

RULE.

Add the product of the height by twice the slope to three times the breadth of base. Multiply the sum by the height, and that product by the length, and divide the product by six.

The calculation of the cubical contents of the mass between stations 1 and 2 will accordingly be thus made:

$$\begin{array}{r} 2 \times 1\frac{1}{2} \times 18 = 54 \\ 3 \times 50 = 150 \\ \hline 204 \end{array} \qquad 204 \times \frac{18 \times 561}{6} = 343332.$$

When these last two conditions are combined (i. e. slopes $1\frac{1}{2}$ to 1 and one height $= 0$) formula [I] becomes, still more simply,

$$\frac{(g + b)\,gl}{2} \quad \ldots\ldots\ldots\ldots \quad [I']$$

FORMULA FOR A SERIES OF EQUAL DISTANCES.

When the cross-sections have been taken at uniform distances apart, (as is usual in the final location of a Road or Railroad, one hundred feet being the customary interval) the calculation of the cubical contents of the successive prismoids may be reduced to a single operation for the whole series, and therefore much shortened, by the use of the symmetrical formula which will be now investigated, and presented in the form of a Rule.

Through the first prismoidal mass of earth, conceive two vertical planes to pass lengthwise, cutting it in the lines in which the side-slopes meet the *base* of the road, (which is the bottom of an excavation, or the top of an embankment) as the lines CG and DH, of Fig. 148. These planes divide the prismoid into a central prism, and two pyramids or frusta. The content of the entire prismoid is expressed, according to formula [G], page 358, by

$$[2s\,(g+h)^2+3b\,(g+h)-2sgh]\times\frac{l}{6}\quad\ldots\ldots\quad[G]$$

This may be decomposed into these two portions:

$$[3b\,(g+h)]\times\frac{l}{6}=\frac{bl}{2}(g+h)\,.\,.\,.\,.\,.\,.\,.\,.\quad[K]$$

$$[2s\,(g+h)^2-2sgh]\times\frac{l}{6}=\frac{sl}{3}[(g+h)^2-gh]\quad\ldots\quad[L]$$

Formula [K] expresses the content of the central prism, and formula [L] that of the two pyramids or frusta. Denoting the end depths (without regarding which is the greater) by h and h', (the former representing the depth at the starting point, and the latter that at the farther end) the formulæ become

$$\frac{bl}{2}(h+h')\,.\,.\,.\,.\,.\,.\,.\,.\quad[M]$$

$$\frac{sl}{3}[(h+h')^2-hh']\quad\ldots\quad[N]$$

Considering now the next prismoid, or following length of excavation, (or embankment) its first depth is seen to be identical with the last depth of the preceding prismoid, i. e. it is h'. Calling the depth at its farther end h'', *the content of its central prism*, by formula [M], will be

$$\frac{bl}{2}(h' + h'')$$

The content of the third length will similarly be

$$\frac{bl}{2}(h'' + h''')$$

and so on for the succeeding portions, l being the same in each.

The sum of any number of these will be

$$\frac{bl}{2}[(h + h') + (h' + h'') + (h'' + h''') + \&c. \ldots .]$$

$$= \frac{bl}{2}(h + 2h' + 2h'' + 2h''' + \&c.)$$

Designating the last depth of the series by H, this expression may be written

$$bl\left(\frac{h}{2} + h' + h'' + h''' + h^{iv} + \&c. \ldots + \frac{H}{2}\right) \ldots . \text{ [O]}$$

Expressed in words, it then gives this

RULE.

To find the cubical contents of *the central prisms*, add together half of the first and last depths, and all the intermediate depths. Multiply their sum by the breadth of base, and that product by the length in feet of one of the equal distances. The last product will be the contents in cubic feet.

The content of the two pyramids or frusta, on each side of the central prism, is for the first length, by formula N,

$$\frac{sl}{3}[(h + h')^2 - hh']$$

For the second length it is $\frac{sl}{3}[(h' + h'')^2 - h'h'']$

For the third length it is $\frac{sl}{3}[(h'' + h''')^2 - h''h''']$; and so on.

For any number of equal lengths, the sum of the contents is

$$\frac{sl}{3}[(h + h')^2 + (h' + h'')^2 + \&c. - (hh' + h'h'' + \&c.)] \ldots \text{ [P]}$$

Expressed in words, it gives this

RULE.

To find the cubical content of *the pyramids or frusta*, square the sum of the first and second depths, the second and third, the third and fourth, and so on, and add these squares together. Multiply the first depth by the second, the second by the third, and so on, and add the products together. Subtract the sum of the products from the sum of the squares. Multiply the difference by the length in feet of one of the equal distances, and that product by the ratio of the side-slopes to unity. Divide the last product by three, and the quotient will be the content in cubic feet.

The sum of the two contents, thus obtained by formulæ [O] and [P], or by the Rules derived from them, will be the total content required.

In the following example, the width of base is 30 feet, the side-slopes 2 to 1, and the equal distances, at which the levels were taken, are each 100 feet. Therefore $b = 30$, $s = 2$, $l = 100$, and h, h', h'' = the successive numbers in the third column of the table. In substituting the values of the quantities in the formulæ they will be more conveniently written under each other.

Station.	Distance.	Depth.
1		$0 = h$
2	100	$2. = h'$
3	100	$4. = h''$
4	100	$3. = h'''$
5	100	$5. = h^{\ \text{v}}$
6	100	$1. = h^{\text{v}}$
7	100	$4. = \text{H}$

The content of the central prism, by formula [O], =

$$30 \times 100 \times \left\{ \begin{array}{r} 0. \\ +\ 2 \\ +\ 4 \\ +\ 3 \\ +\ 5 \\ +\ 1 \\ +\ \underline{2} \\ 17 \end{array} \right\} = 30 \times 100 \times 17\,. = 51000 \text{ cubic feet.}$$

The contents of the pyramids and frusta, by formula **[P]**,

$$= \frac{2 \times 100}{3} \times \left\{ \begin{bmatrix} \begin{Bmatrix} (0+2)^2 \\ +(2+4)^2 \\ +(4+3)^2 \\ +(3+5)^2 \\ +(5+1)^2 \\ +(1+4)^2 \end{Bmatrix} - \begin{Bmatrix} 2 \times 4 \\ +4 \times 3 \\ +3 \times 5 \\ +5 \times 1 \\ +1 \times 4 \end{Bmatrix} \end{bmatrix} \right\}$$

$$= \frac{200}{3} \left\{ \begin{bmatrix} \begin{Bmatrix} 4 \\ +36 \\ +49 \\ +64 \\ +36 \\ +25 \\ \hline 214 \end{Bmatrix} - \begin{Bmatrix} 8 \\ +12 \\ +15 \\ +5 \\ +4 \\ \hline 44 \end{Bmatrix} \end{bmatrix} \right\} = \frac{200}{3} \times 170 = 11333.$$

$51000 + 11333. = 62333$ cubic feet $= 2308.6$ cubic yards $=$ the entire cubical content required.

4. CALCULATION BY MEAN PROPORTIONALS.

A fourth method, called that of "Mean proportionals," is sometimes, though very improperly,employed. It assumes implicitly that the mass is a frustum of a pyramid, *i. e.* that all its sides, if produced, would intersect in one vertex, a supposition which would very seldom be perfectly true. On this assumption the following is the *Rule.*

Add together the areas of the two ends, and a mean proportional between them, (found by extracting the square root of their product) and multiply the sum of these three areas by the length of the frustum, and divide the product by three. The result is always much less than the truth, for it treats as pyramids, or thirds of prisms, the wedge-shaped pieces which are really halves of prisms. It is farthest from the truth when one of the areas $= 0$.

CASE II.—*When the ground is sidelong, i. e., has a transverse slope, "Two-level."*

The cross-section of the ground, at right angles to the direction of the road, has been assumed to be level. But the height of the surface of the ground usually varies considerably within the width to be occupied by the future road, and renders necessary the taking of levels not merely on the centre line, but also on the sides at the points in which the side-slopes, of the cuttings or fillings of the road, would intersect the surface of the ground.

1. When the surface of the ground has the same slope at each end of the mass to be calculated.

On such ground if the centre level, *i. e.*, the height or depth on the centre line of the road, be used to calculate the area of the cross-section, as if the ground were level transversely, the area thus obtained will always be *too small;* the difference being equal to the triangle DQR, in Fig. 149.

Fig. 149.

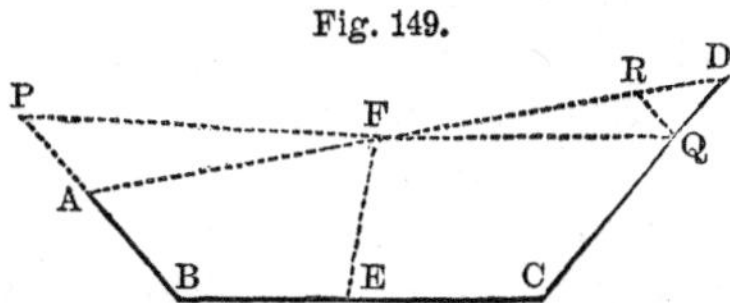

This must be guarded against in calculating the content from a preliminary survey, in which, usually, only one single level is taken along the centre line at each station.

If the average of the extreme heights is taken and used to get the area, as if that were the height of a level, horizontal, transverse section, the result is always *too great.*

There will then be some height, which used as the height of a section level transversely, will produce the *true area.* This is called the *equivalent mean height.* One method for determining the true area is the following:

Fig. 150.

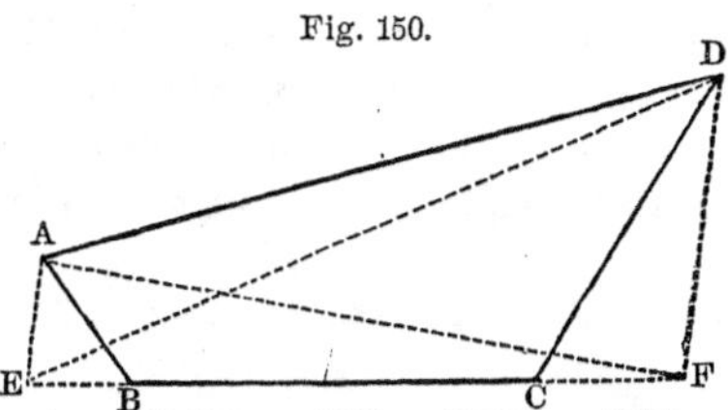

The cross-section ABCD = (EC × ½DF) + (BF × ½AE).

2. When the transverse slope of the ground is not the same at each end of the mass.

In this case the surface of the ground is *warped* or *twisted*, being a hyperbolic paraboloid, but the prismoidal formula still applies, as will now be shown.

THE CALCULATION OF ROAD EXCAVATIONS AND EMBANKMENTS, WHEN THE GROUND IS A WARPED SURFACE.*

When an engineer is laying out a road or railway, he has to determine the amount of earth necessary to be removed in making the "cuts" and "fills" of the road. To do this, his most usual course is to take "cross-sections" or "profiles," of the ground at right angles to the line of road, at convenient intervals, and then to calculate by various methods, commonly near approximations, the volume included between each pair of these cross-sections. The distances apart at which these cross-sections are taken, are determined by the engineer according to the nature of the ground; his aim being that there shall not merely be no abrupt change of height between each pair of these cross-sections, but that the surface from one to the other shall *vary uniformly;* gradually passing, for example, from a small to a great degree of slope, or from a slope to the right into a slope to the left, without any sudden variation at any one place.

The surface fulfilling this condition of varying uniformly, since it is everywhere straight in some direction, is evidently a *ruled surface;* and since the extreme profiles are seldom parallel, it will be a *warped* or *twisted* surface.

Our engineers have been accustomed to consider these surfaces as not admitting of precise calculation, but only of a degree of approximation varying with the nearness of the cross-sections. The object of this paper is to examine the correctness of this position. It will therefore have two parts: firstly, a discussion of the precise nature of the surface; and secondly, an investigation of a formula applying to it.

I. *What sort of a warped surface is the one in question; that is, what is its mode of generation?*

To determine this, we must inquire what the engineer means

---

* This paper was read by Prof. Gillespie before the "American Association for the Advancement of Science," and it has been thought best to insert it without abridgment or alteration.—ED.

when he says that the ground "varies uniformly" from the place at which he stands, and at which he has just taken a cross-section, to the place at which he decides it will be proper to take the next cross-section; whether he means that the ground between the two is straight *cross-wise* or straight *length-wise;* straight at right angles to the direction in which the road runs, or straight in that direction.

Probably few engineers ask themselves this question in so many words; but it would seem that the former conception, or *straightness cross-wise*, is the more likely to be what is meant, for the reason that any deviation from straightness in that direction, at right angles to the line along which we look, is much more easily seen than in the other direction. We can therefore much more readily determine whether the surface of the road is straight or curved from side to side than from end to end; and the surface which we pronounce uniform, is therefore much more likely to be straight cross-wise, than straight length-wise.

Fig. 151.

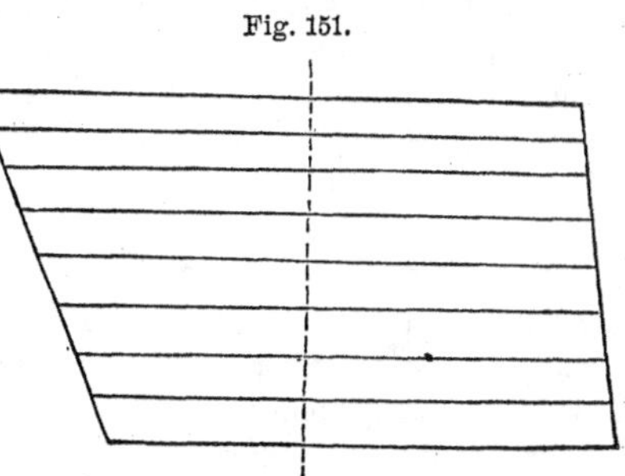

Fig. 152.

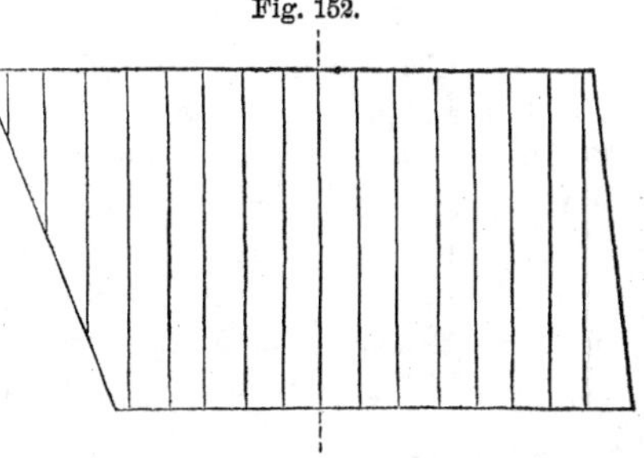

In geometrical language the former surface (which is represented in plan in Fig. 151,) is generated by a straight line resting on the two straight lines which join the extremities of the two profiles, and moving parallel to their planes or perpendicular to the axis of the road. This surface is a "*hyperbolic paraboloid.*"

The latter surface (shown in plan in Fig. 152,) is generated by a straight line resting on the two profiles, and moving parallel to the vertical plane which passes through the axis of the road. It also is a hyperbolic paraboloid, though a different one from the former. The French engineers (*Sganzin* 1, 114; *L'Ecole Centrale, etc.*,) adopt this latter hypothesis. We have seen, however, that the former is the more probable one.

Fig. 153.

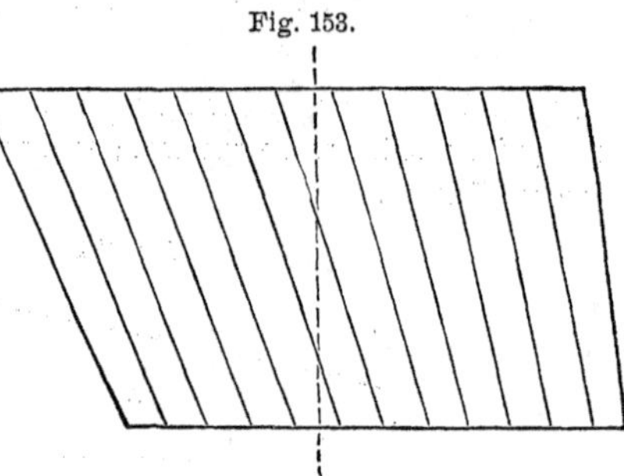

The French hypothesis is farther objectionable on mathematical grounds. As soon as the generating line quits the end lines and rests on the side lines, it has new directrices, and the whole surface generated, is really composed of three different paraboloids; a want of symmetry alone is sufficient to cause the rejection of this system.

Fortunately, the practical difference between the two, is really very slight; for a very small change in the latter hypothesis will make its result identical with that of the former. Conceive the straight line which rests on the two profiles to move on them in such a way as always to divide them *proportionally*, as in Fig. 153. The surface thus generated is identical with that of Fig. 151; as is proven in the higher descriptive geometry.

This last conception is also more probably correct than Fig. 152—even if we suppose the engineer to consider longitudinal straightness,—since he is more likely to extend his imagination from all parts of one profile to the corresponding parts of the other, than in lines perpendicular to the profile on which he stands.*

II. *We will therefore now proceed to investigate the content of a*

* Since the above was written, the author has seen an abstract of the Lectures on Roads, given at "*L'Ecole des Ponts et Chaussées*," (the highest authority on such matters in France, and therefore in the world), in which this last hypothesis is adopted. This removes the only obstacle to the acceptance of the principle which is here advocated.

In the models illustrating the original paper, the surfaces in question were formed by silk threads, representing the generating lines. The identity of the first and third surfaces, and the dissimilarity of the second, were then evident on mere inspection.

solid, bounded on one face by a warped surface generated on the first hypothesis—the other faces being planes.

We will take the case of an excavation; that of an embankment being the same inverted.

We will begin by considering the bottom of the excavation to be level, and its sides to be vertical; and will afterward discuss the more usual form.

Fig. 154.

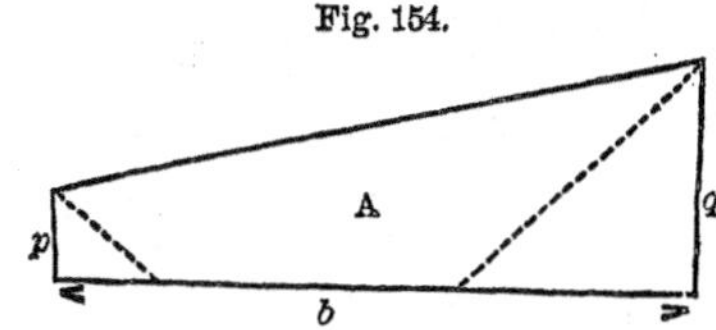

Fig. 155.

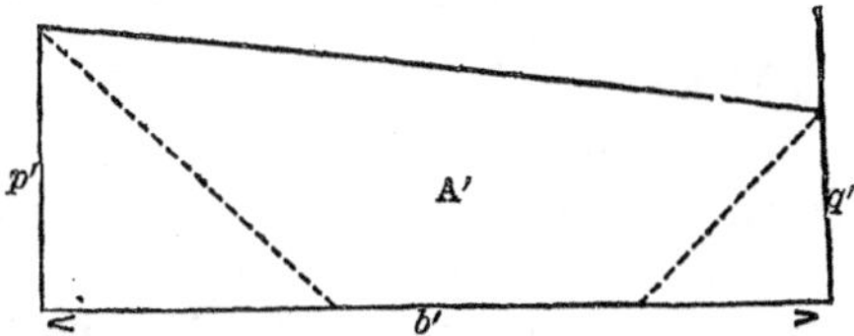

Let A and A′ be the parallel sections at each end of the solid; b and b' their respective breadths; p and q the outside depths of the section A, and p' and q' those of the section A′; and l the length of the solid, measured at right angles to the planes of the sections.

The outside depths are supposed to vary uniformly from p to p' and from q to q'.

Then, at x feet from A, the breadth $= b + \frac{x}{l}(b' - b)$; one outside depth

$$= p + \frac{x}{l}(p' - p); \text{ and the other } = q + \frac{x}{l}(q' - q).$$

The area of that section will therefore be

$$\tfrac{1}{2}\left[b + \frac{x}{l}(b' - b)\right] \times \left[p + \frac{x}{l}(p' - p) + q + \frac{x}{l}(q' - q)\right]. \quad (1)$$

Arranging this expression according to the powers of x, it becomes,

$$\tfrac{1}{2}\Big[b\,(p+q)+\frac{b\,(p'-p+q'-q)+(b'-b)\,(p+q)}{l}x$$
$$+\frac{(b'-b)\,(p'-p+q'-q)}{l^2}\,x^2\Big]$$

The product of this by dx being the differential of the solid, the required volume is,

$$\int\tfrac{1}{2}\Big[b\,(p+q)\,d\,x+\frac{b(p'-p+q'-q)+(b'-b)\,(p+q)}{l}\,x\,d\,x$$
$$+\frac{(b'-b)\,(p'-p+q'-q)}{l^2}\,x^{\,2}d\,x\Big]\quad\ldots\quad(2)$$

Integrating from o to l, we obtain this expression,

$$\tfrac{1}{2}\Big[b\,(p+q)\,l+\tfrac{1}{2}\,b\,(p'-p+q'-q)\,l+\tfrac{1}{2}\,(b'-b)$$
$$(p+q)\,l+\tfrac{1}{3}\,(b'-b)\,(p'-p+q'-q)\,l\Big]$$

Performing the operations indicated and factoring, we finally obtain for the required volume of the solid, this symmetrical formula.

$$\tfrac{1}{6}\,l\Big[(b+\tfrac{1}{2}\,b')\,(p+q)+(b'+\tfrac{1}{2}\,b)\,(p'+q')\Big]^{*}\quad\ldots\quad(3)$$

We now propose to show that the volume given by the preceding formula (3) is the same as would be obtained by applying the familiar prismoidal rule to the given solid.

The area of the section $A=\tfrac{1}{2}\,b\,(p+q)$; and that of the section $A'=\tfrac{1}{2}\,b'\,(p'+q')$.

The area of the section midway between A and B,

$$\tfrac{1}{2}\Big[\Big(\frac{b+b'}{2}\Big)\Big(\frac{p+p'}{2}+\frac{q+q'}{2}\Big)\Big]$$

---

* Two particular cases of this general formula are worthy of special notice.

Let the base of the given solid be a parallelogram,

Then $b=b'$; and formula (3) becomes,

$\tfrac{1}{6}\,l\,[\tfrac{3}{2}\,b\,(p+q)+\tfrac{3}{2}\,b\,(p'+q')]\,l=b\,l+\tfrac{1}{4}\,(p+q+p'+q')=$ The product of the base of the warped surface prism by the arithmetical means of the heights of its four summits.

Let the base be a triangle.

Then $b'=o$, and $p'=q'$; and formula (3) becomes,

$\tfrac{1}{6}l\,[b\,(p+q)+b\,p']=\tfrac{1}{2}\,b\,l\times\tfrac{1}{3}\,(p+q+p')=$ The product of the base by the arithmetical mean of the heights of the three summits.

These two formulæ are also true when the upper surface of the prism is a plane, since a plane is only a particular case of a hyperbolic paraboloid. They thus give a general proof of the well known rules for the content of truncated prisms, which have triangles or parallelograms for bases.

Adding together the areas of A and A′, and four times the middle area, and multiplying the sum by $\frac{1}{6}l$, we obtain,

$$\tfrac{1}{6}l\,(bp + bq + b'p' + b'q' + \tfrac{1}{2}bp' + \tfrac{1}{2}bq' + \tfrac{1}{2}b'p + \tfrac{1}{2}b'q')$$

which can be decomposed into the following:

$$\tfrac{1}{6}l\left[(b+\tfrac{1}{2}b')\,(p+q) + (b'+\tfrac{1}{2}b)\,(p'+q')\right] \quad . \quad . \quad . \quad . \quad (3')$$

This expression is identical with the general formula (3) before obtained.

We thus arrive at the conclusion that *the familiar "prismoidal formula" can be applied with perfect accuracy to such solids as we have discussed, having one of their faces a warped surface generated as in our first or third hypothesis.*

We have thus far been supposing that the road-bed was horizontal, or, in more general terms, that the base of the solid was perpendicular to its ends. The base may, however, make oblique angles with them. Then, to reduce the solid which we have been discussing to this form, we must take from it a wedge-shaped solid, the breadths of whose ends are b and b', and one of whose depths is zero.

But the prismoidal rule also applies to this wedge, and therefore to the solid which remains after it is taken away from our original solid; since all the areas enter the formula only by addition or subtraction, with a common multiplier.

Again, the solids occurring in excavations and embankments usually have sloping sides (as shown by the dotted lines in figures 154 and 155), instead of the vertical sides which we have used in our investigation.

But the solids to be removed to reduce our original solid to this form, are frusta of pyramids, to which the prismoidal formula also applies, and therefore to the new solid in question; for the reasons given in the preceding paragraph.

We will take as an example an excavation of which A and A′ are cross-sections, 100 feet apart. All the dimensions will be in feet. In section A, Fig. 154, let $p = 6$ and $q = 15$. In section A′, Fig. 155, let $p' = 18$, and $q' = 12$. The sections have the side slopes, 1 to 1, shown by the dotted lines. The bottom width of each = 18.

Then, the area of A = 279, and that of A′ = 486. The middle

area, obtained from the mean of the outside depths ($\frac{1}{2} \times (6 + 18) = 12$, and $\frac{1}{2} \times (15 + 12) = 13.5$) is 391.5.

Then the content of the solid by the *prismoidal rule* = 38,850 cubic feet.

The same rule can be applied directly to "*Three-level ground*," *i. e.*, ground given by cross-sections, in which three levels have been taken, viz., one at the centre, and one on each side at the points where the side slopes meet the natural surface. The middle cross-section being obtained from the mean of the levels at each end, the prismoidal rule can be at once applied.

In the case of "*Irregular cross-sections*," in which the inequalities of the surface of the ground have rendered it necessary to take more than these three levels, the rule will still apply after the following preparation. Conceive a series of vertical planes to pass through all the points on each cross-section, at which the transverse slope of the ground changes, and at which, therefore, levels have been taken, and to cut the other cross-section so as to divide the widths of the two *proportionally*.

Then the surfaces between these planes may be regarded as generated on our third hypothesis, and can therefore be calculated by the prismoidal rule; since it has been shown to apply to the surfaces of the first hypothesis, and these are known to be identical with those of the third. Thus, considering the ground on one side of a centre line, let one cross-section have depths of 6.00 in the centre, and 10.00 outside cutting. Let the other end be 8.00 in centre, 12.00 at four feet from centre, and 6.00 outside cutting. Let the half width of road bed be 10 feet, and side slopes 1 to 1. Then the vertical plane passing through the 12.00 level, at 4 feet, a quarter of the whole width $(10 + 6)$, from centre, should cut the other section at one-quarter its width $(10 + 10)$, or 5 feet, from centre. The depth at this point would be $6 + \frac{1}{4}(10 - 6) = 7.00$. This enables us to get a middle area; its depth being $\frac{1}{2}(8 + 6)$ at centre, $\frac{1}{2}(12 + 7)$ at $\frac{1}{2}(4 + 5)$ from centre, and $\frac{1}{2}(6 + 10)$ at the outside cutting.

The prismoidal rule can now be used. A similar preparation for calculating can be applied to cross-sections composed of any number of levels. The labor is much less in practice than it appears in description.

If the views here presented should meet with general acceptance, engineers would be enabled to economize much time and labor, since they would no longer feel themselves under the necessity of taking their cross-sections so near together that the ground between them should be approximately plane, but could take them as far apart as the ground varied uniformly, no matter how much or how far that might be.

It is now proposed to compare the results given by this rule with those obtained by the usual methods, and to establish formulas by which the nature and the amount of the errors which these latter involve can be determined in advance.

A type of the solids in question is represented in Fig. 156, as an excavation seen in perspective. Inverted, it will represent an embankment.

To simplify the investigation, we will conceive the side-slopes to be prolonged till they meet, as shown by the broken lines in the figure. The conclusions at which we may arrive respecting the new solid thus produced, will apply equally well to the original one, since the triangular prism which we imagine added, is common to both the solids discussed, whatever hypothesis we may

Fig. 156.

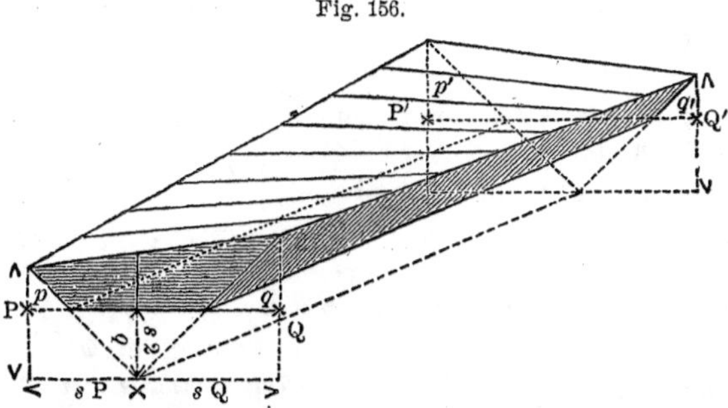

adopt respecting their upper surfaces. The additional depth is equal to the bottom width divided by twice the ratio of the base of the side-slopes to their height, or to $b \div 2\,s$, in the usual symbols. We will suppose the original outside depths p, q, p', q', of the end sections, to be increased by this quantity, and will call these new depths, P and Q for one section, and P′ and Q′ for the other.

Then the area of the triangle which forms one end of the new solid, is the difference between the trapezoid whose parallel sides are P and Q, and the two triangles which have P and Q for their altitudes, and s P and s Q for their bases, and is

$$\tfrac{1}{2}(P + Q) \times (s P + s Q) - \tfrac{1}{2} P \times s P - \tfrac{1}{2} Q \times s Q = s P Q.$$

Similarly, the other end area is $s P' Q'$. The middle section will have the outside depths, $\tfrac{1}{2}(P + P')$ and $\tfrac{1}{2}(Q + Q')$. Consequently its area is

$$s \times \tfrac{1}{2}(P + P') \times \tfrac{1}{2}(Q + Q') = \tfrac{1}{4} s (P + P') \times (Q + Q').$$

The true content of the solid under consideration will then be

$$\tfrac{1}{6} l \left[s P Q + s P' Q' + 4 \times \tfrac{1}{4} s (P + P') \times (Q + Q')\right]$$
$$= \tfrac{1}{6} s l (2 P Q + 2 P' Q' + P Q' + P' Q). \quad . \quad . \quad . \quad . \quad . \quad . \quad (1)$$

We are now prepared to compare with this correct result those given by each of the usual methods of calculation.

I. The method of "*averaging end areas*" will first be examined. This considers the content of the solid to equal the product of the half sum of its end areas by its length; *i. e.*, using the same symbols as above,

$$\tfrac{1}{2} l (s P Q + s P' Q'). \quad . \quad . \quad . \quad . \quad . \quad (2)$$

The excess, if any, of the true content above this, will therefore be obtained by subtracting (2) from (1). It is found, after a little reduction, to be

$$\tfrac{1}{6} s l (P Q' + P' Q - P Q - P' Q'). \quad . \quad . \quad . \quad . \quad (3)$$

The value of this expression is not changed by substituting in it the original depths for the increased depths (owing to its symmetrical character), and it then becomes,

$$\tfrac{1}{6} s l (p q' + p' q - p q - p' q'). \quad . \quad . \quad . \quad . \quad (3')$$

We infer from this formula, that *the true content exceeds the content given by "Averaging end areas"* whenever $p q' + p' q > p q + p' q'$; *i. e., whenever the sum of the products of the pairs of depths (or heights) diagonally opposite to each other, is greater than the sum of the products of those belonging to the same cross-section.* When the former sum is the smaller, then the true result is the smaller. The two sums are the same, and the results therefore equal, only when $p = p'$ or $q = q'$; *i. e.*, when the depths on one or the other side of the solid are the same.

It is so well known, however, that the method of "averaging

end areas" *always* gives more than the true content of a prismoid (such as a tapering stick of timber, a mill-hopper, etc.), that there seems at first glance an apparent inconsistency in the above statement. The difficulty is removed, however, by the consideration that our warped-surface-solid is not a prismoid, although it is to be calculated by the prismoidal rule. A somewhat analogous case is that of a sphere, to which the prismoidal rule also applies, as shown in an ingenious paper by Mr. Ellwood Morris.

II. The method of "Middle areas" will next be taken up. This assumes the content to be equal to the product of the area of the middle cross-section of the solid by its length. This content will therefore be expressed in our symbols thus:

$$\tfrac{1}{4}\, s\, l\, (\text{P} + \text{P}') \times (\text{Q} + \text{Q}'). \quad . \; . \; . \; . \; . \; . \; . \; . \quad (4)$$

Subtracting this from the true content (1), we obtain, after a little reduction, for the excess of the former,

$$\tfrac{1}{12}\, s\, l\, (\text{P Q} + \text{P}'\,\text{Q}' - \text{P Q}' - \text{P}'\,\text{Q}). \quad . \; . \; . \; . \; . \; . \quad (5)$$

For the reasons before given this may be written thus:

$$\tfrac{1}{12}\, s\, l\, (p\, q + p'\, q' - p\, q' - p'\, q). \quad . \; . \; . \; . \; . \; . \quad (5')$$

Comparing this expression with (3′), we see that we have merely to *reverse* the deductions there established; and that this method will give results too small when the preceding method gave them too great, and *vice versa.*

The absolute error, however, will be only half so great; the coefficient in (5′) being only one-half so great as that in (3′).

III. The method of "Equivalent mean heights" (or depths) is now to be examined. It consists (as is well known to engineers) in conceiving the given solid to be transformed in such a way that its top surface shall be a plane, everywhere level crossways at right angles to the length, and that the areas of the ends (which have then become level trapezoids) shall, at the same time, be equivalent to the original areas. The method then assumes that the content of this new solid (which is a true prismoid) is equal to the original content of the real sidelong, warped-surface-solid.

This is the method which it has long been customary to employ when perfect accuracy was desired; and most of the tables and diagrams for sidelong and irregular ground are constructed on this hypothesis. The question of its correctness is therefore an important one.

In Fig. 157, let A B C D be one of the original end areas or cross-sections, and let E F C D be an area equivalent to it, but level on top, if in excavation, as here, or level at bottom, if in embankment. K H, the depth of this new cross-section, is called the "Equivalent mean depth" (or height) of the original cross-section. We have first to obtain an expression for it, in terms of the original side depths.

Fig. 157.

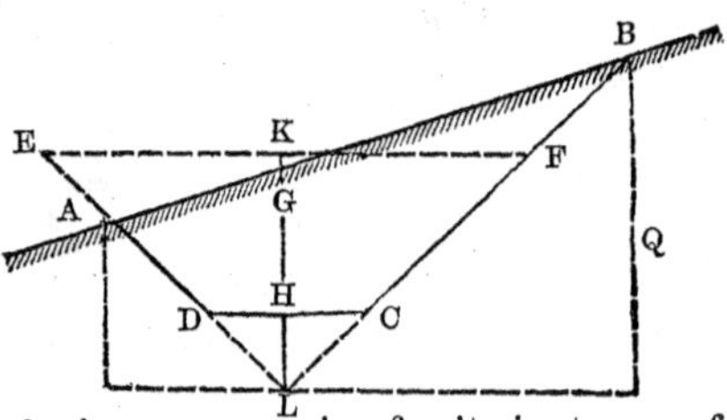

The investigation will be much simplified by the same conception as before, viz., by producing the side slopes till they meet, and calling, as before, the new outside depths P and Q. The height, K L, of the triangle E F L, is what is now wanted. The area of A B L was found on page 375 to be $s\,P\,Q$. Then the area E F L $= \frac{1}{2} \times KL \times EF = s\,KL^2$, being equated with $s\,P\,Q$, we obtain $KL = \sqrt{PQ}$.

The equivalent mean heights for the two end areas will then be $\sqrt{PQ}$ and $\sqrt{P'Q'}$; and the middle equivalent mean height will be their arithmetical mean. The corresponding middle area will be—

$$\tfrac{1}{4}\,s\left(\sqrt{PQ} + \sqrt{P'Q'}\right)^2.$$

Using this middle area and the given end areas in the prismoidal rule, we obtain, as the content of the solid by this method,

$$\tfrac{1}{6}\,l\left[s\,PQ + s\,P'Q' + s\left(\sqrt{PQ} + \sqrt{P'Q'}\right)^2\right]$$

$$= \tfrac{1}{6}\,s\,l\left(2\,PQ + 2\,P'Q' + 2\sqrt{PQP'Q'}\right) \quad . \quad . \quad . \quad . \quad (6)$$

Subtracting this from the true content (1), we find the excess of the latter is, when reduced,

$$\tfrac{1}{6}\,s\,l\left(\sqrt{PQ'} - \sqrt{P'Q}\right)^2 \quad . \quad . \quad . \quad . \quad . \quad . \quad (7)$$

This expression is always positive, whatever the value of P, Q, P′, and Q′, with a single exception, when $PQ' = P'Q$. Hence we have arrived at this result: *The method of "equivalent mean heights" gives contents always less than the true content;* with one exception, viz., when the products of the pairs of heights diagonally opposite to each other are equal.

IV. Some engineers have conceived the surface of the ground lying between two such cross-sections as we have been discussing, to be formed by two triangular planes meeting in a line running diagonally from p to q', or from p' to q (see Fig. 156), and thus forming a ridge or a hollow situated in this line.* But such cases would be abnormal ones, and such ground would not "vary uniformly" between the cross-sections. We will, however, examine this conception, as it will lead us to some interesting results.

We will begin by supposing the solid to be bounded on its sides by vertical planes passing through the outer side-lines of its surface, and to have its base pass through the line in which the prolonged side-slopes would meet, so that the heights of its corners will be P, Q, P′, Q′, as in the preceding discussion.

Let now the diagonal be considered to run from the left-hand corner of the nearest end of the solid to the right-hand corner of the farther end; say, from the height P to the height Q′. We now have to get the middle area. The middle height of the diagonal $= \frac{1}{2}(P + Q')$. The middle width of the left-hand side of the solid $= \frac{1}{2}(sP' + sQ')$, and the middle left-hand height $= \frac{1}{2}(P + P')$. The middle left-hand area is therefore—

$$\tfrac{1}{2} \times \tfrac{1}{2}(sP' + sQ') \times \tfrac{1}{2}(P + Q' + P + P').$$

Similarly we get the middle right-hand area—

$$= \tfrac{1}{2} \times \tfrac{1}{2}(sP + sQ) \times \tfrac{1}{2}(P + Q' + Q + Q').$$

The sum of these two areas gives the complete middle area. From it deduct the areas of the triangles on each side of the original solid. The left-hand one has its height $= \frac{1}{2}(P + P')$, and its base s times that, and the right-hand one has its height $= \frac{1}{2}(Q + Q')$, and its base s times that. Using the middle area thus obtained, with the end areas, in the prismoidal rule, we obtain the content of the solid on the new hypothesis. Its expression may be reduced to the following:

$$\tfrac{1}{3} s l (PQ + P'Q' + PQ') \quad . \quad . \quad . \quad . \quad . \quad . \quad (8)$$

Subtracting this from the true content (1), we get for the excess of the latter.

$$\tfrac{1}{6} s l (P'Q - PQ') \quad . \quad . \quad . \quad . \quad . \quad . \quad . \quad . \quad (9)$$

* See Mr. J. B. Henck's very valuable "Field Book for Railroad Engineers," page 100.

If we next suppose the diagonal to run in the other direction, *i. e.*, from Q to P′, we shall find the excess of the true content then to be,

$$\tfrac{1}{6}sl(\text{P Q}' - \text{P}'\text{Q}) \; . \; . \; . \; . \; . \; . \; . \; . \qquad (10.)$$

Hence, we infer that *the error on either hypothesis is numerically the same; though on one in excess and on the other in defect; but that the true content is the greater when the product of the heights which the diagonal joins is less than the product of the other two heights; and vice versa.**

Some examples will show the practical bearings of the principles which have now been established.

Example 1. We will begin with the solid represented in Fig. 156. It is an exact excavation a hundred feet in length, all the dimensions being given in feet. Its nearer end has the outside cuttings, $p = 6$, and $q = 15$; and its farther end has the outside cuttings, $p' = 18$, and $q' = 12$. The bottom width is 18. The side-slopes are 1 to 1. The areas of the ends are 279 and 486. The middle area, obtained from the mean of the outside depths, is 391.5. Then

---

* This admits of the following geometrical proof:—

Let A B C D be the surface in question. Consider it to be formed by two triangular planes, A B C, A D C, meeting in A C. Conceive also a vertical plane to pass through A C, A′C′, thus forming two truncated prisms. Next consider the surface to be formed by planes meeting in B D, and conceive another vertical plane to pass through B D, B′ D′. Two other truncated prisms are thus formed. Now conceive a plane parallel to A B and D C. It will cut the four planes of the hypothesis in lines parallel to A B and D C, and will thus form a parallelogram II′ I″ I‴. The diagonal II″ divides the lines A D, B C, proportionally (as follows from the similarity of the triangles formed), and is therefore a generatrix of the warped surface which lies between the two pairs of planes. But this diagonal of course bisects its parallelogram; the same is true of any other generatrix; consequently the surface which they form is everywhere midway between the surfaces of the two pairs of truncated prisms, and is therefore equal to half their sum.

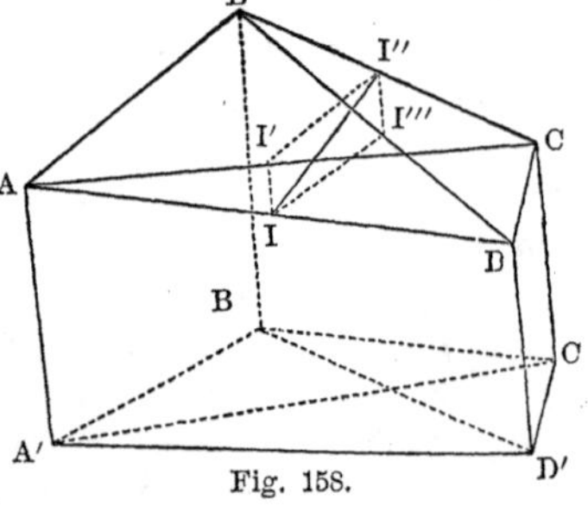

Fig. 158.

the true content of the solid, by the prismoidal rule, is 38,850 cubic feet.

Applying to this example the method of "Averaging end areas," we get a content of 38,250 cubic feet, or 600 cubic feet too little. It is too little, because the sum of the products of the depths diagonally opposite to each other is greater than the sum of the products of the depths belonging to the same cross-section. The precise deficiency is given directly by formula (3').

The method of "Middle areas" gives 39,150 cubic feet, or 300 cubic feet too much; in accordance with formula (5').

The method of "Equivalent mean heights" comes next. The formula on page 377. gives the "equivalent mean heights" of the two sections as 9.97366 and 14.81176 feet. Their mean gives a "middle area" = 376.65. The corresponding content = 38,860 cubic feet. The deficiency is 990 cubic feet. The same is given in advance by formula (7); since we have (adding $6 + 2s = 18 + 2 = 9$ to the original depths), $P = 15$, $Q = 24$, $P' = 27$, and $Q' = 21$; whence,

$$\tfrac{1}{6} \times 1 \times 100 \left(\sqrt{15 \times 21} - \sqrt{27 \times 24}\right)^2 = 990.$$

The method of imaginary "Diagonals" gives 33,300 cubic feet, if we suppose the diagonal to run from p to q'; *i. e.*, from 6 to 12, thus forming a hollow; or 44,400 cubic feet, if it runs from p' to q; *i. e.*, from 15 to 18, thus forming a ridge. The deficiency in the former case is 5550 cubic feet; and the excess in the latter case is the same; conformably to formulas (9) and (10).

Example 2. Conceive the outside depths of the farther end of this solid to be interchanged, so that 12 may be on the left, and 18 on the right. The true content will then be 37,950 cubic feet.

But "Averaging end areas" still gives the same as before, viz., 38,250 cubic feet. It was less than the true content in the former case, but it is now more, in accordance with formula (3'). The "Middle area" method gives 37,800, or too little, while before it gave too much; this result being still in accordance with formula (5'). "Equivalent mean heights" give the same as before, and therefore still too little.

Example 3. Conceive the depth q', of the solid of Example 1, to be changed from 12 to 15, all the other dimensions remaining the same. The new end area is 567, and the true content becomes 42,300 cubic feet. But $q = q'$. Therefore, according to the principles established on page 378. the method of "Averaging areas" should give the same result, and it does so. So too with the method of "Middle areas." The method of "Equivalent mean heights," however, still gives too little, because $P \times Q'$ is not equal to $P' \times Q$. On making the calculation (the equivalent heights being 9.97366 and 13.21475), we get a content = 41,600 cubic feet, or 700 cubic feet too little; and formula (7) gives the same result.

Example 4. In another warped surface solid, let one end area have depths of 15 on the left and 5 on the right, and the other end be 5 on the left and 15 on the right. Let the breadth of road bed be 20 feet, and the side slopes 2 to 1. The true content will be 38,333 cubic feet. The "Averaging method" gives 35,000 cubic feet; too little by formula (3'), because

$$5 \times 5 + 15 \times 15 > 15 \times 5 + 15 \times 5.$$

The "Middle area" method gives 40,000 cubic feet, an error in excess of half the amount of the preceding deficiency. "Equivalent mean heights" give 35,000 cubic feet; not enough, because $P \times Q'$, or 20×20 (adding $20 \div 2 \times 2$ to the given depths) is not equal to $P' \times Q$, or 20×20.

Example 5. Reverse one of these sections so that both may be 15 on the left, and both be 5 on the right. The surface is then a plane, and the solid is a prism with a uniform section of 3500 square feet. For this solid all the methods give the same content; and this is a final corroboration of our formulas. The "Averaging" method is now correct, because $p = p'$, each being 15, or because $q = q'$, each being 5. The "Middle area" method is correct for the same reason. The method of "Equivalent mean heights" is now correct, because now $P\,Q = P'\,Q$.

The method of "Equivalent mean heights" which the preceding investigation most particularly affects, seems to have been introduced by Telford, and has since been adopted without question by most writers (the present one included), *when perfect accuracy was desired.* The difficulty has been the want of any better standard than itself with which to compare its results. But if the positions which the writer endeavored to establish in the first part of this paper be accepted as correct, this method should be at once and entirely abandoned—since its errors are not of the kind which balance each other in the long run, but are always on the same side—since they are committed too with a belief of its perfect accuracy, and therefore in the most important and delicate cases—and since they may sometimes be of serious moment, the deficiency of the first example given being more than 2½ per cent. of the whole amount; no trifling item in a class of work which on some railroads is counted by millions of yards.

CASE III.—"*Three level*" *ground.*

This is ground whose surface is such as may be fairly represented by the centre "level" and the outside "levels," *i. e.*, heights or depths taken on the centre line, and at the "outside cuttings" or "fillings," which are the tops, or bottoms, of each side slope.

The areas are obtained by dividing them into triangles. See Figs. 160 and 164.

A common mode of calculating is that of "cross averaging," *i. e.*, taking one-fourth the sum of the outside heights and twice the middle one, and using the average as if it were the height of a level trapezoidal section. The areas thus obtained are always *too great.* For Fig. 160 it gives 74.82 instead of 74.64.

The "Equivalent mean height" is often used, and there are tables for this, but this always gives *too small* a content.

The *true* content is given by the *Prismoidal Rule.* See page 359.

The height of the ground above the grade line of the road on the centre line is called the "centre cutting;" and the heights at the intersection of the side-slopes of the cuttings with the ground on each side of any station are called the "right cutting" and "left cutting;" abbreviated into C. C. ..... R. C. ..... L. C.

In embankments, the corresponding heights are called "centre bank," "right bank," and "left bank;" usually written C. B. ..... R. B. ..... L. B.

For greater accuracy, these cross-sections should be taken at every chain or less. If an abrupt change in the level of the ground requires a levelling between these regular stations, it is called an "intermediate" one.

The following table presents various examples of irregular cross sections The slopes are assumed to be 2 to 1, and the width of the road to be 20 feet.

Station.	Distance	L. C.	C. C.	R. C.	L. B.	C. B.	R. B	End Areas Excavation.	End Areas Embankment.
1		0	0	0				0	
2	100	2.0	2.0	2.0				48.	
3	100	3.0	2.6	3.4				74.64	
4	100	3.0		2.0				62.	
Inter.	60	1.0	0	0	0	0	0	5.	0
5	40	0	0	0	0	0	2.0	0	10.
6	100				3.0	4.0	6.0		121.
7	100				0	0	0		0

We will proceed to sketch and note each cross-section, writing each height vertically in its appropriate place, and show how its area is obtained by dividing it into triangles, of which the base and height are known.

At station 1 the cutting begins, with an area = 0.

Fig. 159.

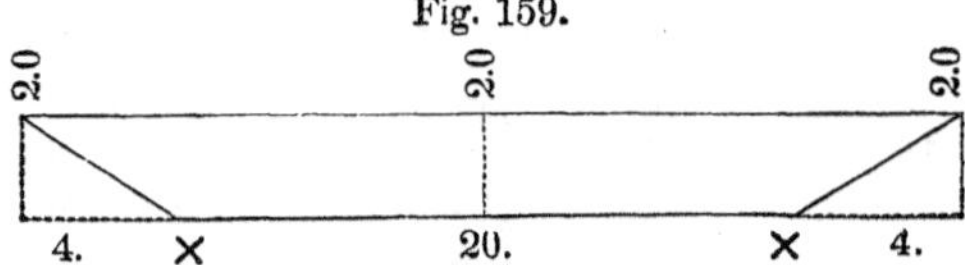

At station 2, Fig. 159, the section is of uniform depth, and its area is simply $(20 + 2 \times 2) \times 2.0 = 48$.

Fig. 160.

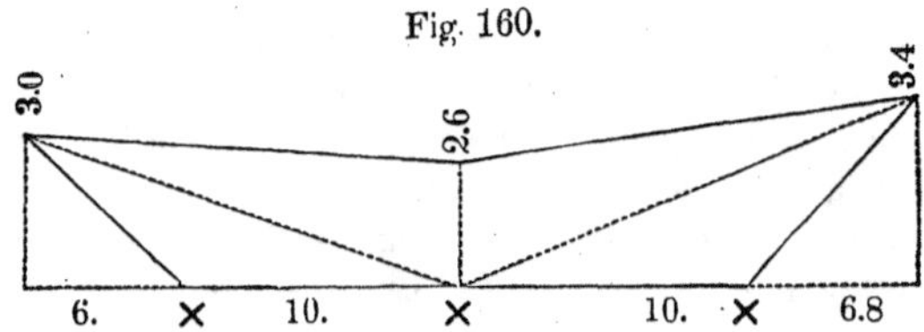

At station 3, Fig. 160, the lower left-hand triangle $= \frac{10 \times 3}{2} = 15.$

The lower right-hand triangle $= \frac{10 \times 3.4}{2} = 17.$

The two remaining triangles $= \frac{2.6 \times (6+10+10+6.8)}{2} = 42.64$

The entire area therefore - - - $= 74.64$

Fig. 161.

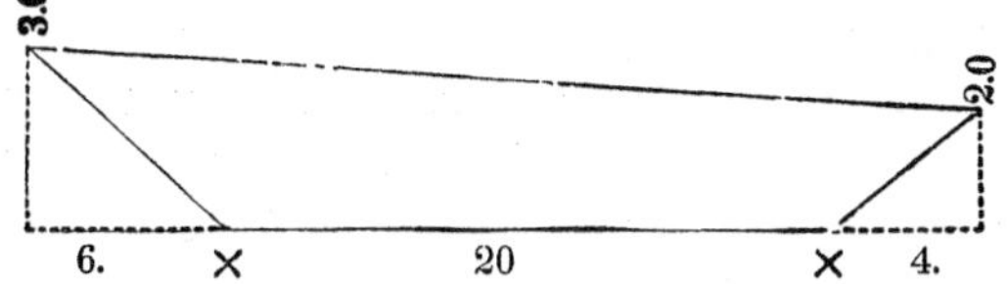

At station 4, only two levels were thought necessary, viz. those of the outside cuttings, without the centre one. To find the area, consider the figure as a trapezoid, minus the right-angled triangles at each end.

$$\text{Trapezoid} = (6 + 20 + 4) \times \frac{2. + 3.}{2} = \quad 75.$$

$$\text{Left-hand triangle } \frac{6 \times 3}{2} = \quad -9$$

$$\text{Right-hand triangle } \frac{4 \times 2}{2} = \quad -4$$

$$\overline{-13} \qquad -13.$$

$$\text{Area of cross-section, - - - } \quad 62.$$

A simple algebraic expression for this area may be found thus: call the breadth of base b, the outside cuttings d and e, the ratio of side-slopes to unity s. The area will be

$$\frac{(b + sd + se)\,(d + e)}{2} - \frac{sd^2}{2} - \frac{se^2}{2} = b\,\frac{d + e}{2} + sde.$$

The above example would then be $20 \times \frac{5}{2} + 2 \times 3 \times 2 = 50 + 12 = 62.$

Fig. 162.

1.0

10. × 10.

Between stations 4 and 5, at 60 feet from the former, an intermediate cross-section was made necessary, by the cutting "running out" on one side. The area, Fig. 152, is only the single triangle $\frac{10 \times 1.0}{2} = 5.$

At station 5, 40 feet farther, the cutting entirely runs out, and its area at that point becomes 0. The embankment had commenced

Fig. 163.

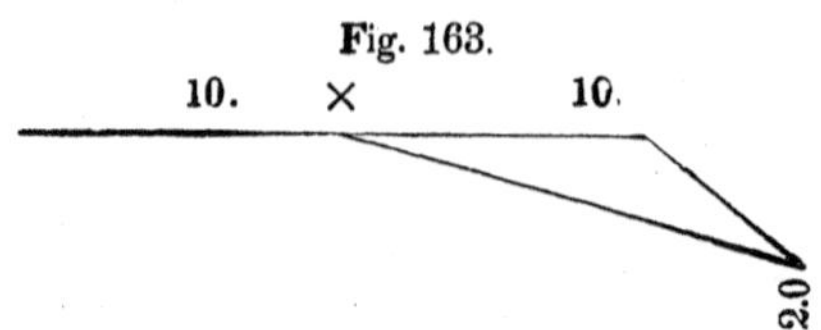

with area 0, at the preceding intermediate station, and at this station its area, Fig. 153, is $\frac{10 \times 2}{2} = 10$.

At station 6, the cross-section resembles that at station 3, in-

Fig. 164.

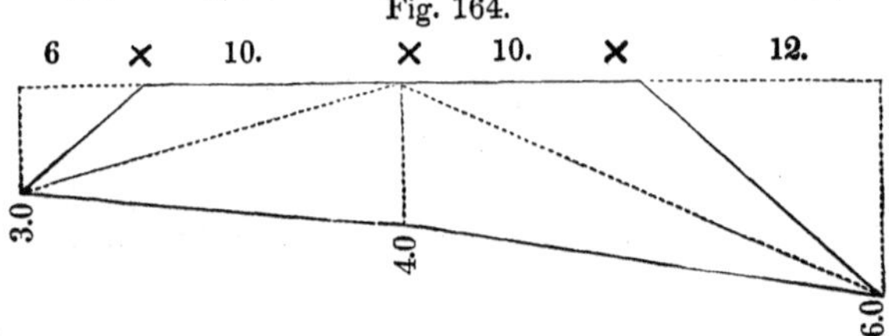

verted, and is calculated in the same manner by division into triangles, as is shown in Fig. 154.

Left-hand triangle - $= \frac{10 \times 3}{2}$ $= 15.$

Right-hand triangle - $= \frac{10 \times 6}{2}$ $= 30$

Two remaining triangles $= \frac{4 \times (6 + 10 + 10 + 12)}{2}$ $= 76.$

Entire area, - - - - - $= 121.$

At station 7, the embankment runs out, and the area $= 0$.

MEAN HEIGHTS.

To apply the prismoidal formula to cases of irregular cross-sections, it is necessary to calculate the mean heights of these cross-sections, to be subsequently averaged together to find the middle height, which produces the middle area. The following *problem* is therefore to be solved: Given the area of any irregular section, required the mean height which would produce the same area, the base and slopes remaining the same.

Fig. 165.

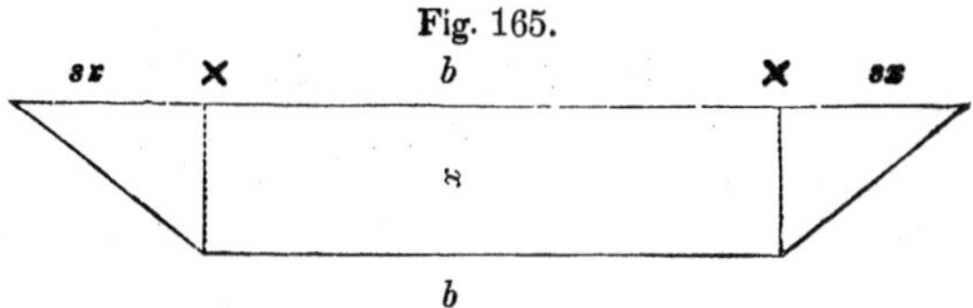

Let a represent the given area ; b the breadth of base or road-bed ; s, the ratio of side-slopes to unity ; and x the mean height required.

Then $a = sx^2 + bx$; by solving which equation we obtain

$$x = \sqrt{\left[\frac{a}{s} + \left(\frac{b}{2s}\right)^2\right]} - \frac{b}{2s}.$$

In all the preceding examples, $\frac{b}{2s} = \frac{20}{2 \times 2} = 5.$

At station 3, (p. 365) $a = 74.6 \therefore x = \sqrt{\left(\frac{74.6}{2} + 5^2\right)} - 5 = \sqrt{62.3} - 5 = 7.89 - 5 = 2.89.$ If this mean height be verified, it will be found to produce the original area. Thus substituting it in the above expression for a, we obtain $2 \times 2.89^2 + 20 \times 2.89 = 74.6$.

A similar process will give the mean heights for the remaining cross-sections. They may then be employed, as were the uniform heights in the original examples, to find the middle heights, and thence the middle areas required by the prismoidal formula; or as the values of g and h in the easier formulæ, which have been therefrom deduced.

In most cases, it will be sufficiently accurate to take only three levels, viz., at the centre, and at the foot, or top, of each side slope. The "*Equivalent mean height*" can be then obtained directly by a remarkably simple expression, without previously calculating the area. Let c = the cut or fill at the centre, and p and q the outside cuttings or fillings. Find the expression for the area, and put it equal to $sx^2 + bx$, as above, and the following expression will be obtained for the value of the mean height:

$$x = \sqrt{\frac{(sp + sq + b)(b + 2sc) - b.}{2s}}$$

When the "distances out" are given, calling them d and d', the above expression becomes

$$x = \sqrt{\frac{(d + d')(b + 2sc) - b.}{2s}}$$

Case IV.—*Irregular ground.*

This is ground, such that its cross-section requires more than three heights to be taken in order to represent its transverse profile correctly.

Usually, the area of such a cross-section is considered to be divided into triangles, whose bases and perpendiculars are known, and are always horizontal and vertical, and the sum of their areas gives that of the whole cross-section.

The triangles are usually taken in pairs, as far as possible, the vertical heights being taken as the bases, and the horizontal distances as the perpendiculars. The *sum* of the products are divided by 2 instead of dividing each product separately.

For example, in Fig. 166, commencing on the left, the small triangle has 5 for a base and 4 for a perpendicular. The next vertical, 5.4 is a common base for the triangles whose perpendiculars are 13 and 6. The next pair has a common base of 8.3, and the sum of the perpendiculars is 16. So on for the whole cross-section.

Fig. 166.

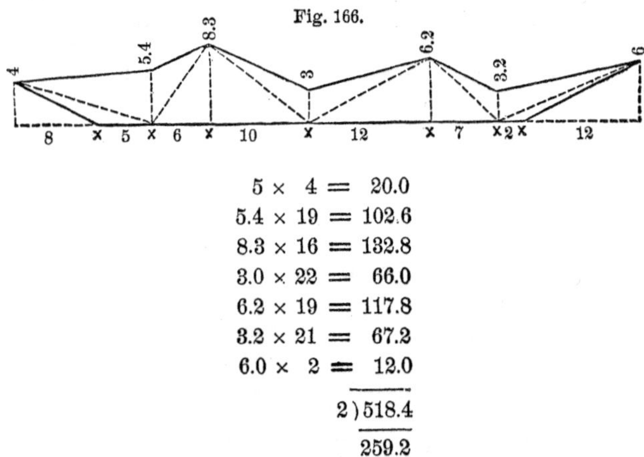

$$
\begin{aligned}
5 \times 4 &= 20.0 \\
5.4 \times 19 &= 102.6 \\
8.3 \times 16 &= 132.8 \\
3.0 \times 22 &= 66.0 \\
6.2 \times 19 &= 117.8 \\
3.2 \times 21 &= 67.2 \\
6.0 \times 2 &= 12.0 \\
\hline
2\,)\,&518.4 \\
\hline
&259.2
\end{aligned}
$$

These end areas are then USUALLY averaged to get the content of the mass between them. The correct method is given on p. 358

Sometimes it is impossible to take the second set of levels (those on the finished work) exactly over, or under, the first set. Then find the area of such cross-sections above some common datum and take their difference. The corresponding levels might be found by proportion.

When the ground is very irregular and great accuracy is required, its surface may be divided into rectangles or squares, and levels taken at each corner of these before the cutting or filling is made. The original base lines are axes of ordinates, and are carefully preserved. After the work has been done, levels are again taken at the same points. Then the difference of the two sets of levels, taken at these points, will be the depth of the cutting, or height of the filling. The content can then be calculated, either by combining the successive cross-sections, or by the method of truncated prisms.

When the ground is very irregular in plan and in heights, as in the case of foundation pits, etc., the method of cross-sections cannot be conveniently or completely applied. Then the mass of earth which is to be removed (or added) must be conceived to be divided by various vertical planes into prisms generally truncated, or pyramids, and calculated by the familiar rules of mensuration.

CASE V.—*Excavation and Embankment on Curves.*

Since the distances are measured along the centre line of a road, on curves as well as on straight lines, the calculation of the contents will not be correct when the ground is not level transversely. When the cross-sections are taken at right angles to the chords of the curves, as is usual, the content will be too great on the concave side of the curve, and too little on the convex side. The two balance each other only on level ground.

If the sections be measured at right angles to the tangents at the points where they are taken, the results will be more nearly correct.

The theorem of Guldinus applies here, *i. e.*, "The content of any body of revolution equals its generating cross-section, multiplied by the length of the path passed over by its centre of gravity."

The following formula for the correction, in excavation on curves, is from Henck's Field Book, Art. 130:

Let c = centre height, h = greatest side height, h' = least side height, d = greatest distance out, d' = least distance out, b = breadth of road-bed, and R = radius of curve, to find the correction, C.

$$C = \left[\tfrac{1}{2}\, c\,(d - d') + \tfrac{1}{4}\, b\,(h - h')\right] \times \frac{100\,(d + d')}{3\,R}$$

This correction is to be added when the highest ground is on the convex side of the curve, and subtracted when the highest ground is on the concave side.

TABLES

FOR CALCULATING EXCAVATION AND EMBANKMENT.

The TABLES at the end of this volume are extracted from those of Sir John Macneill, referred to on page 358. The numerals at the top and side of each table represent the depths or heights of the cutting or filling at its ends. The numbers in the body of the table indicate the number of cubic yards for the corresponding depths, and for a longitudinal distance of 1 foot. Thus, if the slopes of a given cutting be $1\frac{1}{2}$ to 1, the base 20 feet, the depths at the two ends 2 and 5 feet, and the distance between them 100 feet, find in TABLE I. the numeral 2 in the side column ; follow out the horizontal line corresponding to it till it meets the vertical column under the numeral 5 in the top line. At the intersection is 3.31, the cubic yards for a distance of 1 foot. Multiply this by 100, and the product is the number of cubic yards required.

The use of such Tables is limited by the inconvenience of making them voluminous enough to embrace every variety of slope, base, and depths, (though the fractional numbers wanting may be interpolated) but in the cases to which they apply, they unite the advantages of greatly lessened labor, and increased accuracy.

If much work is to be done for any base and side slope, not found in the tables, labor is saved and accuracy increased by calculating one for them.

APPENDIX B.

LOCATION OF ROADS.

1. *Planning the Route.*

THE true bearing or azimuth of one place from another, when the latitude and longitude of each are given, can be found by spherical trigonometry. But if the two places be very distant, the "rhumb" or loxodromic line between them, *i. e.*, the line having the same bearing throughout its length, will not be the shortest distance. The arc of the great circle passing through the points must then be adopted. Its bearing, *i. e.*, the angle which it makes with the meridians which it crosses, will be constantly changing. Thus, if one point be due west from another, the east and west line, *i. e.*, the parallel connecting them, is not the shortest line between them. For example, calling San Francisco due west from St. Louis, the shortest line between them by an arc of a great circle is about 9 miles shorter than the due west line following the parallel of latitude, and runs about 70 miles north of this parallel. A similar difference exists for every other line except a due north and south line.

For these reasons, in planning a long route, the position of points, situated on the arc of a great circle connecting the extremities, should be determined in advance by calculating their latitude and longitude. It would usually be impossible to follow this line perfectly, but it should be approximated to as far as possible, as is done for a straight line for short distances, as on page 82.

Other considerations cause the line of a railroad to deviate from the shortest line, as in common roads, in order to obtain good grades, moderate cuttings and fillings, and to pass through certain ruling points on the line.

It is usually best to follow the valleys of the water-courses lying nearest in the direction of the required line, and in passing from one valley to another to select that pass which can be reached by the most uniform grade.

2. *Reconnoissance and Preliminary Survey.*

For long distances this may be executed by determining the latitude and longitude of the ruling points with a sextant and chronometer, and determining the heights by a barometer.

For shorter distances the reconnoissance is conducted as explained on pp. 81 to 86, for common roads. More care, however, is necessary, owing to the greater expense in building, sustaining, and working a railroad. In the preliminary survey of the London and Birmingham Railroad, Robert Stephenson walked over the ground twenty times—a distance of one hundred and twelve miles.

3. *Survey and Location.*

The transit party usually goes ahead, and consists of a chief, a transitman, two flagmen, two chainmen, and one or more axmen, according to the country.

The line is marked out by the transit party, by placing a small peg in the line at every hundred feet. These pegs are driven so that their tops are nearly to the surface of the ground, and then large stakes are driven near them, to aid in finding the former in retracing the line. The number of the station is marked on the large stake with red chalk. Sometimes the larger stakes are placed in the line, and the small "level pegs" are only driven at every five hundred feet.

"Reference points" are also located along the line at important points, so that if the stakes in the line be lost, the exact point can be found again.

Fig. 167.

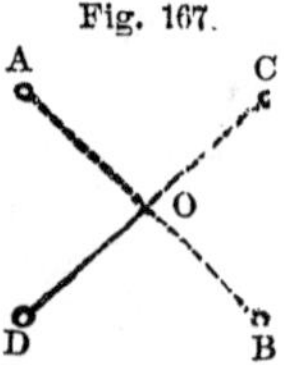

Let O be the point whose position it is desired to fix. Select four points as, A, B, C, and D (as permanent as possible), in such positions that the lines AB and CD will intersect at O. Now if the stake at O be lost, it can be replaced by finding the intersection of the two lines. The reference points, A, B, C, and D, should be at such distances from the line as not to be disturbed in building the road. Any of the methods for determining the position of a point can be used; as, rectangular, angular, or polar co-ordinates, but the one here given is generally used.

The number and size of the openings (drains or culverts) required to pass the water-courses under the road should be carefully noted and abundant room given; also the length and height of bridges.

The geological formation of the country should be examined, the nature of the surface noted, and generally everything which can affect the cost of the construction and maintenance of the road.

A common form for the "Transit Notes" is the following: the left-hand page is ruled into five columns, which are headed as follows:

Station.	Def. Ang.	Degree of Curve.	Bearing of Tangent.	Remarks.

In the first column is placed the number of the station; in the second, the deflection of each tangent from the preceding one; in the third, the degree of the curve connecting the tangents; in the fourth, the bearing of each tangent; and in the fifth, memoranda of the things spoken of in the preceding paragraph.

On the right-hand page of the Transit Field Book, plot the line approximately in the field, and on it sketch the topography, the hills, valleys, and water-courses, as nearly as possible in their true places.* This page should be ruled in squares.

The notes should be commenced at the bottom of each page, so that when holding the book in the hand and looking along the line, the line in the book will have the same direction as the one on the ground.

The "leveller" follows the transit party, and takes the heights of the small pegs which they have set. He is assisted by a "rod-man." "Cross-levels" are also taken, to a greater or less width, according to the ground, in order to determine what will be the

* On sketching ground by various methods, see "Gillespie's Levelling, Topography, and Higher Surveying." Part IV.

effect, in the cuttings and fillings, of moving the line to the right or the left in order to improve its grade or curvature.

The most perfect preliminary location of a line would be made by first making a topographical survey, and getting contour lines over all the surface near the proposed line. This can be done very rapidly with a level provided with extra "stadia hairs."

Preliminary surveys being completed, plots and profiles are made; curves put in so as to obtain the line of easiest curvature; a grade line put on the profile so as to nearly equalize the cutting and filling, and at the same time get easy grades. (See p. 154.)

Approximate grade contour lines may be obtained from the cross-sections thus: Knowing at each station the relative heights of the ground, find where a horizontal line, passing through the grade point at each station, perpendicular to the line of the road, would intersect the surface of the ground.

This is most easily done, if the slope is taken in degrees, by a traverse table. Opening the table to the degree of the slope, calling the depth of the cut or fill the departure, and finding the latitude corresponding to it, which will be the distance of the required intersection to the right or left. Mark on the plot the places of these points of intersection, and draw a line through these. This will be a line on which there will be no cutting or filling, and may be called a grade contour line.

The located line should approach this as nearly as other considerations (curvatures, etc.) will allow. It should be a compromise between this line and the straight line.

Grades in railroads should be grouped by bringing the steep ones together and obtaining some uniformity of them over such a length of road as would be worked by the same engine; because no one engine can advantageously work easy grades and steep ones.

4. *Comparison of Lines.*

The various lines which have been surveyed and estimated, between two termini, are now to be "equated." One line may be straight, but have many grades; another level, but have sharp curves, or be longer than the former; and so on.

In order to ascertain the most economical line, we must determine what additional distances the curves or grades are equivalent to.

Then the sum of these distances being added to the measured distance of the several lines will give their equated lengths. These are to be used in calculating the cost of working the road and the additional capital to which this is equivalent.

In the following example, to equate for grades, we will call each 24′ ascent equivalent to 1 mile additional in length, and will consider 1000° of curvature equivalent to 1 mile of distance. To obtain the data for the later, the length of each curve in chains is multiplied by the degree of the curve, and the sum of all these products gives the total curvature of the line. Thus:

No. of curve.	Radius.	Degree.	Length.	Total curvature.
1	1146	5°	1000	50°
2	5730	1°	4600	46°

Example.—Line A has 5 miles 24′ grade, 6 miles 12′ grade, and total curvature = 4000°. Line B has 10 miles 48′ grade, 10 miles 24′ grade, and total curvature = 1000°. The expenses of maintaining the road varies with the travel on it. Call it $1000 annually per nile of actual length, and find the equivalent capital at 6 per cent. The expense of working also varies with the trafic. It is proportional to the equated length. We will call it $2000 per mile of this. Find its equivalent capital. The last column is the sum of the three preceding columns.

Name of Line.	Measured Length.	Equated Length.	Estimated cost of construction	Capital for maintaining.	Capital for Working.	Total Capital.
A.	100	112	$4000000	1666666.66	3733333.33	9400000.01
B.	90	121	$3800000	1500000.00	4033333.33	9333333.33

Final Location.—The best line having been chosen, it is then to be staked out. For grade book, see p. 146. Its columns 3, 4, 5, and 6 may be omitted. The side stakes for construction are set as on page 457. The estimates are made as for common roads, with the addition of the new items of rails, ties, etc., etc.

APPENDIX C.

RAILROAD CURVES.

A RAILROAD curve is a portion of the road curved horizontally, so as to form an arc, usually circular, terminating at each end in straight portions which are tangent to it.

A railroad curve is "*determined*" when its *starting point*, its *radius*, and its *length* are known. When these have been obtained, points in the curve can be fixed in various ways. Such points are angles of a polygon whose sides are chords of the desired arcs, and approximately coincide with them.

Usually these chords are chains of one hundred feet, and the angle in degrees which each one subtends at the centre, is called the "*degree*" of the curve. It equals the angle of deflection of each of these chords from the preceding one. The relation between this angle and the radius is important.

Approximately, and sufficiently near for the usual curves, the angle of deflection in degrees = 5730 ÷ radius in feet

Precisely : Sine of half the degree $= \dfrac{\text{chord}}{\text{twice radius}}$.

The subject is divided into two parts :—

PART I.

GENERAL PROBLEMS ON CURVES; or how to determine a curve so that it shall fulfill certain conditions, *e. g.*,

A. To unite two given tangents.

B. To start from a given tangent and pass through a given point.

C. To unite a given tangent line and a given curve, etc., etc.

PART II.

METHODS OF RUNNING CURVES WHEN DETERMINED; *i. e.*, methods for fixing points in them; and transformations of the formulas of Part I. to suit these different methods.

PART I.

GENERAL PROBLEMS

CASE A.—TO UNITE TWO GIVEN TANGENT LINES.

Fig. 1.

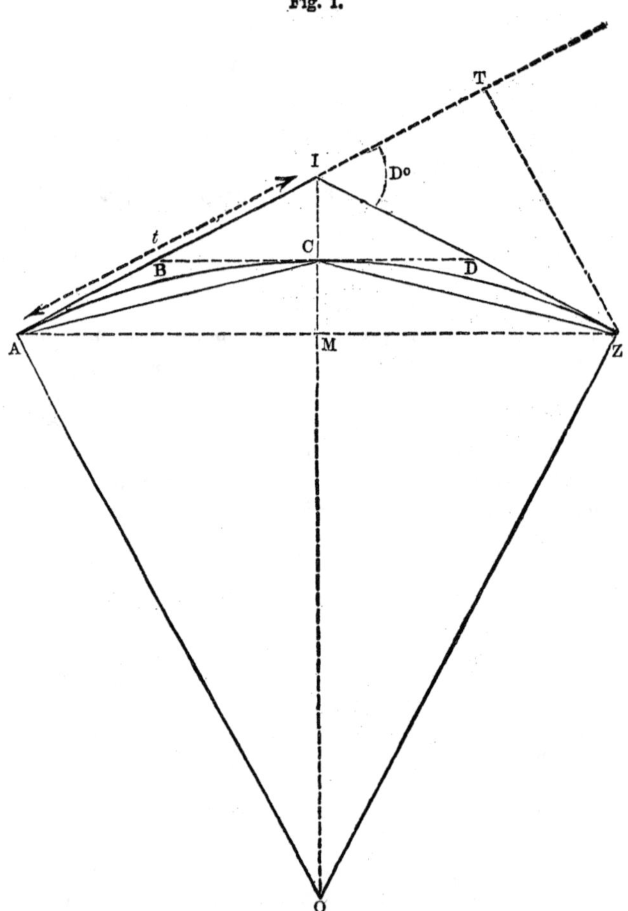

In all the figures the starting point of the curve is lettered A, and its terminus Z. I is the point of intersection of the tangents,

and D° is the angle of deflection of one from the other. O is the centre of the arc.

Its radius is therefore $OA = OZ = r$. The equal tangents are, $AI = IZ = t$.

$$AOZ = ZIT = D°, \quad IAZ = IZA = \tfrac{1}{2}AOZ = \tfrac{1}{2}D°.$$

Problem I

Given two intersecting tangents, and also the starting point, A, *on one of them, to find radius and length of curve.* Fig. 1.

Graphically, on a plot. Set off IZ = AI. At A and Z draw perpendiculars to the respective tangents, and their intersection will be the centre required. It will also be in the line bisecting the angle, AIZ.

When the lines are given on the ground. Set off IZ = AI. Measure AZ; mark its middle point, M, and measure IM. Then from the similar triangles, AMI and AMO,

$$OA = \frac{AI \times AM}{IM}.$$

Trigonometrically. When the lines are given by their angle of deflection, then from the right-angled triangle, OAI,

$$OA = AI \cot. \tfrac{1}{2}TIZ = t \cot. \tfrac{1}{2}D°.$$

The length of curve $AZ = OA\,\frac{AOZ}{57.3} = r\,\frac{D°}{57.3}$.

Problem II.

Given two tangents, and also the desired radius, OA, *to find the starting point, and length of curve.*

Graphically on a plot. Draw parallels to the tangents at a distance = radius. Their intersection will be the centre. It will also be in the line bisecting the angle I.

By calculation, when the lines are given on the ground, Fig. 2, measure equal distances from I to P and Q. Measure PQ; mark its middle point, S, and measure SI. Then from the similar triangles, AOI and PSI,

$$IA = \frac{OA \times SI}{SP}.$$

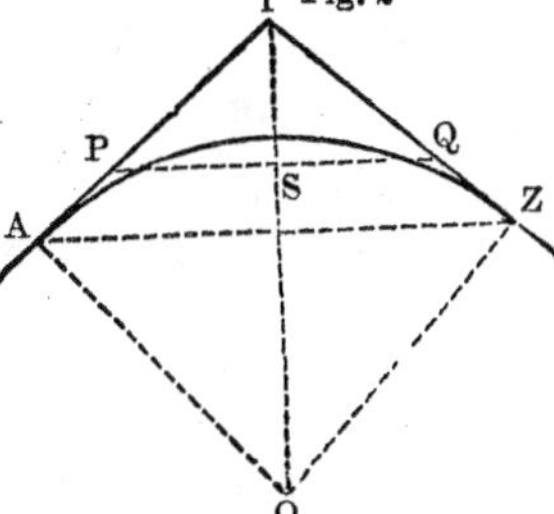

Otherwise, by trial, Fig. 3 Run from a random point, A′, till the tangent at the end of the curve is parallel to second tangent, I Z, as U V at V. Measure V Z parallel to first tangent, A I; move the starting point that distance and repeat.

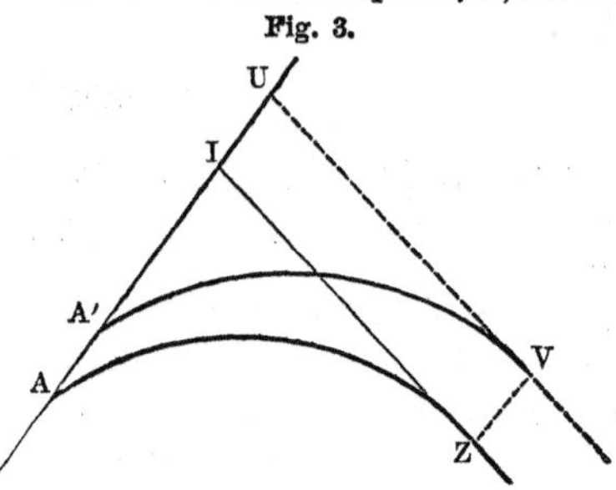

Fig. 3.

When the lines are given by their angle of deflection, we have,

$$\text{I A} = \text{O A} \tan. \tfrac{1}{2}\text{D}^\circ. \qquad \text{Length of curve} = r\frac{\text{D}^\circ}{57.3}.$$

PROBLEM III.

Given two intersecting tangents, and also the distance I C *from their intersection to a point through which the curve must pass. Required to find the starting point,* A, *radius, r, and length of curve.* Fig. 1.

Graphically. Draw a line bisecting A I Z. Through C (at the given distance measured on the line) draw a perpendicular to it, meeting the tangents in B and D. Set off C B from B to A, and its equal C D, from D to Z. Perpendiculars at A and Z will intersect in the centre, which will also be in the bisecting line.

By calculation. In triangle, A C I,

$$\sin.\text{ I A C} : \sin.\text{ A C I} :: \text{I C} : \text{A I} = \text{I C}\frac{\sin.\text{ A C I}}{\sin.\text{ I A C}}.$$

Now,

$$\sin.\text{ A C I} = \sin.\text{ A C O} = \cos.\text{C A Z} = \cos.\tfrac{1}{2}\text{I A Z} = \cos.\tfrac{1}{4}\text{A O Z} = \cos.\tfrac{1}{4}\text{D}^\circ$$

and,

$$\sin.\text{ I A C} = \sin.\tfrac{1}{2}\text{A O C} = \sin.\tfrac{1}{4}\text{A O Z} = \sin.\tfrac{1}{4}\text{D}^\circ.$$

Hence,

$$\text{A I} = \text{I C}\frac{\cos.\tfrac{1}{4}\text{D}}{\sin.\tfrac{1}{4}\text{D}} = \text{I C} \cot.\tfrac{1}{4}\text{D}^\circ.$$

To find the radius,

$$\text{O A} = \text{A I} \cot.\text{ A O I} = \text{A I} \cot.\tfrac{1}{2}\text{D} = \text{I C} \cot.\tfrac{1}{4}\text{D} . \cot.\tfrac{1}{2}\text{D}^\circ.$$

Hence,

$$\text{O A} = \text{I C} \cot.\tfrac{1}{4}\text{D}^\circ \cot.\tfrac{1}{2}\text{D}^\circ.$$

Conversely,

$$\text{I C} = t . \tan.\tfrac{1}{4}\text{D}^\circ. \quad \text{And I C} = r . \tan.\tfrac{1}{4}\text{D}^\circ . \tan.\tfrac{1}{2}\text{D}^\circ.$$

Problem IV.

Given two intersecting tangents, and also a point, P, *through which the curve must pass, to find the starting point,* A, *radius,* r, *and length of curve.* Fig. 4.

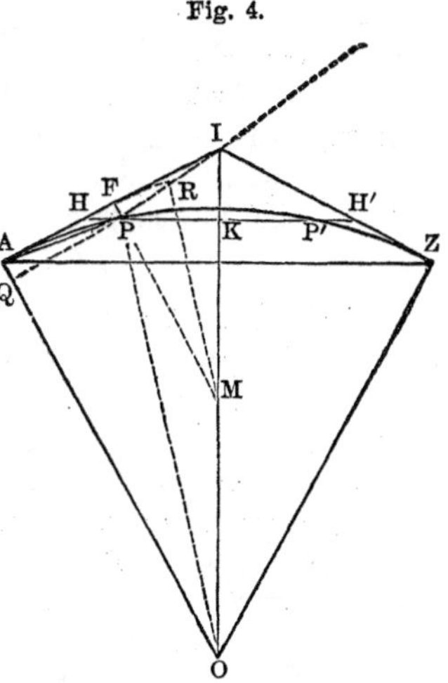

Graphically, on the ground. Bisect angle at I. Through P draw a perpendicular to bisecting line, intersecting tangents at H and H′. Construct a mean proportional between H P and P H′. It equals H A, since H A $=\sqrt{(\text{HP} \times \text{HP}')}=\sqrt{(\text{HP} \times \text{PH}')}$. This gives A, and, therefore, gives I A, and thence A O, by Problem I.

On a plot. Draw the bisectrix of the angle, I. Join I P. Through P draw a perpendicular, M F, to the nearest tangent. With M as a centre and M F as a radius, describe an arc cutting I P in R. Join F R and M R. Draw P O parallel to R M, and P A parallel to R F. O will be the required centre, and A the starting point.

By calculation. Measure or calculate the rectangular co-ordinates, I F and F P, of the given point. Then we get,

$$\text{FA} = \text{FP}, \cot. \tfrac{1}{2}\text{D}^\circ \pm \sqrt{[(\text{FP} \cot. \tfrac{1}{2}\text{D}^\circ + \text{IF})^2 - \text{IF}^2 - \text{FP}^2]}$$

$$\text{IA} = \text{IF} + \text{FA}, \text{ and } \text{OA} = \text{IA} \cot. \tfrac{1}{2}\text{D}^\circ.$$

Analytically. Given angle, A I Z, and the point, P, by rectangular co-ordinates from I, P F and F I, to find F A, etc.

$$\text{AF} = \text{QP} = \sqrt{(\text{OP}^2 - \text{OQ}^2)} = \sqrt{[\text{AO}^2 - (\text{AO} - \text{FP})^2]}$$
$$= \sqrt{(2\,\text{AO}, \text{FP} - \text{FP}^2)}.$$

By Prob. VI. (or by mensuration),

$$x = \sqrt{(2\,r\,y - y^2)}, \text{ or } \text{AF} = \sqrt{(2\,\text{AO} \times \text{FP} - \text{FP}^2)}.$$

By Prob. I $\text{AO} = \text{AI}, \tan. \tfrac{1}{2}\text{AIZ} = (\text{AF} + \text{FI}) \tan. \tfrac{1}{2}\text{I}.$

Substituting A O in the above;

$$AF = \sqrt{[2\,FP\,(AF + FI)\tan.\tfrac{1}{2}I - FP^2]}.$$

$$AF^2 = 2\,FP, AF\tan.\tfrac{1}{2}I + 2\,FP, FI\tan.\tfrac{1}{2}I - FP^2.$$

$$AF^2 - 2\,FP, \tan.\tfrac{1}{2}I, AF = 2\,FP, FI, \tan.\tfrac{1}{2}I - FP^2.$$

$$AF = FP, \tan.\tfrac{1}{2}I \pm \sqrt{(2\,FP, FI\tan.\tfrac{1}{2}I - FP^2 + FP^2\tan.^2\tfrac{1}{2}I)}$$

$$AF = FP, \tan.\tfrac{1}{2}I \pm \sqrt{[(FP, \tan.\tfrac{1}{2}I + FI)^2 - FI^2 - FP^2]}.$$

Note to Case A, *Probs. I., II., III., and IV.*

When the intersection, I, of the tangents is inaccessible, D° and *t* must be calculated. Fig. 5. From A, run one or more lines to meet the other tangent at some point, as W. Then the desired angle at I is obtained by subtracting the sum of all the interior angles

Fig. 5.

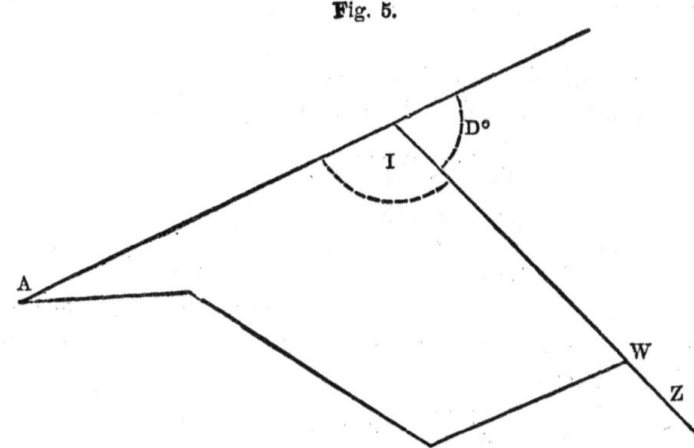

from two right angles, taken as many times, less two, as the figure has sides. When the lines are run by traversing, the reading from W to Z at once gives D°. A W is calculated by latitudes and departures. Then I A, in the triangle A I W is calculated by trigonometry.

Case B.—To start from a given tangent and pass through a given point.

Problem V.

Given this tangent and point, Z, *to find radius, length of tangent, and length of curve.*

Fig. 6.

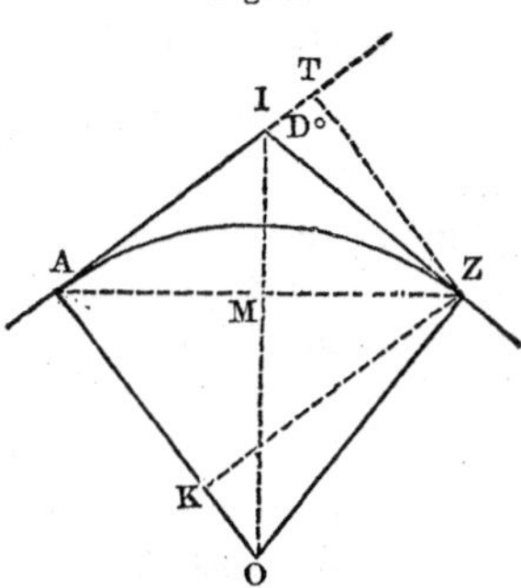

Graphically on a plot. At Z make an angle A Z I = Z A I. Draw perpendiculars at A and Z, and their intersections is the centre. *Otherwise,* draw perpendicular at A, and make angle A Z O = Z A O.

By calculation. There are two sub-cases, according to what the data are.

1. When the point is given by its polar co-ordinates, as T A Z and A Z,

$$\text{We have, } r = \frac{\text{A Z}}{2 \sin. \text{T A Z}}, \text{ and } t = \frac{\text{A Z}}{2 \cos. \text{T A Z}}.$$

$$\text{Length of curve} = 2\, r\, \frac{\text{T A Z}}{57.3}.$$

Angular length of curve = A O Z° = 2 T A Z.

Conversely. Given radius and T A Z, to find A Z. A Z = 2 r sin. T A Z.

2. When the point is given by its rectangular co-ordinates, viz., A T = x, and T Z = y.

$$r = \frac{x^2 + y^2}{2\,y}, \text{ and } t = \frac{x^2 + y^2}{2\,x}.$$

$$\text{Length of curve} = 2\, r\, \frac{\text{I A Z}}{57.3}; \quad \tan. \text{I A Z} = \frac{\text{T Z}}{\text{A T}} = \frac{y}{x}.$$

Angular length of curve, A O Z = 2 T A Z.

The direction of the final tangent at Z, *i. e.*, its deflection T I Z from the first tangent = A O Z.

Note.—To calculate the rectangular co-ordinates of a point of a curve from various data.

1. Given the polar co-ordinates,

$$x = \text{A T} = \text{A Z} \cos. \text{T A Z}, \; y = \text{T Z} = \text{A Z} \sin. \text{T A Z}.$$

2. Given the angle of deflection of the tangents, and radius of curve,

$$x = \text{A T} = \text{O A} \sin. \text{D}, \; y = \text{T Z} = \text{O A}(1 - \cos. \text{D}) = 2\, \text{O A}(\sin. \tfrac{1}{2}\text{D})^2.$$

3. Given radius and length of tangent.
Find D°, having tangent

$$\tfrac{1}{2}\ \text{A I Z} = \frac{r}{t};\ \text{and D} = 180^\circ - \text{A I Z}.$$

Then, $x = t(1 + \cos. \text{D}^\circ)$, and $y = t.\ \sin.\ \text{A I Z} = t.\ \sin.\ \text{D}^\circ$.

4. Given the radius of curve and its length,

$$\text{A O Z} = \text{D}^\circ = \frac{\text{length of curve} \times 57.3}{\text{radius}}.$$

Then apply the second case.

Finding the rectangular co-ordinates of the end of a curve, is equivalent to finding how far the curve will depart from its first tangent, and what point of that tangent its extremity will be opposite to.

Problem VI.

Given this tangent and point, as in Prob. V., *and also the radius, to find the starting point, length of tangent, and length of curve.* (Fig. 6.)

When the point is given by polar co-ordinates, change them to rectangular co-ordinates by the preceding formulas, *i. e.*, find T Z and the position of T.

Then,

$$\text{T A} = \sqrt{(2\ \text{A O} \times \text{T Z} - \text{T Z}^2)};\ \text{or, } x = \sqrt{(2\,r\,y - y^2)}.$$

Length of I A and of curve are as in the last problem.

Conversely. Given radius and tangent, to find T Z.

$$\text{T Z} = \text{O A} - \sqrt{(\text{O A}^2 - \text{A T}^2)};\ \text{or, } y = r - \sqrt{(r^2 - x^2)}.$$

Case C.—Given a tangent line and a curve already run.

Problem VII.

Given the radius, r, and length, l, of a curve; required the radius, r′, of another curve, A Z′, or A′ Z′, which shall start from the same tangent, and pass at a given distance, Z Z′, from the end of the first curve.

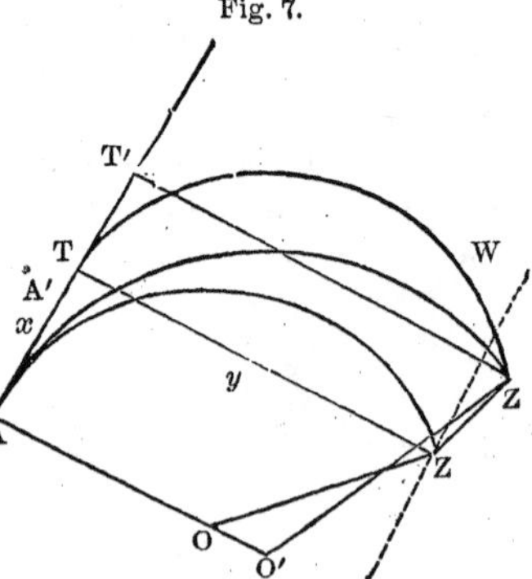

Fig. 7.

Precisely. Fig. 7.—Find the rectangular co-ordinates of Z, and then of Z′, and then apply Problem V., Case 2.

Conversely. Given the two curves, A Z and A Z′, or A′ Z′,

to find the distance apart, Z Z′. When they start from the same point, A;

$$\text{Z Z}' = \sqrt{[(x \backsim x')^2 + (y \backsim y')^2]}.$$

When they do not, add A A′ (with proper sign) to x', and use the sum for x' in the formula; x' and y' are the co-ordinates of Z′.

To find the *direction* of the distance, Z Z′, *i. e.*, the angle Z′ Z W, we have,

$$\sin.\ \text{Z}'\,\text{Z W} = \frac{y - y'}{\text{Z Z}'}, \text{ and } \cos.\ \text{Z}'\,\text{Z W} = \frac{x' - x}{\text{Z Z}'}.$$

Reckoning this angle around from Z W, to the left, as is usual, the trigonometric signs will determine the quadrant in which Z Z′ lies, and therefore its absolute direction.

Approximately. When the curves are of the same length, of large radius, and do not diverge far.

For the general problem. When the curves start from the same point, A,

$$r' = r\,\frac{\text{A Z}^2}{\text{A Z}^2 \pm 2\,r\,\text{Z Z}'},$$

using the plus sign when the curve A Z′ passes farther from the tangent than does A Z, and *vice versa.*

When they start from different points, A and A′,

$$r' = r\,\frac{(\text{A Z} \pm \text{A A}')^2}{\text{A Z}^2 \pm 2\,r\,\text{Z Z}'}.$$

Conversely. Given the two curves to find their distance apart. When they start from the same point, A,

$$\text{Z Z}' = \text{A Z}^2\,\frac{r - r'}{2\,r\,r'}.$$

When they do not,

$$\text{Z Z}' = \frac{r(\text{A Z} + \text{A A}')^2 - r'\,\text{A Z}^2}{2\,r\,r'}.$$

PROBLEM VIII.

Given the radius, r, and the length, l, of a curve; required the radius, r', of another curve, which shall start from the same tangent at A′, *and meet the first curve at a point,* Z.

Precisely. Find the rectangular co-ordinates of Z, and then apply Prob. V., Case 2.

Approximately.

$$r' = r\left(\frac{\text{A Z} \pm \text{A A}'}{\text{A Z}}\right)^2.$$

This is from the approximation in Prob. VII., by making **z x = 0.**

PART II.

To run the Curves.

FIRST METHOD.

By "tangential angles;" i. e., angles of divergence from tangents.

From the starting point set off, with a transit or a compass, equal diverging angles, each subtended by equal chords; *i. e.*,

Fig. 8.

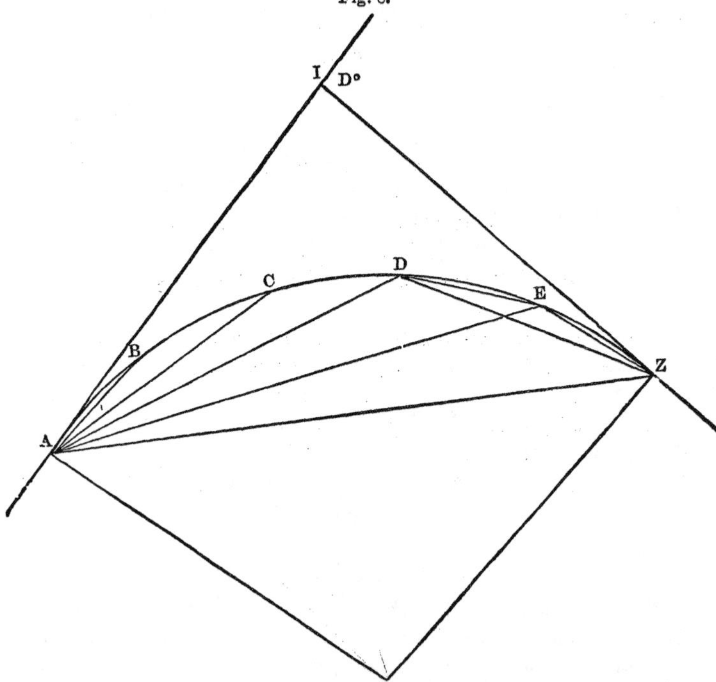

chains. After determining *n* points, go to the last one, sight back to the first one, and deflect from the chord, Z A, an angle equal to that already turned; *i. e.*, I A Z. You are then pointing in the tangent Z I, which may be prolonged as a tangent, or used to continue the curve as at first. This is the method most commonly used.

For example; let it be required to run a four-degree curve for five stations, commencing at A, Fig. 8. Then the angle to be turned off each time is two degrees. This is called the "tangential angle," and is represented by δ.

Set up the transit at A, with the telescope pointing toward I. Turn 2° to the right, and fix the point B in the line of sight at a distance from A of 100 feet. Then turn 2° more to the right, and fix the point C in the line of sight at a distance of 100 feet from B. So on for any number of stations, turning 2° each time, and fixing the station in the line of sight at a distance of 100 feet from the preceding one. To get on the tangent at Z, set up the transit at Z, with the telescope pointing to A. Turn to the right 10° (the number of degrees deflected from A I), and the telescope will then be pointing to I, along the tangent I Z. It frequently happens that the entire curve cannot be run from A. Suppose it is desired to make a changing point at D, set up at D with the telescope directed toward A. Turn to the right 6° (the whole number of degrees deflected from the tangent), and the telescope will then be pointing along the tangent at D, and the curve can be prolonged in the same manner as when starting at A.

When the "tangential angle" contains some odd seconds, keep account of them, and when they amount to a minute, add it in.

When a curve does not come out just right, *i. e.*, to some point Z′, instead of Z, some engineers, instead of running it over until it does, will move Z′ to Z, and the other stakes a distance proportional to the square of their distance from the starting point. This is a tolerable approximation.

NOTATION.

δ° = "tangential angle" by which the curve is run.

$2\delta^\circ$ = "degree" of curve = the angle subtended at the centre by a chord of 100′.

c = length of one of the equal chords, usually 100 feet.

n = number of chords in the curve.

r = radius of the curve.

Chord and radius must be in the same unit of measure.

$$\text{IAZ} = n\,\delta^\circ. \quad \text{D}^\circ = \text{AOZ} = 2\,n\,\delta^\circ = 2\,\text{IAZ}.$$

The length of the curve in n chords $= \dfrac{\text{AOZ}^\circ}{2\,\delta^\circ} = \dfrac{\text{D}^\circ}{2\,\delta^\circ}$

If n have a *fraction*, the curve will end with a "*sub-chord*," *i. e.*, a similar fraction of a whole chord.

Fundamental Theorems.

$$\text{Sin. } \delta^\circ = \frac{c}{2\,r}; \qquad r = \frac{c}{2 \text{ sin. } \delta^\circ}.$$

With hundred feet chords, approximately,

$$\delta^\circ = \frac{2865}{r} \qquad r = \frac{2865}{\delta^\circ}.$$

This is near enough for curves of large radius.

For any other chord, c′, and a corresponding tangential angle δ'°.

$$\text{sin. } \delta'^\circ = \text{sin. } \delta^\circ \frac{c'}{c}.$$

By this formula long chords may be used when more convenient.

For short chords, approximately,

$$\delta'^\circ = \delta^\circ \frac{c'}{c}.$$

A curve whose chords are of an equal length, and δ has odd minutes and seconds, may be run with δ'° in even minutes, by using another chord, c', given by, $c' = c \dfrac{\text{sin. } \delta'}{\text{sin. } \delta}$.

PROBLEMS.

The enunciations are as in the "General Problems," Part I., only substituting "Tangential Angle" for "Radius."

Case A.

Problem I.

Given two intersecting tangents, and also the starting point on one of them, i. e., given D° *and t, to find* δ°. (Fig. 1.)

$$\text{sin. } \delta^\circ = c\,\frac{\text{tan. } \frac{1}{2}\,\text{D}^\circ}{2\,t}, \text{ and angular length of curve} = 2\,n\,\delta^\circ.$$

PROBLEM II.

Given two intersecting tangents, and also the tangential angle, i. e., given D° *and* $\delta°$, *to find t.* (Fig. 1.)

$$t = c\,\frac{\tan.\ \frac{1}{2}\,D°}{2\ \sin.\ \delta°}.$$

PROBLEM III.

Given two intersecting tangents, and also the distance from the vertex to a point through which the curve must pass, i. e., given D° *and* I C, *to find t and* $\delta°$. (Fig. 1.)

$$\sin.\ \delta° = c\,\frac{\tan.\ \frac{1}{4}\,D°\ \tan.\ \frac{1}{2}\,D°}{2\,I\,C},\ \text{and}\ t = I\,C.\ \cot.\ \tfrac{1}{4}\,D°.$$

Conversely,

$$I\,C = c\,\frac{\tan.\ \frac{1}{4}\,D°.\ \tan.\ \frac{1}{2}\,D°}{2\ \sin.\ \delta°},\ \text{and}\ I\,C = t\ \tan.\ \tfrac{1}{4}\,D°.$$

PROBLEM IV.

Given two intersecting tangents, and also a point through which the curve must pass, i. e., given D° *and the co-ordinates of* P, *to find* $\delta°$ *and* A I. (Fig. 4.)

Find F A and A I as in General Problem IV.

$$\sin.\ \delta° = c\,\frac{\tan.\ \frac{1}{2}\,D°}{2\,A\,I}.$$

CASE B.—TO START FROM A GIVEN TANGENT, AND PASS THROUGH A GIVEN POINT.

PROBLEM V.

Given this tangent and point, and also the starting point. (Fig. 6.)

1. Given A Z and I A Z, to find $\delta°$ and I A.

$$\sin.\ \delta° = c\,\frac{\sin.\ I\,A\,Z}{A\,Z},\ \text{and}\ A\,I = \frac{A\,Z}{2\ \cos.\ I\,A\,Z}.$$

Approximately,

$$\delta° = \frac{I\,A\,Z}{A\,Z\ (\text{in chains})};$$

Conversely,

$$A\,Z = c\,\frac{\sin.\ I\,A\,Z}{\sin.\ \delta°}.$$

2. Given A T $= x$, and T Z $= y$, to find δ° and A I.

$$\text{Sin. } \delta^\circ = \text{C} \frac{y}{x^2 + y^2}, \text{ and A I} = \frac{x^2 + y^2}{2x}.$$

Conversely,
$$\text{A T} = x = \text{C} \frac{\sin. 2n\delta^\circ}{2 \sin. \delta^\circ},$$

$$\text{T Z} = y = \text{C} \frac{(\sin. n\delta)^2}{\sin. \delta};$$

or,
$$y = \text{C} \frac{1 - \cos. 2n\delta^\circ}{2 \sin. \delta^\circ}$$

Problem VI.

Given the tangent and point, as in Problem V., *and also δ° and* T Z, *to find* A T. (Fig. 6.)

$$\text{A T} = x = \sqrt{\left(\text{C} \frac{y}{\sin. \delta} - y^2\right)}.$$

Conversely. Given δ° and A T to find T Z.

$$\text{Radius} = \frac{\text{C}}{2 \sin. \delta^\circ}; \text{ then T Z} = y = r - \sqrt{(r^2 - x^2)}.$$

Case C.

Problem VII.

Given δ°, n, and Z Z′, *to find δ'°.* (Fig. 7.)

Accurately, as in General Problem VII.

Approximately, Z Z′ being in feet,

$$\delta'^\circ = \delta^\circ \pm \frac{4 \text{ Z Z}'}{7 n^2}, \text{ and conversely, Z Z}' = (\delta^\circ \backsim \delta'^\circ) \frac{7 n^2}{4}.$$

When the curves start at a distance apart = A A′ (in chains),

$$\delta'^\circ = \frac{n^2 \delta \backsim \frac{4}{7} \text{ Z Z}'}{(n + \text{A A}')^2}, \text{ and Z Z}' = \tfrac{7}{4} [n^2 \delta \backsim (n + \text{A A}')^2 \delta'].$$

Problem VIII.

Given δ°, n, and A A′, *to find δ'°.* (Fig. 7.)

Accurately, as in General Problem VIII.

Approximately,
$$\delta'^\circ = \delta^\circ \left(\frac{n}{n + \text{A A}'}\right)^2.$$

SECOND METHOD.

By "chord angles," i.e., angles of deflection from chords. (Fig. 9.)

Turn $\delta°$ and fix B as in the first method. Set up at B, and turn $2\delta°$ from A B prolonged, and fix C at a distance of 100 feet from B,

Fig. 9.

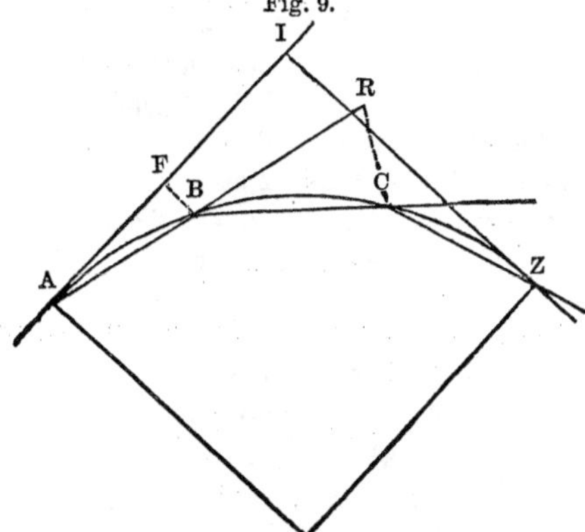

and so go on any number of stations, turning $2\,\delta°$ from each chord produced. To get into the tangent at any point, turn $\delta°$. Find $\delta°$ as in the first method. The defect of this method is the frequent setting of the instrument.

THIRD METHOD.

With two transits and no chain. (Fig. 10.)

Set the transits at A and Z. Turn the telescope of the former to I, and that of the latter to A. Then deflect equal angles in the same direction (to the right), and set stakes at the intersections of the corresponding lines of sight. The principle is that the vertices of a series of equal angles, constructed on the same chord, will all lie in the arc of a circle. Given the chord, the angle to be turned is found as in the first method.

Fig. 10.

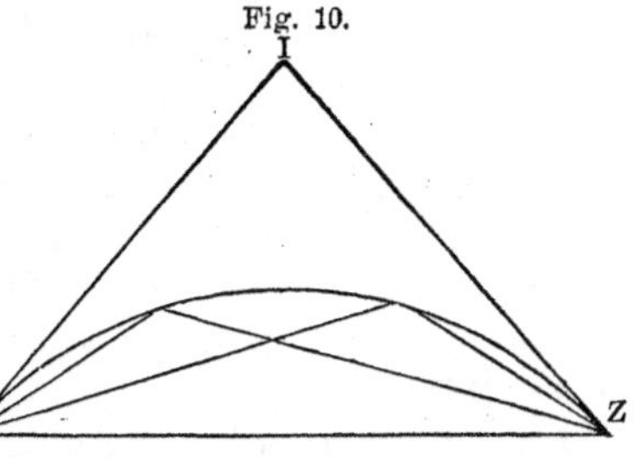

FOURTH METHOD.

With a sextant; reflecting the angle in a segment. (Fig. 1.)

Given A Z and I A Z; either on the ground or calculated. Set the sextant (or other reflecting instrument) to the supplement of I A Z. Move about till poles at A and Z (seen, one by direct vision, and the other by reflection) appear to coincide. Drop a plumb line from the eye, and it will fix one point of the curve. Repeat this at as many points of the curve between A and Z as are desired. The principle is, that the angle between the tangent and chord at any point of a circle is equal to the angle inscribed in the segment, and equal to the supplement of the angle inscribed in the original segment.

FIFTH METHOD.

By versed sines. For the method of running the curve, see pp. 140 and 141. (Figs. 64 and 65.)

Let $v =$ versed sine D E,

$c =$ chord A E = E F.

$$v = \frac{c^2}{2r}.$$

$$\text{When } c = 100,\ v = \frac{5000}{r};\ \text{and } r = \frac{5000}{v}.$$

$$\text{By Prob. I. } v = \frac{c^2}{2\,\text{AI}\tan.\frac{1}{2}\,\text{AIZ}}.$$

For a sub-chord c', the versed sine $v' = v\left(\frac{c'}{c}\right)^2$.

Hence, when $c' = \frac{1}{2}c$, $v' = \frac{1}{4}v$, and so on.

To find, approximately, intermediate points,

$$\text{ED} = \text{AE}\ \sin.\ \text{EAD};\ \text{or } v = c\ \sin.\ \delta^\circ.$$

Approximately, $v = \frac{7}{4}\,\delta^\circ$ and $\delta^\circ = \frac{4}{7}\,v$.

SIXTH METHOD.

By deflection distances from chords produced, or double versed sines. (Fig. 9.)

Let d represent one of the deflection distances, as C R. Then $d = 2v = \frac{c^2}{r}$.

A B is prolonged until $BR = AB$. The first station, B, is set by a "tangent deflection," F B, from the tangent, A I. $FB = \frac{1}{2} RC$; *i. e.*, a tangent deflection is half a chord deflection.

SEVENTH METHOD.

By offsets from tangents. (Fig. 11.)

Fig. 11.

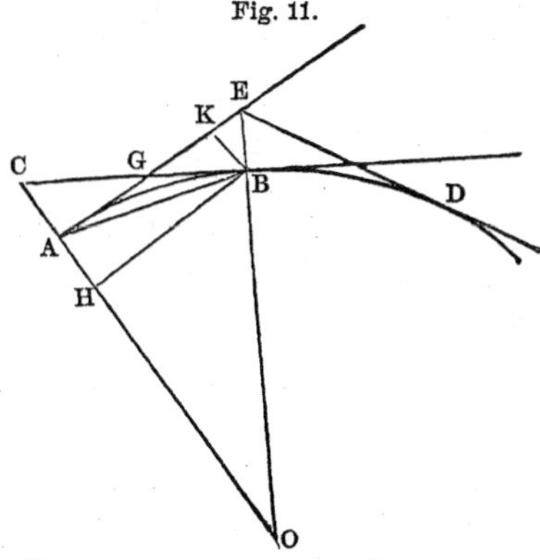

1. *Exactly.* Given radius, O A, and distance on tangent, A K, to find offset from tangent, K B.

$$KB = AH = r - \sqrt{(r^2 - AK^2)}.$$

2. *Approximately.*

$$AH : AB :: AB : 2\,OA; \therefore AH = \frac{AB^2}{2\,OA}.$$

Calling $AK = AB$ (which it is approximately), we have,

$$AH = \frac{AK^2}{2\,OA} = \frac{\text{tangent}^2}{2 \times \text{radius}}.$$

When tangent $= \frac{1}{10}$ radius, the error of the approximation is .000013 radius $= \frac{1}{77000}$ radius.

When tangent $= \frac{1}{4}$ radius, error $= .00051\, r = \frac{1}{2000}\, r$.

" " $= \frac{1}{2}$ " " $0.009\, r = \frac{1}{111}\, r$.

Required, length of tangent which will make the chords A B, etc., even chains.

$$\text{Sin.}\, \tfrac{1}{2} AOB = \frac{AB}{2\,OA} \quad . \quad AK = HB = OA \sin. AOB.$$

When the offsets become too long to be set off with accuracy or ease, or when the tangent deviates from the desired curve so far as to fall on impracticable ground, "*auxiliary tangents*" may be used.

From A points have been fixed to B. A new tangent at B is desired. From A set off a distance, A C, to be calculated by a formula given below. C B produced is the desired new tangent. Set off from it offsets as before. If the ground prevents A C being set off, set off A G, and produce G B as before.

To get a new tangent at D, set off B E = A C, and E D produced is the third tangent required.

Or, set off K E on first tangent prolonged, and prolong E D as before. So on for other points.

$$\text{A C} = \frac{\text{O B} \times \text{B K}}{\text{O H}} = \frac{\text{rad.} \times \text{offset}}{\text{rad.} - \text{offset}}.$$

$$\text{A G} = \frac{\text{O B} \times \text{B K}}{\text{B H}} = \frac{\text{rad.} \times \text{offset}}{\text{tangent}}.$$

$$\text{K E} = \frac{\text{H B} \times \text{K B}}{\text{H O}} = \frac{\text{tan.} \times \text{offset}}{\text{rad.} - \text{offset}}.$$

These formulas are derived from the similar triangles, C A G, E B G, C H B, B H O, B C O, B K E, and B K G.

EIGHTH METHOD.

By Ordinates to Chords.

Fig. 12.

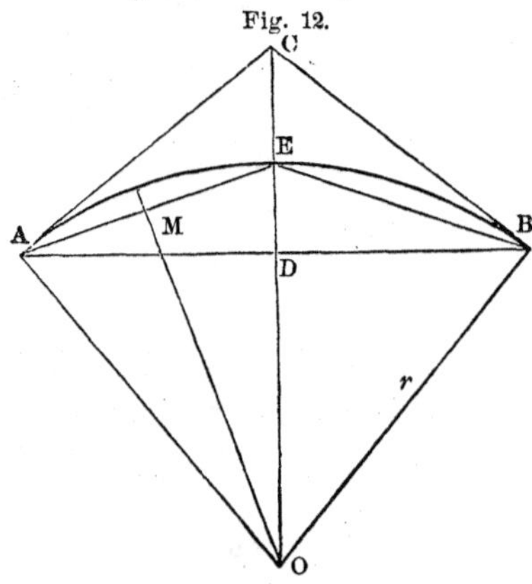

1. *To find the middle ordinate* (*m*).

A. Given, the tangents on the ground.

AE bisects CAD. Hence, DE : EC :: AD : AC.

And $$\text{DE} : \text{DE} + \text{EC} :: \text{AD} : \text{AD} + \text{AC};$$

or, $$\text{DE} : \text{DC} :: \text{AD} : \text{AD} + \text{AC}.$$

Hence, $$\text{DE} = \frac{\text{AD} \times \text{CD}}{\text{AD} + \text{AC}} = m.$$

Also, $$\text{DE} : \text{EC} :: \text{AD} : \text{AC} :: \cos.\ \text{CAB} : 1.$$
$$\text{DE} : \text{DE} + \text{EC} :: \cos.\ \text{CAB} : \cos.\ \text{CAB} + 1.$$

Hence, $$m = \text{DE} = \frac{\cos.\ \text{CAB}}{1 + \cos.\ \text{CAB}}.$$

B. Given, the radius and chord.

$$\text{OA}^2 = \text{AD}^2 + \text{DO}^2 = (\tfrac{1}{2}\text{C})^2 + (\text{OE} - \text{DE})^2, \text{ or, } r^2 = \tfrac{1}{4}\text{C}^2 + (r - m)^2$$
$$2\,r\,m = \tfrac{1}{4}\text{C}^2 + m^2.$$

Hence, $$m = r - \sqrt{(r^2 - \tfrac{1}{4}\text{C}^2)}.$$

Approximately. In the equation
$$2\,r\,m = \tfrac{1}{4}\text{C}^2 + m^2,$$
neglecting the last term in the second member (which is very small in railroad curves), we have,
$$m = \frac{\text{C}^2}{8\,r}.$$

C. Given, radius and chord of half the arc $= \text{C}'$.

From the similar triangles, EAD and AOM, we have,
$$\text{AO} : \text{AM} (= \tfrac{1}{2}\text{AE}) :: \text{AE} : \text{ED}.$$

Hence,
$$\text{ED} = \frac{\text{AE}^2}{2\,\text{AO}}, \text{ or, } m = \frac{\text{C}'^2}{2\,r}.$$

D. Given, radius and tangent.

$$\text{DE} = \text{DC} - \text{CE} = \text{DC} - (\text{OC} - \text{OE})$$
$$= \text{DC} - \left[\sqrt{(\text{AO}^2 + \text{AC}^2)} - \text{AO}\right] = \text{DC} + \text{AO} - \sqrt{(\text{AO}^2 + \text{AC}^2)}.$$

Hence, $$\text{DE} = m = \text{DC} + \text{AO} - \sqrt{(\text{AO}^2 + \text{AC}^2)}.$$

E. Given, the "tangential angle."

$$\text{ED} = \text{AD}\ .\ \tan.\ \text{EAD}.$$

Hence, $$\text{ED} = m = \tfrac{1}{2}c\ .\ \tan.\ \tfrac{1}{2}\delta°.$$

2. *To find intermediate ordinates.* (Fig. 13.)

Let D G, the distance of the foot of the required ordinate from the middle chord be represented by f.

Fig. 13.

The required intermediate ordinate,

$$\text{FG} = \text{FH} - \text{GH},$$

we will find FH and GH in turn.

$$\text{FH}^2 = \text{HL} \times \text{HI} = (\text{OL} + \text{OH})(\text{OL} - \text{OH})$$

$$= (r + f)(r - f) = r^2 - f^2.$$

Hence, $\text{FH} = \sqrt{(r^2 - f^2)}$.

$$\text{GH}^2 = \text{AK}^2 = \text{IK} \times \text{KL} = (\text{OI} - \text{OK})(\text{OL} + \text{OK})$$

$$= (r - \tfrac{1}{2}c)(r + \tfrac{1}{2}c) + r^2 - \tfrac{1}{4}c^2.$$

Hence, $\text{GH} = \sqrt{(r^2 - \tfrac{1}{4}c^2)}$.

Then, $\text{FG} = \sqrt{(r^2 - f^2)} - \sqrt{(r^2 - \tfrac{1}{4}c^2)}$.

Approximately. (Fig. 14.)

Fig. 14.

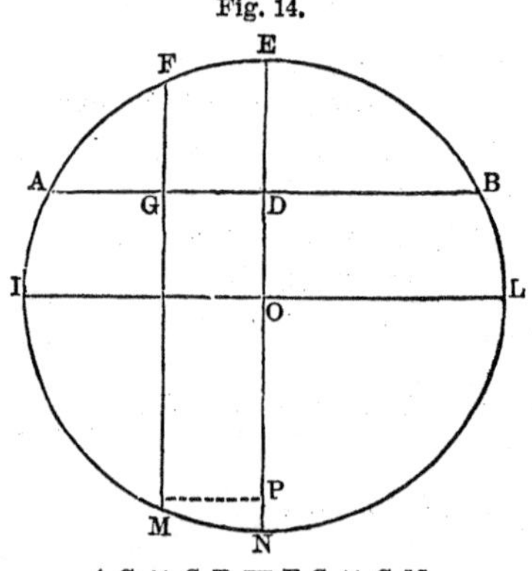

$$\text{AG} \times \text{GB} = \text{FG} \times \text{GM}.$$

$$\text{FG} = \frac{\text{AG} \times \text{GB}}{\text{GM}}.$$

$$\text{GM} = \text{DP} = \text{EN} - (\text{ED} + \text{PN}).$$

Omit the subtrahend, as very small compared with E N, and we have,

$$\text{G M} = \text{E N} = 2\,r, \text{ and } \text{F G} = \frac{\text{A G} \times \text{G B}}{2\,r}.$$

COMPOUND CURVES.

A single arc of a circle uniting two tangents, must meet them at equal distances from their point of intersection or vertex. If it be required to unite two tangents by a curve meeting them at unequal distances from the vertex, a compound curve must be employed, composed of two or more arcs of circles of different radii.

A fundamental condition is that, the centres of the two adjoining arcs and their point of meeting, must lie in the same straight line; since these arcs must have a common tangent at their point of meeting.

An infinite number of pairs of curves would satisfy the preceding conditions, consequently, another condition must be introduced. This may be that one radius shall be given, or that the difference of the two radii shall be a minimum; or their ratio a minimum, etc.

Problem I.

It is required that the ratio of the two radii shall be a minimum.

In this case the common tangent will be parallel to the line A Z.

Analytically. Let t and t' be the two tangents, I A and I Z; r and r' the corresponding radii; φ' and φ the angles comprised by the curves, or their angular lengths; and α the angle A I Z. Put

$$m = \sqrt{(t^2 + t'^2 - 2t \,.\, t' \,.\, \cos.\, \alpha.)}$$

Then,

$$r = \frac{m}{2\,t' \,.\, \sin.\, \alpha} \,.\, (m + t - t').$$

$$r' = \frac{m}{2\,t \,.\, \sin.\, \alpha} \,.\, (m - t + t').$$

Fig. 15.

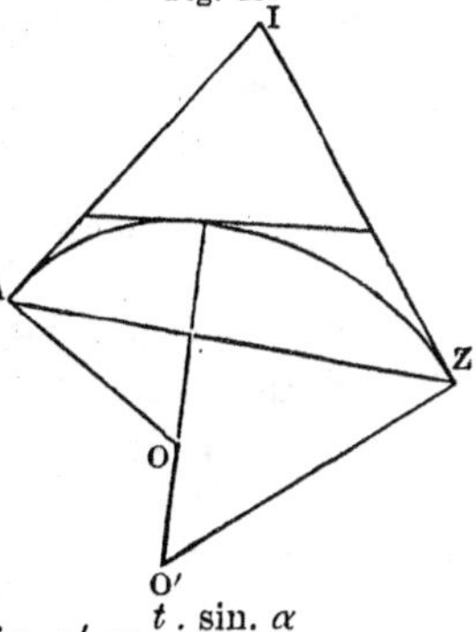

$$\sin.\, \varphi = \frac{t' \,.\, \sin.\, \alpha}{m}. \qquad \sin.\, \varphi' = \frac{t \,.\, \sin.\, \alpha}{m}.$$

PROBLEM II.

It is required to make the difference of the two radii a minimum.

$$r = t \,.\, \tan. \tfrac{1}{2}\alpha + \frac{t - t'}{2 \cos. \tfrac{1}{2}\alpha}.$$

$$r' = t' \,.\, \tan. \tfrac{1}{2}\alpha - \frac{t - t'}{2 \cos. \tfrac{1}{2}\alpha}. \qquad \varphi = \varphi' = 90 - \tfrac{1}{2}\alpha.$$

PROBLEM III.

When one radius is given to find the other.

By construction. Draw perpendiculars at A and Z. Set off the given radius on each of them, from A to some point, O, and from Z to some point, P. Join O P, and bisect it by a perpendicular. This perpendicular meets the perpendicular from Z in O′. O and O′ are the desired centres, and the two curves will unite on the line through these points.

*Analytically.** Let A Z, the angles I A Z and I Z A, and the radius A O $= r$, be given, to find the second radius r'.

From A run a curve with the first radius to D, where the tangent, D E, becomes parallel to Z I. The line, Z D, prolonged will meet the curve at the common tangent point, C.

Fig. 16.

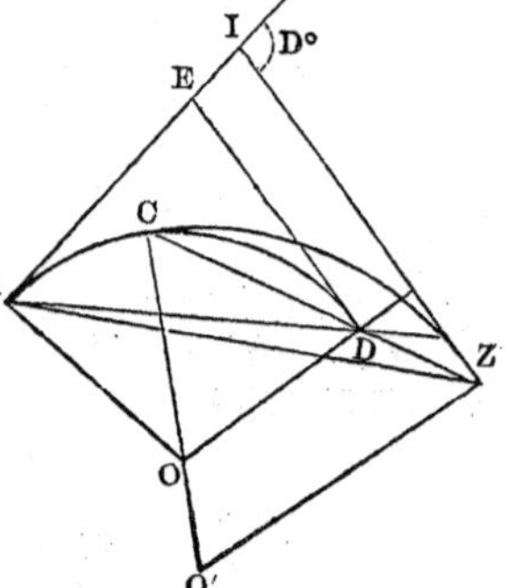

$$r' = r + \frac{\text{Z D}}{2 \sin. \text{C Z I}}.$$

When the angle at Z is greater than the angle at A, the formula becomes,

$$r' = r - \frac{\text{Z D}}{2 \sin. \text{C Z I}}.$$

In the field the point, D, may be found by laying off the angle, I A D $= \frac{1}{2}$ (I A Z + I Z A), and measuring the distance, A D $= 2r \,.\, \sin. \frac{1}{2}$(I A Z + I Z A).

REVERSED CURVES.

If the two branches of the curve, instead of both lying on the same

* From Henck's Field Book.

side of the common tangent, as in "Compound Curves," are on opposite sides, it is called a "Reversed Curve."

Fig. 17.

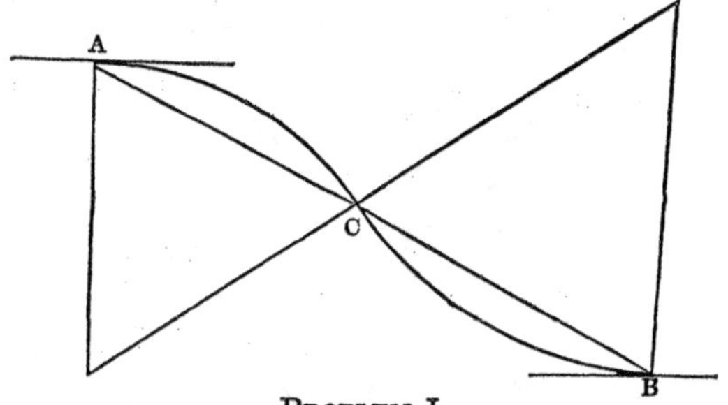

Problem I.

When the tangents are parallel, and both curves are to have the same radius, to find the radius, r.

When the tangents are parallel, the point of reversed curvature is on the line joining the tangent points; and if the two radii are equal, it is the centre of the line.

Let the distance A B, Fig. 17, be represented by a, and the perpendicular distance between the tangents, by d. Then we have,

$$r = \frac{a^2}{4d}.$$

Problem II.

When the tangents are parallel, and one radius, r, is given, to find the other radius, r′.

$$r' = \frac{a^2}{2d} - r;$$

$$\text{A C} = \frac{2rd}{a}, \text{ and, B C} = \text{A B} - \text{A C}.$$

Problem III.

When the tangents are not parallel, and both curves are to have the same radius, r, the tangent points being given. (Fig. 18.)

$$r = \text{B Y} \frac{\sin. \text{B P C}}{\sin. \text{C} + \sin. \text{C}'}.$$

$$\text{Sin. } \mathrm{BPC} = \tfrac{1}{2}(\cos.\ \mathrm{ABY} + \cos.\ \mathrm{BYZ});$$
$$\mathrm{C} = 270° - \mathrm{BPC} - \mathrm{ABY};$$
$$\mathrm{C}' = 270° - \mathrm{BPC} - \mathrm{BYZ}.$$

Fig. 18.

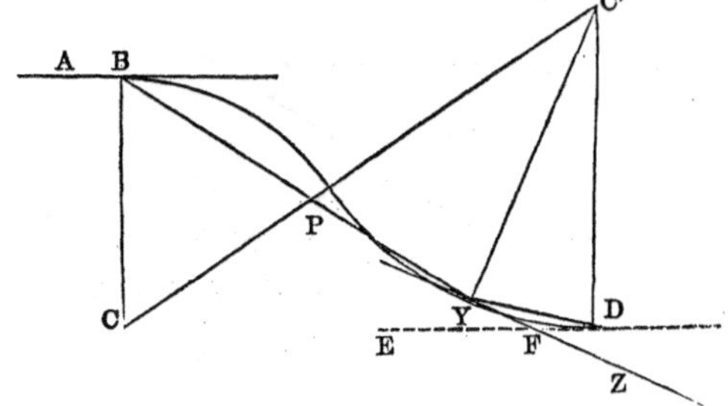

Problem IV.

When the tangents are not parallel, and one of the tangents, r, is given, to find the other tangent, r′.

Run with the given radius to some point, D, where the tangent, D E, is parallel to A B, and then apply Prob. I. To find D, lay off Z Y D $= \frac{1}{2}$ E F B, and make Y D $= 2\,r$ sin. $\frac{1}{2}$ B.

To lay out a compound, or reversed curve, run to the point of common tangency of the two branches of the curve, by one of the methods given. At that point get on the tangent, and then run out the remaining branch in the usual manner.

APPENDIX D.

ESTIMATION.

General Principles.

We have to determine: 1st, The cost of the raw material; 2d, Time employed in working on it; 3d, Price of a unit of this time.

1. *First Cost.*

This varies with the locality, demand, etc. The waste in shaping a material must be allowed for. In common cut stone it is about $\frac{1}{10}$; in converting round timber to hewn, about $\frac{1}{5}$.

2. *Time.*

The time in which an average workman will perform a certain amount of work of any kind, is called the "*Constant*" of labor for that work, it being nearly the same for all times and places. To get the cost of that work it is only necessary to multiply this constant by the price of labor for that unit of time.

APPLICATION TO ROAD-MAKING.

A. *Constants for Excavation.*

The table on page 126 gives for one cubic yard of excavation, previously loosened, including throwing, for common earth, 1.25 hours to .83 of an hour; loose and light earth, 1.25 hours; mud, 1.43 hours to .62 hour; clay and stony earth, 2.5 hours; rock, after blasting, 4.5 hours. Other experiments for "hard pan," 4.2 hours; compact sand, .43 hour.

The table on page 128 gives for excavating earth, and loading it into barrows, a constant of .42 hour, and for excavating and loading it into horse-carts, .8 hour.

Cole's Erie Canal Experiments give .47 hour for barrows. A man has shovelled into a wagon at the rate of .48 hour. The constants from the bottom of page 125 are for shovelling into a cart, earth previously loosened. Gravel and clay, 1 hour; loam, .83 hour; sandy earth, .71 hour; average for all, .6 hour.

Excavating clay tow-path and depositing it behind the bank, 1.8 hours.

B. *Barrow Work.*

Constant for wheeling 1 cubic yard 100 feet:

1. Page 126 gives, for common earth, from .5 to .3,—average, .4 hour.

2. Page 128 gives, average .44 hour.

3. Birmingham Railroad, page 128, gives .34 hour.

4. Morin says one man can wheel 400 lbs. (= 3 cubic feet), 20,000′ in 10 hours. Constant per .45 hour. (No return.)

5. Gauthey says removing 1 cubic metre a distance of 30 metres, takes .5 hour: constant per yard for 100′ = .38 hour.

6. Erie Canal work with barrows holding $\frac{1}{18}$ cubic yard; wheelers travel 250′ per minute, on a level run; delays starting, etc., $\frac{3}{4}$ minute. Constant per yard for 100′ = .275 hour.

Work of loading into a barrow and wheeling common earth, the following length of run-way, for a day of 10 hours, on the Erie Canal enlargement:

Length, in feet, of run-way.	Cubic yards.	Constant, in hours, per yard.	Cost per yard, at $1 per day.
50	16	0.625	6.25
100	14	.71	7.1
150	12	.833	8.33
200	10	1.00	10.00
300	8	1.25	12.5
400	7	1.428	14.28
500	6	1.67	16.7

The above table gives a constant of .24 hour for each hundred feet after starting.

C. *Wagon Work.*

Average performance, on 10 miles of the Erie Canal, of the men in loading wagons, including picking, and loss of time in waiting for wagons to come in, etc., was, per day of 12 hours: sand and loam, 15 cubic yards, constant, .8 hour; clay and gravel, 12 cubic yards, constant, 1 hour; hard pan, 4 yards, constant 3 hours; stiff clay (earth of 1½ men.), 10 yards, constant, 1.2 hours.

Six men can work to best advantage in loading wagons. They can fill a wagon containing $\frac{2}{3}$ cubic yard in 2 minutes. Unloading and other delays take 3 minutes, and 100 feet at each end for a turning space should be added to the hauling distance. The horses travel at about the rate of $2\frac{1}{2}$ miles per hour, or 220′ per minute.

Work done in a day of 10 hours, by a pair of horses, working on a level road, with a common wagon:

Distance in feet....	100	200	500	1000	1500	2000	3000	5000
Cubic yards per day	61	53	40	28	21	17	12	8
Constant per yard..	0.16 h.	0.19 h.	0.25 h.					

Up a slope of 100′ per mile ($= \frac{1}{53}$) only $\frac{3}{4}$ as much can be drawn. Up a slope of 260′ per mile (1 in 20), only $\frac{1}{2}$ as much can be drawn.

D. *Railway Work.*

Work done on a rail track with horses kept constantly moving, and hauling on a level three loaded cars, containing $1\frac{1}{2}$ yards each, at $2\frac{1}{2}$ miles per hour.

Distance in feet	1000	2000	3000	4000	5000	10000	20000
Cubic yards, per day.	225	127	90	69	56	29	15

LIMITING DISTANCES.

A limiting distance for any mode of transportation is that at which that mode becomes more expensive than some other mode.

The first means of moving earth very short distances is by throwing it with the shovel, which can be done 12′ horizontally and 5′ vertically. For twice that distance two men may throw twice, and so on. The scraper is cheaper for more than 12′.

For long distances and heavy work, rails should always be laid on which one horse can draw several cars, which can be dumped where desired.

There is a certain distance at which the various modes of transport become successively more expensive than some other. This limit is best found by putting tabular results of experiments in a diagram.

The average of many experiments give the following limiting distances. Men shovelling to 24′ in two throws.

Scraper, ...	thence to 100′
Barrow, ...	" " 200
One-horse cart, . .	" " $\frac{1}{3}$ mile
Two-horse wagons, . .	" " $\frac{2}{3}$ "

Railroad with horses thence to $1\frac{1}{2}$ miles.

Railroad with locomotive for greater distances.

The horse railroad should be used for less distances, if the amount to be moved is large, which will also effect the preceding limits.

APPENDIX E.

TUNNELS.

WHEN the depth of an excavation passes beyond a certain limit, it becomes cheaper to tunnel. To determine when to change from open cutting to tunnel: let e equal the cost of excavation per cubic yard, t equal the cost of the tunnel per running foot, b equal the base of the excavation, s equal the ratio of the side slopes, and x the unknown depth at which the costs of excavating and tunnelling are equal. Then we have:

Cross-section of excavation $= x\,(b + s\,x)$,

Contents for one running foot $= x\,(b + s\,x)$,

Cost of running foot of excavation $= \dfrac{x\,(b + s\,x)}{27} \times e$.

" " tunnel $= t$,

Hence,

$$\frac{x\,(b + s\,x)}{27} \times e = t,$$

$$x = -\frac{b}{2\,s} \pm \sqrt{\left(\frac{27\,t}{e\,s} + \frac{b^2}{4\,s^2}\right)}.$$

A somewhat greater depth than that deduced from the formula would be arrived at before beginning to tunnel, because of the uncertainty and delays of tunnel work.

Dimensions.—Width from 24 to 30 feet (for double track), height from 18 to 25 feet.

The Mount Cenis tunnel is 26¼ ft. wide, 20 ft. 8 in. above the rails, and 7 miles 1044 yds. long. There is no shaft. The depth of tunnel below the summit of the mountain is one mile.

The dimensions adopted for the numerous tunnels on the Central Pacific Railroad was: width 16 ft., height 19, consisting of a rectangle at the bottom 16 × 11, and a semicircle at the top, 16 ft. in diameter.

The Hoosic tunnel is to be 24500 ft. long.

Laying them out.—The centre line is first set out on the surface of the ground with great accuracy. Then the line is carried into the adits and down the shafts.

Construction.—The work is commenced with a "heading" or "driftway" about 6 ft. square, sometimes at the bottom, and sometimes at the top. In solid rock it is better to carry in the heading at the top. This driftway is afterward enlarged to the full cross-section. Tunnels in earth or loose rock are lined with timber or masonry.

When the tunnels are long, shafts are sunk at convenient places in order to expedite the work. From the bottom of the shaft the excavating is carried on both ways, the earth being raised in buckets.

The usual dimensions for shafts are from 9 to 12 ft. in diameter. If rectangular, about 8 × 12 ft. It is recommended by some to carry the shaft down at the side of the tunnel, instead of over the centre. Sometimes they are impracticable, as at Mount Cenis.

When the material to be excavated is rock, blasting becomes necessary. See page 160.

Nitro-glycerine is extensively used for rock blasting. Much more rapid progress can be made with it than with powder. The drills used are smaller, fewer holes are required, and the rock is broken into smaller pieces. It is much less expensive, and if manufactured on the ground where it is used, and handled with proper care, it seems no more dangerous than powder.

At Mount Cenis the drilling was done by machinery, worked by air, which was compressed by water power near the tunnel. The compressed air, after doing its work, was discharged from the machine and served for ventilation.

The alignment inside the tunnel is secured by wooden plugs, inserted into drill holes in the roof. The exact centre line is marked by tacks, driven into the plugs, to which a piece of cord is fastened.

Progress.—This depends on the rapidity with which the "heading" or "driftway" can be pushed forward, as the "bottom" can be taken out much more rapidly. It varies from 2 ft. per day in hard granite, to 10 or 12 ft. in soft rock and earth. The slowness of the work is due to the lack of room, and the disadvantage of working against the face of the rock.

Cost.—In the United States tunnels cost from $2.00 per cub. yd. in soft slate to $7.00 in hard graywacke.

On the Baltimore and Ohio Railroad the average cost per cub. yd., without counting the shafts, was $2.60, the length being from 100 to 1200 ft. With the same material, excavations in the open cutting cost about one-fourth as much.

The Bergen tunnel for the Erie Railroad is 4300 ft. long, 23 ft. high, and 28 ft. wide. It cost about $1,000,000.

The Summit tunnel on the Central Pacific Railroad was 1659 ft. long with one shaft, and cost about $15.00 (gold) per cub. yd. with powder, and $10.00 with nitro-glycerine.

In England tunnels for single track usually cost from $35.00 to $75.00 per running foot. Some have cost as high as $150.00 per running foot.

In earth the mere excavation is a small part of the expense. In one English tunnel the cost of excavating was only about one fourth as much as the propping and arching.

The principal difficulties met with in tunnelling are want of ventilation and drainage. Headings can be driven but a few hundred feet before artificial ventilation becomes necessary. The air being confined is soon rendered impure by the respiration of the men and the smoke of the lamps; and after each blast the smoke of the powder would make it impossible to continue the work for some time. By forcing air through pipes into the heading, the smoke is at once driven out and pure air supplied to the men.

When a tunnel enters a hill on an "up grade" there is no difficulty about drainage but when the work is on a "down grade," or from the bottom of a shaft, the water which collects in the working must be lifted in buckets, or pumped out.

APPENDIX F.

BRIDGES.

General Principles.—A body is to be supported over an inaccessible space. Its weight is a force acting vertically. It can be supported only by the action of two oblique forces, or pieces supporting it by their resistance to compression or extension; *i. e.*, by acting as struts or ties. For example, Figs. 1 and 2.

Fig. 1.

Fig. 2.

Every method of sustaining a material point in space may be reduced to these two. A beam combines an infinite number of pairs of struts and pairs of ties.

A series of points are usually supported, as in Figs. 1 and 2, at such distances apart that a beam resting on two of them will support the load between them. The combinations supporting the several points form a truss, the beam being nature's truss.

Weights to be supported.—The greatest possible load is a crowd of men. Equal 70 pounds per square foot. A drove of cattle is 40 pounds per square foot. A double row of the heaviest loaded wagons, with horses, gives 600 pounds per running foot. Calling each row six feet wide, we have 50 pounds per square foot. A heavy freight train weighs half a ton per running foot. A row of engines weighs one ton per running foot. If the track be double, of course the weight and the strain will be double.

The weight of the bridge itself, is the first thing to be determined in proportioning a bridge. The weight of a good, single-track wooden bridge, per running foot of span = 0.3 ton gross (invariable), + 0.15 ton for a span of 150 ft. (the latter item increasing as the square of the span), + 0.3 ton for the increase of the weight by velocity. This is Trautwine's rule. Thus for a span of 150 feet, we will have: weight of bridge = 150 × (0.3 + 0.15 + 0.3) = 150 × 0.75 = 112½ gross tons = 1630 pounds per running foot.

For two hundred feet span, the second item is obtained thus: $150^2 : 200^2 :: 0.15\,T : 0.222\,T$. Hence, weight per running foot is 0.3 + 0.222 + 0.3 = 0.822 T = 1850 pounds.

An Erie Canal farm bridge of six open panels of 72 feet span, contains 10 cubic feet of timber per running foot. One of 48 feet span contained the same.

A double road bridge over the canal of 90 feet span contains 22 cubic feet of timber per running foot. A railroad bridge of four panels of 30 feet span contains 20 cubic feet per running foot. One of Long's bridges, 100 feet long, contains 42,000 feet B. M., or 35 cubic feet per running foot. A McCallum's bridge of 200 feet span contains 50 cubic feet of timber per running foot, or averages 1500 to 2000 pounds per running foot.

Iron truss bridges weigh from 1000 to 2000 pounds per running foot. Wood weighs from 30 to 60 pounds per cubic foot—average 40 pounds. Cast iron weighs 450 pounds, and wrought iron 480 pounds per cubic foot.

In calculating bridges, assume some approximate weight for the bridge, and after determining the necessary sizes (and consequently weights) of the different parts, the calculations should be again made, using the weight of the bridge just found.

Classification.—Bridges will be here classified; First, As to their material: As wood, stone, iron, or brick, and these subdivided; Secondly, As to the manner in which their points are supported, viz., Trabeate, Arcuate, or Suspension.

WOODEN BRIDGES.

Trabeate.

The simplest form of a bridge is a plank. Next to it is a pair of timbers with planks crosswise upon them. For a hasty bridge, two trunks of trees with smaller trunks laid across them.

In calculating bridges of this kind, the timbers are considered to be supported at both ends and loaded uniformly with the greatest load that can come upon them. To determine the weight any given beam in this condition will bear, or the size necessary to bear a given weight, we have the formula,

$$W = 2s\frac{bh^2}{L}.$$

In which W represents the breaking weight, b and h the breadth and height in inches, L the length in feet, and s a coefficient found by experiment. For common American timber s varies from 300 to 700; for white pine it is 410; for white oak, 580; for tamarack, 300; for hemlock, 380; for red pine, 510; for white ash, 600; for hickory, 700. The safe load may be taken at $\frac{1}{5}$ the breaking weight.

When a span becomes too long for a single beam, "corbels" are used, as in Fig. 3. The strain on them is like that of a beam fixed at one end and loaded uniformly. They may be strengthened by struts.

Fig. 3.

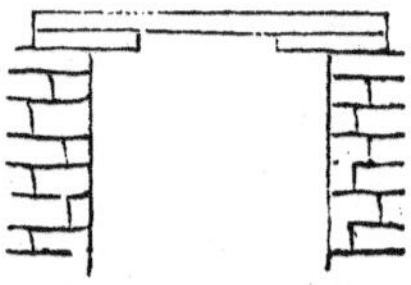

A series of such bridges resting on simple piles, is known as "*pile bridging.*" It is the usual way in which railroads cross shallow waters. When high above the water they are called "*trestle work*" bridges.

Fig. 4 shows the arrangement of the piers for the railroad pile bridge across the South Platte. The piers are placed sixteen feet apart. There are four piles about one foot in diameter in each pier; the middle ones being 5 feet apart, and the outside ones 4½ feet. Sway braces, 3″ × 10″ and 14 feet long, are bolted to the piles.

Fig. 4.

The greatest trestle bridge is the Portage High Bridge. It is 800 feet long and 190 feet high. Contains 1,600,000 feet B. M. of wood, and 109,000 pounds of bolts.

Comparative cost of trestle work and earth work. For five feet high, trestle work costs four times as much as earth work. The Portage Bridge cost only a quarter that of earth work. The cost of earth work increases nearly as the square of the heights. The cost of trestle work increases *about* as the $\frac{3}{2}$ power. Making the heights the abscissas, and the cost the ordinates, the curve for the earth work will be a common parabola, and for trestle work a semicubical parabola.

CLASSIFICATION OF BRIDGE FRAMES.

CLASS I. *The oblique pieces all resist compression.*

1. The weight is transferred to the abutments directly.
2. The weight is transferred to the abutments indirectly; as Long's, Howe's, etc.
3. Combinations of sub-classes 1 and 2; as Latrobe's.

CLASS II. *The oblique pieces all resist extension.*

1. The weight is transferred to the abutments directly; as Bollman's, etc.
2. The weight is transferred to the abutments indirectly, being conveyed to the oblique ties by vertical posts; as Linville's, etc.
3. Combinations of sub-classes 1 and 2; as Fink's.

CLASS III. *The oblique pieces are alternately compressed and extended.*

The weight is transferred to the abutments indirectly; as Warren's, etc.

Arched Bridges are polygonal modifications of Class I.

Suspension Bridges are polygonal modifications of Class II.

CLASS I. *The oblique pieces all resisting compression.*

1. When the weight is transferred to the abutments directly, as in Figs. 5, 6, 7, etc.

Fig. 5.

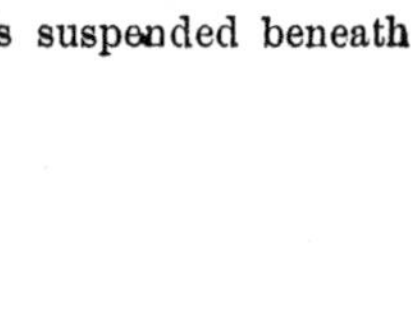

In Fig. 5, the middle point being supported, twice the span can be obtained with the same strength of beam. One-half the weight is borne by the struts. In calculating the strain on a pair of struts, as A B and A′ B, the effect is the same, whether the weight rests directly on the top of the struts or is suspended beneath them.

CALCULATIONS.

Construct a parallelogram, having for one diagonal a line B F, representing the number of units in the weight, and having its sides parallel to the struts. These sides will represent the strains on the beams to which they are parallel. Let t represent the thrust in the direction of their length, h the horizontal thrust, and W the weight. Then we have,

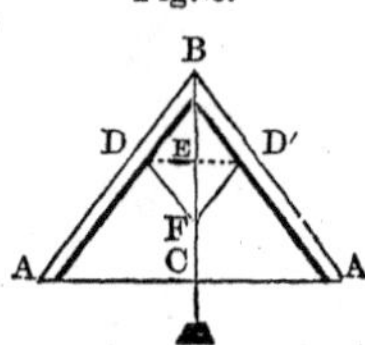

$$W : t :: BF : BD;$$
$$:: 2BE : BD,$$
$$:: 2BC : BA.$$

Hence, $t = \frac{1}{2} W \frac{BA}{BC} = \frac{1}{2} W \frac{\text{length}}{\text{rise}} = \frac{1}{2} W \text{ cosec. } A.$

By similar triangles,

$$BE : DE :: BC : AC. \qquad \frac{1}{2} W : h :: \text{rise} : \frac{1}{2} \text{ span}.$$

Hence, $h = \frac{1}{4} W \frac{\text{span}}{\text{rise}} = \frac{1}{2} W \text{ cot. } A.$

Any change in the obliquity of the beams increases or diminishes the strains, as can readily be seen from the formulas. The length of the beams has no effect on the stresses, but the strength of a beam decreases as its length increases.

NOTE.--Safe loads for wooden posts, whose crushing load is 6000 lbs. per square inch.

Ratio of length to side.	Safe stress in lbs. per sq. inch.
1	900
10	800
15	690
20	560
25	420
30	340
35	280
40	225
45	175
50	130
55	105
60	85
65	70
70	60

If the post is not square, take the ratio of the length to the smallest dimension.

Two points may be supported in space, as in Fig. 7. Each pair has one long and one short strut; we will calculate one pair separately. Let the pair at the left be represented by Fig. 8.

Fig. 7.

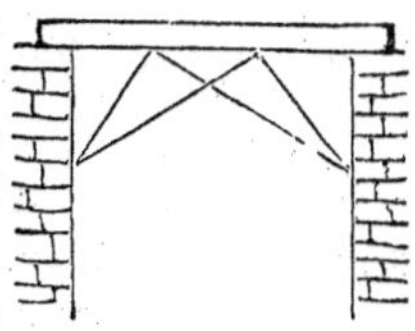

Let the wt. = 1500 lbs. Let each inch of the diagonal represent 500 lbs. Let the angle of the beams = 100°, the one making an angle of 60° with the vertical, and the other an angle of 40°. Required the strain on the two beams, A B and A C. (The Fig. is not drawn to scale.)

Fig. 8.

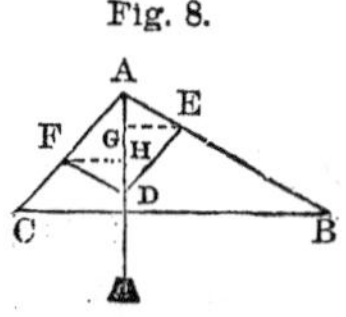

1. *Graphically.* Set off on the vertical 3″, and complete the parallelogram as before. The proportions of the wt. borne by B and C are, respectively, A G = D H and A H = D G. When B and C are at different heights, to find the portion of the wt. borne by each, draw the lines, not horizontal, but parallel to B C.

2. *Trigonometrically.*

$$\text{W} : \text{strain on AB} :: \text{AD} : \text{AE} :: \sin.\text{AED} : \sin.\text{ADE}.$$

Hence, $$\text{strain on AB} = \text{W}\,\frac{\sin.\text{ADE}}{\sin.\text{AED}}.$$

Again, $$\text{W} : \text{strain on AC} :: \text{AD} : \text{AF} :: \sin.\text{AFD} : \sin.\text{ADF}.$$

Hence, $$\text{strain on AC} = \text{W}\,\frac{\sin.\text{ADF}}{\sin.\text{AFD}}.$$

Substituting these values in the formula, we have,

$$\text{strain on AB} = 1500 \times \frac{\sin.\ 60^\circ}{\sin.\ 80^\circ} = 1319;$$

$$\text{strain on AC} = 1500 \times \frac{\sin.\ 40^\circ}{\sin.\ 80^\circ} = 979.$$

Resolving these strains into their vertical and horizontal components, we shall find that the horizontal pressure of one of the beams exactly equals that of the other, whatever be the difference of their inclination to the vertical, and that the sum of the two vertical components equals the whole weight. The numerical calculation of these components is made as before trigonometrically.

When the span and heights of the struts, and hence their lengths are given, we have more simply: the weight supported at either end $= W \times \frac{\text{non-adjacent segment}}{\text{whole span}}$.

That is the Weight supported at $B = W \times \frac{KC}{BC}$;

Weight supported at $C = W \times \frac{BK}{BC}$;

Thrust on AB = weight on $B \times \frac{AB}{AK}$;

" " AC = weight on $C \times \frac{AC}{AK}$.

Horizontal thrust = weight on $B \times \frac{BK}{AK}$ = weight on $C \times \frac{KC}{AK}$.

Instead of supporting the two points by two pairs of struts, as in Fig. 7, we may employ two struts, and a straining beam, Fig. 9. The calculations are the same, as for Fig. 12.

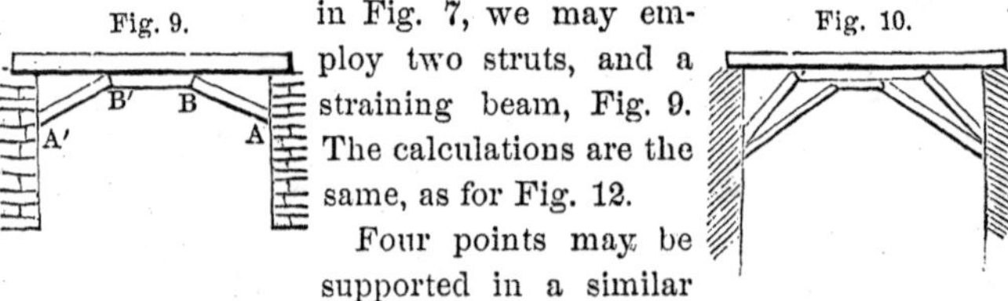

Fig. 9. Fig. 10.

Four points may be supported in a similar manner by two pairs of struts and two straining beams, as in Fig. 10.

This principle may be extended to great spans, but long timbers are hard to get and are weaker. A bridge at Wettingen, on this plan, has a span of 397 feet.

In many cases, supports from below may be objectionable, as exerting too much thrust against the abutments, and being liable to be carried away by freshets, etc. The beams must in such cases be strengthened by supports from above, as in the following class.

2. When the weight is transferred to the abutments indirectly, being conveyed to the oblique struts by vertical ties.

The simplest form of this class is a pair of struts, Fig. 11. The calculations for this are the same as for Fig. 5.

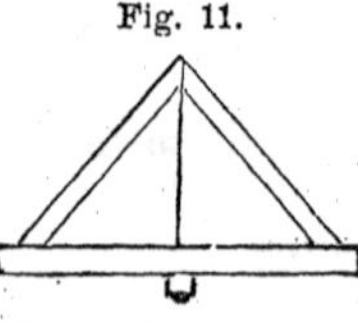
Fig. 11.

Bridges built on this plan can be used on railroads for spans of from 10 to 25 feet. For longer spans (say up to 35 feet) two struts and a straining beam may be used, Fig. 12.

Calculations. With loads W and W′ on C and C′, the stresses are analogous to Fig. 6. Consider the upward reaction at A and A′, which are each equal to W. Then, the triangle, A B C, gives stress on A C $= W \frac{AC}{BC}$; horizontal stress $= W \frac{AB}{BC}$. All the stresses are the same as if the struts, A C and A′ C′, were produced upward to meet, and the whole load (W + W′) placed at that point, as in Fig. 6.

Fig. 12.

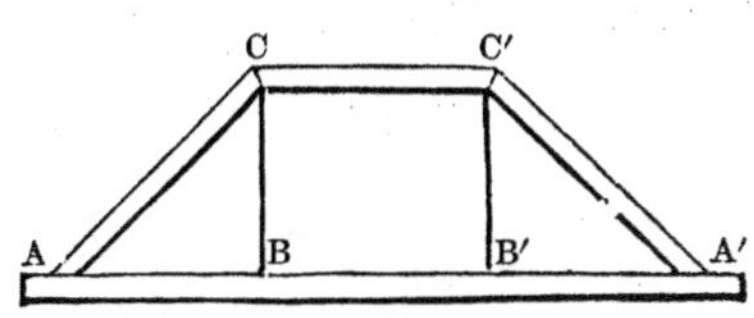

Suppose the load on A A′ to be uniform, and supported by rods, B C and B′ C′; A A′ not to be continuous, but to be divided at B and B′. This corresponds to the weight on C and C′ in Fig. 12. If the beam, A A′, be continuous and level, then $\frac{11}{30}$ W is on B C and B′ C′, and $\frac{4}{30}$ on A and A′. This is safest to take, being greatest, though it is not generally done. In either case, calling W′ the load on B C or B′ C′, then the horizontal thrust on C C′ and the pull on A A′ $= W' \frac{AB}{BC} = W' \frac{1}{3} \frac{AA'}{BC} = \frac{1}{3} W' \frac{\text{span}}{\text{rise}} = \frac{1}{9} W \frac{\text{span}}{\text{rise}}$. The thrust on C A and C′ A′ $= W' \frac{AC}{BC}$. If a similar load were on top, the stresses would be the same.

When a bridge of this form is reversed the stresses remain the same except that the former stresses of compression have become extension, and *vice versa.* This arrangement may be extended to any number of panels. It is preferable for materials, like wood and wrought iron, because the shortest pieces are exposed to compression. A large number of points may be supported in the same way.

Suppose a uniform load, W, on a beam, A A′, Fig. 13. Each post or vertical tie supports $\frac{1}{4}$ W = W′. The struts, E B and E B′ resist a stress $= \frac{1}{2} W' \frac{BE}{CE}$, as found

Fig. 13.

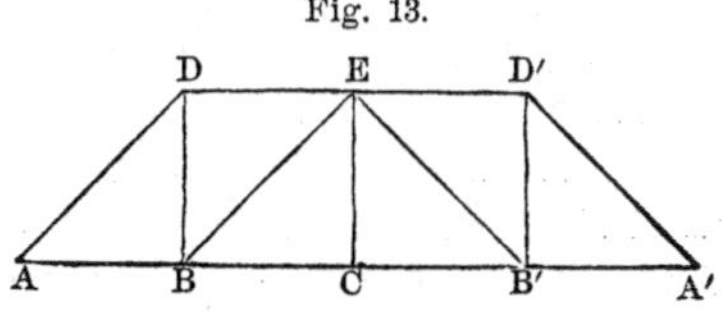

from Fig. 6. They produce a horizontal strain at E and on B B′ $= \frac{1}{2} w' \frac{BC}{CE}$. The weight on D or D′ $= w' + \frac{1}{2} w' = \frac{3}{2} w'$ ($\frac{1}{2} w'$ being transferred from E). Consequently, D A and D′ A′ have a stress $= \frac{3}{2} w' \frac{AD}{BD}$. They produce a horizontal strain on D D′ and A A′ $= \frac{3}{2} w' \frac{AB}{DB}$. The horizontal strain B B′ = the sum of the two horizontal strains $= (\frac{1}{2} w' + \frac{3}{2} w') \frac{1}{4} \frac{\text{span}}{\text{rise}} = \frac{1}{2} w \frac{\frac{1}{4}\text{ span}}{\text{rise}} = \frac{1}{8} w \frac{\text{span}}{\text{rise}}$. So proceed for any number of panels. It will be found that the strains on the posts and on the struts increase in a direct ratio to the distance from the centre. Their strength and size should, therefore, be increased in the same ratio. The strains on the top and bottom beams increase from the ends to the middle, but not in a direct ratio. The increase is most rapid as you proceed from each end, and becomes less rapid on approaching the centre or middle. It is analogous to that of a solid beam, in which latter case the relative increase is indicated graphically by a parabolic curve.

The usual formula for the horizontal stress on a frame, caused by a uniform load ($\frac{1}{8} w \frac{\text{span}}{\text{rise}}$), supposes the weight to be uniformly applied at the ends of the struts, as well as distributed uniformly over the roadway. This is the case in frames which have an even number of panels; but is not so with those of an uneven number. For example, with three panels, the horizontal stress $= \frac{1}{9} w \frac{\text{span}}{\text{rise}}$; for five panels $\frac{1}{8\frac{1}{3}} w \frac{\text{span}}{\text{rise}}$; for seven panels $= \frac{1}{8\frac{1}{6}} w \frac{\text{span}}{\text{rise}}$; and generally for n panels (n being uneven) $= \left(\frac{1}{8} - \frac{1}{8n^2}\right) w \frac{\text{span}}{\text{rise}}$.

We now see that all truss bridges are composed of three systems or sets of pieces: 1. Chords or stringers, horizontal, or nearly so. 2. Ties or posts, vertical, or oblique. 3. Struts.

With these three elements bridges may be constructed of several hundred feet span, and bear safely a load uniformly distributed: but unless very heavy they will not bear safely a partial load.

Counter-bracing. A bridge uniformly loaded tends to assume the form indicated by dotted lines in Fig. 14, in which the rectangular panels become rhomboids. This tendency is resisted by the struts which must be compressed or broken before this tendency can be carried out.

Fig. 14.

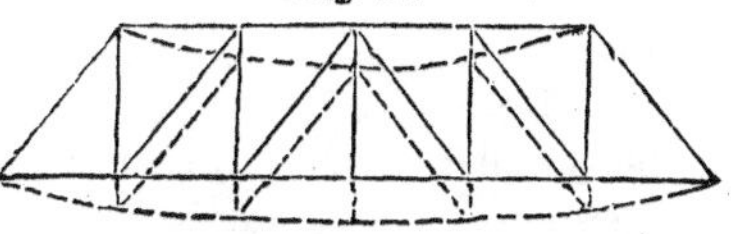

But let a passing load be at some point, C, of the bridge, being supported finally by the points A and B. The directions of its pressure are C A and C B, and the force, C A, tends to make the bridge rise at D, and to assume the form shown by the broken lines in Fig. 15. The struts do not resist this action, for they now occupy the long diagonals. This tendency must be resisted by fastening the ends of the struts to the chords, or putting tie-rods beside them, or as is most usual by *counter-braces*, *i. e.*, braces placed in the other diagonals of the panels. The bridge cannot now rise as indicated in Fig. 15, without breaking or bending these counter-braces.

Fig. 15.

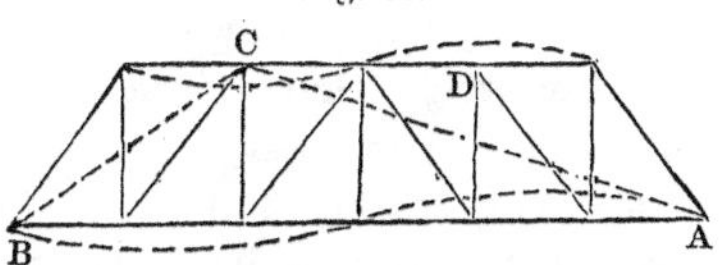

This counter-bracing, therefore, checks the up-and-down vibrations of a bridge, and renders it stiff against *passing* loads, while the main braces give it strength to bear uniform loads. In very heavy bridges their weight may render counter-bracing unnecessary.

The strain on a counter-brace equals the greatest weight which can ever press upon any point of the bridge, multiplied by the length of the counter-brace, divided by its height. In a railroad bridge this greatest weight would be the load on a pair of drivers of an engine. On a common road bridge it would be the greatest load between a pair of posts. This system of counter-braces was first fully carried out by Colonel Long.

LONG'S BRIDGE.

The joints and fastenings are simple, the strain on the timber is direct, and any piece can be easily removed and replaced. All the principal pieces are of timber. In very long spans, struts (called arch-braces) are placed under the ends, and a roof truss in the middle of each truss, or a pair of struts and a straining beam along each side.

The peculiarity of Long's bridge is in the mode of keying the counter-braces. They are keyed or wedged so strongly that the string-pieces are constantly pressing against them, and when a load comes on the bridge its only effect is to relieve the counter-braces from the pressure against them and to transfer it to the main braces.

Thus there is no more strain on the bridge when fully loaded than when unloaded; only the strain is on different parts. The effect of this mode of keying is the same as if the string-pieces had been originally curved upward, or arched, and then brought down straight by weights hung to them, the counter-braces then wedged tight, and finally the weights removed.

A load now coming on the bridge puts it in the same condition as it was before these imaginary weights were removed, *i. e.*, it takes the strain off the counter-braces. There is, therefore, a constant pressure which makes the bridge very stiff.

HOWE'S BRIDGE.

In this an iron rod replaces the vertical post. These bridges are very generally used, but are not durable. The expansion and contraction of the rods strain the bridge out of shape and require constant screwing up. Extra struts at the end are usually added, sometimes extending to the abutment under the bridge. An improved form of angle block is now used to prevent crushing the lower chord by the nut.

McCALLUM'S BRIDGE.

Its peculiarity is that the upper chord is arched. The ends are also strengthened by struts, or "arch-braces" (so called), thrusting against the abutments. This bridge is very stiff, but uses much timber. Altogether it is one of the very best railroad bridges. Its counter-braces are adjusted by screws. Sometimes iron rods are added near the ends, so as to suspend that part of the bridge.

LATROBE'S BRIDGE.

In this bridge two systems are combined; viz., that of long struts, transferring the weight directly; and that of struts and tie-rods, transferring the weight indirectly. The advisability of any such combination is questionable, owing to the impossibility of so adjusting the two that they shall bear their exact proportions of the load.

Class II.—*The oblique pieces all resist extension.*

This is rarely used for wooden bridges; chiefly for iron bridges. Hall's and Pratt's bridges belong in this class. The principle is good; the shorter pieces being compressed in which way timbers resist most advantageously.

In calculating the stresses on oblique ties, we apply the same formulas as for struts. The strain is now one of extension instead of compression. For a pair of ties, the horizontal strain produced by a weight, $w = \frac{1}{4} w \frac{\text{span}}{\text{depression}}$, and the strain on each tie $= \frac{1}{2} w \frac{\text{length}}{\text{depression}}$.

Class III.—*The oblique pieces are alternately compressed and extended.*

The weight is transferred to the abutments indirectly.

In the preceding forms the ties were vertical and the struts inclined. In Fig. 16 both ties and struts are inclined. The stresses, however, follow similar laws. With a uniform load, w, such as its own weight, the vertical strain increases uniformly from the middle, where it equals zero toward the end where it equals $\frac{1}{2}$ w. At x' from end, or x'' from middle, it equals $w \frac{x''}{\text{span}}$. The strain on any diagonal whose middle is x'' from the middle of the bridge $= w \frac{x''}{\text{span}} \times \frac{\text{length of diagonal}}{\text{depth of panel}}$. The horizontal strain at the centre $= \frac{1}{8} w \frac{\text{span}}{\text{rise}}$. At any point x' from middle or x'' from end,

Fig. 16.

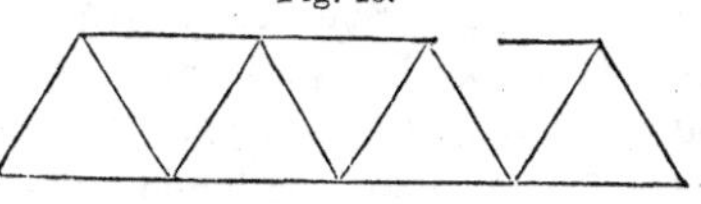

it $= \frac{w}{2} \times \frac{x' (\text{span} - x'')}{\text{span} \times \text{depth}}$, diminishing from the centre to the ends in the ratio before shown by a parabolic curve.

When the loads are applied along the top or bottom, or along both. The distance x' and x'', in the preceding formulas, are measured to the tops of the diagonals, when the load is attached to the bottom of the beam; to the lower ends, if it be on the top; and to their middle, if the load be equally on the top and bottom, as its own weight.

Bridges of this form are called "Triangular girders," or "Warren's," or "Neville's." If the number of oblique pieces be doubled, then each sustains half the above strains.

TOWN'S LATTICE.

This is a lattice of common plank. It is easily made, but, though strong, is deficient in stiffness. The material is not advantageously disposed, too much of it being near the "neutral axis." It soon gets loose and sags, or twists sidewise, *i. e.*, buckles. It is sometimes strengthened by long struts and straining beams, or by arch ribs. To calculate a lattice bridge, consider the truss a solid beam with holes cut out of it where the spaces in the lattice are.

MISCELLANEOUS DETAILS.

1. *The ratio of height to length.* This is important. The most economical is $\frac{1}{6}$. Short spans, requiring great strength, may have $\frac{1}{4}$. In long spans this would give the wind too much hold, and the sides would twist or buckle. Then for great strength use $\frac{1}{6}$ or $\frac{1}{7}$, while for moderate length and stress $\frac{1}{10}$ may do.

2. *Horizontal braces or sway-braces.* They are to prevent lateral flexure. The greatest possible strain on them is the wind, which operates as a uniform load. They are shown in plan in Fig. 17.

Fig. 17.

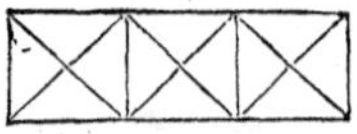

3. *Stiffening the sides.* When the roadway is on top ("Deck bridges") use transverse vertical bracing, extending from the top chord of one truss across to the bottom chord of the other truss. When the road is not on top, extend the needle-beams beyond each side of the bridge, and brace the top chord from it; otherwise make gallows frames.

4. *Wedging up the ends of the lower chords.* This produces an initial strain of compression, which the stress of the load must overcome before it begins to bring a strain of extension upon this lower chord. The lower chord then acts somewhat as an arch. An objection is that it makes the strength of the bridge depend upon the resistance of the abutment.

5. *Double roadway.* In important bridges it is best to have each track separate to prevent a one-sided strain.

6. *Durability.* An uncovered wooden bridge is seldom safe for more than eight or ten years. If covered, sided, and well painted, it may last thirty or forty years. Some have been used fifty or sixty years.

WOODEN ARCH BRIDGES.

A beam resting on two supports, sustains a load by the compression of its upper fibres, and the extension of its lower fibres. If we confine the ends of the beam by immovable obstacles, these will be substitutes for the tension of the lower fibres, which may therefore be removed without lessening the strength of the beam, as may also the extreme portions of the upper fibres. So too a board laid on two supports will bear a certain weight. Bend it up and confine its ends and it will bear a much greater weight. This principle may be adopted in building bridges of considerable span. Strong, cheap bridges may be made by forming an arch of planks. One such, with a span of 130 feet, rise 14 feet, was formed of 3″ plank in 15 layers and 30″ wide. Three locomotives on it caused a deflection of only ⅜″. The roadway may pass either over the top, resting on posts and struts, or be at the springs and thus act as a tie-beam, being suspended from the arch. It is then called a "Bowstring" bridge.

Perhaps the strongest and cheapest form of bridge, where abutments can be obtained, would be a parabolic arch, increasing in cross-section from crown to spring, according to stress, and stiffened by counter-bracing. The counter-braces may be wedged down, as in Long's bridge, and thus made very stiff, as well as strong.

Double or parallel arches are always bad. Suppose the "neutral axis" to pass near the middle of the lower arch rib. Only half the strength of the timber is used, being the upper portion of the upper arch rib, and the extreme portion of the lower arch rib. The Erie Railroad Cascade Bridge was built on this plan. Span, 275 feet; rise, 45 feet.

Combination of an arch and truss. This is much used, and its expediency is advocated by some eminent engineers. There are, however, grave objections. It is impossible so to combine them that the arch and truss shall each bear its due share of the pressure. One will give way before the pressure comes on the other. One of the best combinations is Burr's bridge. The relative stress on the arch and truss, of a combination, may be so adjusted by set-screws as to throw any desired portion of the stress upon either the arch or the truss; but this ratio will be changed by every passing load, and by every change in the temperature.

If an arch be used, and the abutments will allow, it is best to depend for the whole strength upon it, and to employ a truss merely to stiffen it.

Wooden Suspension Bridges. Wood is rarely employed in this way, notwithstanding its greater strength to resist extension than compression, because of the loss of material caused by the necessary bolts and straps. A bridge on this plan was built by Burr across the Mohawk at Schenectady, N. Y., in 1808, and is still (1871) in use.

Lavé's Bridge. This is a combination of a wooden arch and suspension bridge. Fig. 18. A timber is sawn nearly through lengthwise, its ends confined, and the middle portions are wedged apart. It will now bear a much greater load than before. Two timbers may be thus combined. For great spans, the upper and lower portions may be formed by splicing timbers. The principle is good. It is recommended for military bridges.

Fig. 18.

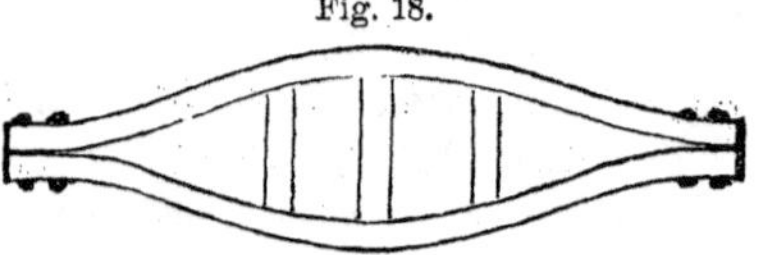

Wooden bridges have been extensively used in this country, on account of their cheapness; timber being plenty and capital limited. They are, however, faulty from their elasticity and consequent vibration, and their perishable nature.

Iron bridges are employed with great success, and their use is increasing. They have the requisite rigidity; and although the first cost is greater than for wooden bridges, their imperishable nature, if well cared for, renders them, in a majority of cases, most economical.

IRON BRIDGES.

They are divided, like wooden, into Trabeate, Arcuate, and Suspension. The stresses, strains, and calculations are, of course, the same for them as for wooden bridges; only using the experimental coefficient of strength for iron, cast or wrought, instead of that for wood.

CLASSIFICATION.

I. TRABEATE.

1. *Cast iron girders.*
   Simple girders. Built girders. Trussed girders.
2. *Wrought iron girders.*
   I-shaped beams. Box or tubular girders.
3. *Wrought iron truss work.*
   Post's, Fink's, Bollman's, Whipple's, Rider's, Heath's, etc., etc.

I. TRABEATE IRON BRIDGES.

1. CAST IRON GIRDERS.

Relative strength of cast iron beams. Fig. 19 (*a*) is a cross-section of a beam made by Boulton & Watts in 1801. It was improved by Fairbairn in 1825, the vertical rib being made thinner and the lower flange thicker (*b*). Tredgold's beam (*c*) has equal upper and lower flanges. The strongest form is Hodgkinson's (*d*), the lower flange being six times the upper one. The relative strength of these beams, Hodgkinson's being taken as unity, is: Boulton & Watts', 0.51; Fairbairn's, 0.75; Tredgold's, 0.62; and Hodgkinson's, 1.

Fig. 19.

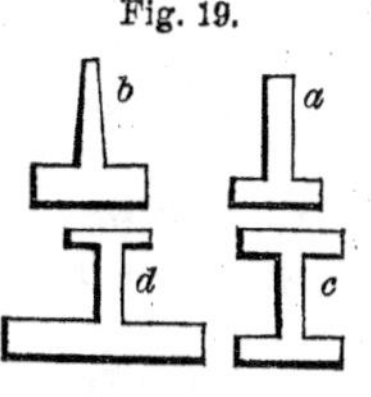

For short distances, a single girder may be used for each rail. For 30 or 40 feet, use two girders on each side, with a timber between them to carry the rail.

The greatest possible load should not exceed one-sixth the breaking weight. The test load should be about twice the greatest load, or about one-third the breaking weight. The deflection under the permanent load should not be more than $\frac{1}{600}$ of the length. A "camber" of 1 in 300 should be used. One girder 76 feet long has been cast.

Built girders. For spaces too long for simple girders, built girders are used, fitted closely at the joints with flanges there bolted together. Spans of 120 feet have been thus crossed.

Fig. 20. Fig. 21.

Trussed girders. Cast iron beams sometimes have wrought iron tension rods applied to them, as in Figs. 20 and 21, with the object of strengthening the lower flange; the two rods helping it to resist extension. The rods are tightened by screws or wedges, so as to have any amount of initial tension in advance; but it is difficult so to adjust the two, that each shall bear its share of the strain; and even if this adjustment were once made it would be altered after any strain, owing to the different "*sets*" of wrought iron and cast iron. Since for respective stresses equal to $\frac{1}{3}$ breaking weight for each (say 5 tons per square inch for cast iron and 15 tons for wrought), the elongation for wrought iron is $2\frac{1}{2}$ times that of cast, and its *set*, 10 times as great as that of cast. This adjustment, and with it the strains coming on each, would also vary with every change of temperature, since wrought iron expands with heat more than cast iron. The combination is therefore bad. Cast iron is never *safe* for girders; wrought iron should be used.

Wrought iron bridges. The resistance of wrought iron for railroad bridges is safely 8600 pounds per square inch, or about $\frac{1}{7}$ of its breaking weight. For common road bridges, 11,400 pounds. These are safe limits. In England, 11,400 is used for railroads, and 18,000 for cast iron. The greatest possible load for an iron railroad bridge is in Austria called 2800 per running foot for each track; in Russia, 1600; in France, 2700; in England, 2300 to 3400.

The proper trial load may be from 40 to 80 pounds per square foot of roadway, according to the probabilities and importance of the bridge.

In France, iron railroad bridges are by law tested thus: For spans under 64 feet, 3300 pounds per running foot; and for spans over that, 2640 pounds per running foot is used; but the load in this last case must be at least 200,000 pounds.

Wrought iron resists extension much more than compression, therefore the compressed parts of wrought iron beams (the upper flange of a beam supported at both ends) should be nearly as 2 : 1. They are usually made nearly the same, since for small strains its resistances are about the same.

Bridges of I-*shaped beams.* Up to $2\frac{1}{2}$ feet, a single rail would answer. From that to five feet, double rails, bolted together by the lower flange. A common form for a wrought iron girder is shown in Fig. 22. The dimensions will, of course, depend on the span. The usual ratio of depth to span is about 1 to 14. Parabolic girders have been used.

Fig. 22.

Box or tubular girders. The ultimate tenacity of plate girders with double riveted covering plates, is 45,000 pounds per square inch of cross-section. The ultimate resistance to crushing is 36,000 pounds per square inch. One such bridge has a span of 150 feet, the girders being 12 feet high and 3 feet wide.

Another, of two tubular beams, of 170 feet span, weighed 130 tons, gross, and cost $100 per ton, equals $76 per foot. Another, of one girder of 76 feet span, cost $100 per ton, and $42 per foot.

When the beams are small and liable to give way by bending, use the formula for wrought iron posts. When the thickness of the plates is not less than $\frac{1}{30}$, the diameter of a square tube, the ultimate resistance of it to buckling or bending is 27,000 pounds per square inch of the cross-section. Fig. 23 shows the common form of tubular girders. For small spans each line of rail rests on a tube. For greater spans, each line of rails

Fig. 23.

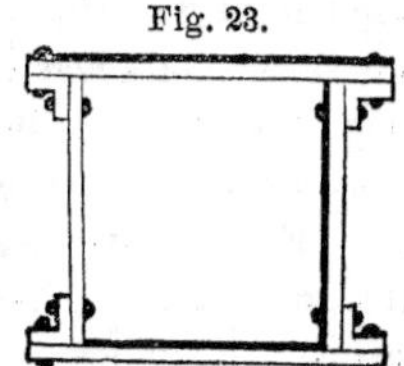

will have a pair of them. For still greater spans, the roadway may go through the tube. For example, the Britannia bridge.

The *Britannia tubular bridge*, over the Menai Straits, has two spans each of 460 feet, and two of 250, its total length being 1500 feet. Its tubes are 30 feet high and 14 wide. Its top and bottom are cellular, being composed of two parallel sheets, 18 inches apart, and connected by cross-plates which form a series of square cells or tubes. The material is boiler iron, from $\frac{3}{8}$ to $\frac{3}{4}$ inch thick, in sheets united by two million rivets, and stiffened by sixty-five miles of angle iron. Heavy trains daily cross it, with scarcely perceptible vibration. But its cost, $2,500,000, must always render it more a subject of admiration than of imitation.

The *Victoria bridge* at Montreal is on the same plan. The centre span is 330 feet, and 12 spans on each side, each 242 feet. The plates of tubular bridges *should* vary in thickness in the same ratio as the chords and braces of truss bridges.

TRUSS WORK.

Rider's truss. This is Long's bridge in iron.

Heath's strut truss. This is built of sheet iron, stiffened by T irons.

NOTE.—For a discussion of the comparative merits of the Fink, Bollman, Jones, Murphy-Whipple, Post, Triangular, and Linville trusses, see Col. Merrill on "Iron Truss Bridges for Railroads."

Triangular girders. On this plan is Brunel's "Crumlin Viaduct." It has 10 spans of 150 feet each. Each span composed of nine equilateral triangles 15 feet high. Piers 200 feet high, of cast iron columns strongly tied together.

The best angle for the struts and ties is 45°. Depth usually $\frac{1}{12}$ to $\frac{1}{15}$ the length.

Lattice bridges. In a good one of six spans, each 90 feet in clear, the height was 10 feet. The angles were 45°. Width of upper and lower stringers was 10″. Thickness $2\frac{3}{8}$″, made of three bars superimposed. Lattice bars $3\frac{1}{2}$″ broad and $\frac{5}{8}$″ thick. Distance apart from centre to centre 13″. Riveted at every crossing. Distance from rivet to rivet was 18″. The objection to these bridges is, that they are liable to buckle. There is considerable competition between the advocates of these and boiler-plate bridges.

II. Arcuate Iron Bridges.

1. *Cast iron arch.* This is the strongest of all forms of cast iron bridges. Whipple's arch truss is one of this class.

An arch formed of cast iron tubes, through which the water passes, serves as both a bridge and conduit on the Washington Aqueduct. Span 200′, rise 20′, diameter of tube 4 feet.

Wrought iron arches are usually of the bow-string form.

The steel arch bridge across the Mississippi, at St. Louis, is to have three spans, the middle one being 515 feet.

Lave's form. The greatest one is Brunel's Saltash bridge. It has two spans of 445 feet each.

SUSPENSION BRIDGES.

Various plans are proposed for stiffened suspension bridges for railroads; among them are these:

1. Adding a heavy and stiff platform.
2. Connecting a truss with the chain. (Niagara.)
3. Making the chain itself a truss. (See Latham, plate II.)
4. Suspending many points of the platform directly from the piers. (Dredge's plan.)
5. Sustaining the bridge and load as in Bollman's bridge, the rods themselves being supported by a chain. (Ordish's plan.) See Latham, plate VI.
6. Applying stay rods. (Niagara.)

Comparison of a suspension bridge and a girder. Suppose them each of 400 feet span and 40 feet deep. The weight of a chain of proper strength would be about 260 tons. The weight of a girder of equal strength would be about 900 tons. Under a stationary load, the former would deflect about twice as much as the latter. Under a moving load, such as would cause a wave of 2 feet on the former and 3 inches on the latter, the shock to the structures would be 128 times as great on the suspension bridge as on the other. Also oscillations tend to accumulate on the suspension bridge.

Sections to give the chains. The French government rule is this: On trials, apply 40 pounds per square foot of platform. The tension shall not exceed for bar iron $\frac{1}{3}$, and for wire $\frac{1}{4}$ the breaking weight, which is a tension corresponding to about 17,000 pounds and 26,000 pounds per square inch respectively. The strain of the unloaded bridge is about $\frac{1}{2}$ this.

Roebling allows 7 wires of $\frac{1}{8}''$ diameter for each ton of maximum tension, equivalent to 320 pounds per wire, or $\frac{1}{3}$ the breaking weight. The constant load is $\frac{1}{5}$ breaking weight. The vertical suspending rods are loaded to only $\frac{1}{10}$ breaking weight, being exposed to shocks. The weight of the cable increases as the square of the span.

Possible length. They might be built of one mile length or span. For example, a No. 10 wire will support safely a strain of 500 pounds, its breaking weight being three or four times that. Such a wire suspended over a span of 4000′, with a versed sine of 500′, would have a tension of only 212 pounds, or $\frac{1}{8}$ breaking weight. Such a wire would bear its own weight across a span of three miles, with a versed sine of $\frac{1}{12}$ that.

The East River Suspension Bridge, connecting New York and Brooklyn, is to have a single span of 1600 feet.

STONE BRIDGES.

The *bridges* necessary on railroads, when of stone, present peculiar difficulties in their construction. This is owing to the frequently unavoidable *flatness* of the arches (a characteristic which it is not easy to unite with sufficient strength, both in reality and in appearance), and to the *obliquity* with which they often cross other roads, and which compels the employment of "skew-arches," which require more than ordinary skill in both the engineer and the builder.

MOVABLE BRIDGES.

I. *Turning bridges.*
    1. Turning on one end.
    2. Turning on the centre, or pivot bridges.

II. *Lifting bridges.*

III. *Sliding bridges.*
    1. Raise or lower one end of the draw, and shove it back on rollers.

2. Shove the roadway sideways, to make room to shove the draw back.
3. Shove the draw sideways and then run it back.

IV. *Floating bridges.*

1. Boat bridges.
2. Pontoon bridges.

APPENDIX G.

SPECIFICATIONS.

In making out specifications for the execution of any work, everything should be plainly expressed and nothing left to be inferred.

SPECIFICATIONS FOR GRADING.

1. Description of the Work.

This usually refers to the maps and plans defining the centre line, cross-section, true grade, and sub-grade. Grade is the top of the bank, as completed and ballasted. Sub-grade is from one to two feet below this. It is the top surface of the earthwork before the ballast is put on.

2. Preliminary Work.

Clearing. All trees, logs, brush, and other vegetable matter to be removed from the ground on which the banks are to be placed.

Grubbing. All stumps and large roots to be grubbed out, the entire width of the work.

Mucking. All soft earth to be removed down to two feet below sub-grade.

3. Excavation.

All the dimensions should be given, *i. e.*, width, side slopes, etc. Also the distance below grade to which the excavation is to be made to allow for ballasting. Ditches must be cut along the top of the slope to protect the slopes of the cut. Their size to depend on how much will be required of them.

Classification. One railroad divides the material only into earth and rock, the former including everything except rock in ledges or boulders measuring more than 10 cubic feet.

Another road divides it into earth (including "hard-pan" and stones less than 1 cubic yard), loose rock (comprising detached stones of 1 cubic yard and over), and solid rock (embracing all rock in ledges).

Another road has also three classes: solid rock, or that requiring blasting; soft or rotten rock, requiring the bar and pick, but not blasting; and earth. Detached stones less than 3 cubic feet come in the last class; and those between 3 and 20 cubic feet in the second class.

On the Erie Canal enlargement the classes were: common earth, hard-pan, quicksand, slate rock, and solid rock.

How measured and paid for. Excavation is sometimes paid for in the cut, and sometimes in the bank.

The *average haul* should be named; also the distance beyond which extra hauling is paid for. On one road this limit was 1500 feet, and beyond that the contractor was paid $\frac{3}{4}$ ct. per yard per hundred feet. On another road the "haul" was 1000 feet, and the contractor received 1 ct. per yard for every additional hundred feet.

Usually the grade is so established as to make the "cut" and "fill" nearly balance, and the whole work is measured in the cut and is paid for as excavation only, unless the "haul" exceeds the limit named in the specification, in which case the extra hauling is paid for. In case the cut does not quite make the fill, the extra material is measured in the "borrowing pit," so that all earthwork is measured in the excavation.

4. Embankment.

Dimensions. This includes width, side slopes, etc.

Material. No soft mud, muck, or vegetable matter allowed in the bank.

Subsidence. In making high banks in the usual manner, allowance must be made for settling, and the banks be made so much higher originally. The following has been used :—

For banks		5	feet high,		3	inches
"	"	10	"	"	5	"
"	"	20	"	"	8½	"
"	"	28	"	"	10	"
"	"	35	"	"	11	"
"	"	40	"	"	12	"

For intermediate heights allowance is made in the same proportion.

An embankment should never be carried up to any piece of masonry, as a bridge abutment, by dumping from the top of the bank in the usual way; but should be wheeled in and rammed in layers.

5. BALLAST.

The kind of ballast to be used, and the thickness, must be named. If of gravel, the quality; if of broken stone, the kind of stone and the size of the pieces. It is measured on the finished work.

6. DETAILS.

The position, size, and slope of the ditches. Providing for the passage of roads, both public and private, and of water-courses. Protecting banks from the action of water, by rip-rap, slope walls, piles, etc. Extra excavations, as foundation pits for bridges, stations, etc. Location of spoil banks and borrow pits.

SPECIFICATIONS FOR MASONRY.

A full description of the work should be given, accompanied by the requisite drawings. It should also be distinctly stated what are the requirements for the first, second, third, and fourth class masonry; *i. e.*, the size of the stones to be used, manner of laying, arrangement of headers and stretchers, kind and amount of dressing, thickness of mortar joints, quality of cement, etc.

CLASSIFICATION ON THE CROTON AQUEDUCT.

1. *Cut stone.* "This means that a tooled draft $1\frac{1}{2}''$ wide shall be cut on the face and joints, so as to bring the stone into the proper lines and angles. The face that shows is to be axed down fair and even. The beds and end joints to be dressed so as to be laid to a joint not more than $\frac{3}{16}''$. The rear beds and joints to be dressed parallel."

2. *Well-hammered work.* "The stone is to be taken 'out of wind' and dressed with hammer, pick, and points, so as to admit of being laid to a compact joint, not more than $\frac{3}{8}''$. The stones are to hold their full size for half their length from front to rear, and on the rear to be at least $\frac{2}{3}$ as wide and $\frac{2}{3}$ as thick as on the front. The face is to be fair but not very smooth."

3. *Rough-hammered work.* "This means that the stones are to be dressed and formed with so much regularity, as will admit of their being laid in a compact and substantial manner and so as to make good lined work." See pp. 186 and 187.

The hydraulic cement used should be fresh-ground. The usual proportion of sand and cement for cement mortar, is one part of cement to two of clean sharp sand. When lime is used in the mortar, the usual proportions are, one part of cement, two of lime, and five of sand.

RAILROAD RESISTANCES.

[NOTE TO PAGE 265.]

The *axle friction* is directly as the radius of the axle, and inversely as that of the wheel. Let w' equal the weight resting on the wheels, and r and r' the radii. Then the axle friction equals $f w' \frac{r}{r'}$; in which $f =$ the coefficient of friction $= .05$ to $.017$, and $\frac{r}{r'} = \frac{1}{13}$ to $\frac{1}{20}$. As a mean we have $.035\, w' \times \frac{1}{15} = .0023\, w'$ or about $\frac{1}{400}\, w'$.

The *rolling friction* at the circumference of the wheel equals $f'\, w$, w being the whole weight, and f' averaging .001, that is, about, .001 of the whole weight, or about one-half the axle friction.

Both combined equals, approximately, $\frac{1}{300}$ of the whole weight. The fraction $\frac{1}{280}$ is often used for the friction and the other resistances at very low speeds, at which they are very small.

The friction on railroads has usually been determined by letting cars run down a steep inclined plane, succeeded by a level or an ascent, until they are stopped by friction.

Let $w =$ the weight of the car, $h =$ the vertical descent of the inclined plane, $h' =$ the vertical ascent of the succeeding plane, $x =$ the distance of the descent, $x' =$ length of the ascent, and $f =$ the coefficient of friction.

Then the "*work*" accumulated in descending $= w\, h$. The work done before the car comes to rest $= f w\, (x + x') + w\, h'$. Equating these two expressions, and reducing, we get, $f = \frac{h - h'}{x + x'}$. If the second plane be level, this becomes $f = \frac{h}{x + x'}$. If the second plane desend, $f = \frac{h + h'}{x + x'}$.

Recent experiments indicate that the friction is not entirely independent of the extent of the surface or the velocity, but that it increases somewhat with them ; and under great pressures it increases somewhat faster than the weights, owing to abrasion taking place.

The *resistance of the air* is found thus : A velocity of one mile per hour $= \frac{44}{30}$ ft. per minute. Then the formula, $s = \frac{v^2}{2g}$, becomes for this speed, $s = (\frac{44}{30})^2 \div 2 \times 32 = 0'.0334$. A column of air of this height, and a base of one square foot, weighs $1 \times 0.0334 \times 0.08 = 0.0027$ pounds.

[NOTE TO PAGE 269.]

A simple formula by D. Gooch for the resistance on railroads is this : The resistance of the train in pounds per T (ton) $= 6 + .03\ (v' - 10)$; in which v' is the velocity in miles per hour That is, 6 lbs. per T + 0.3 lbs. per T per mile per velocities beyond ten miles per hour. For less velocities omit the second term. For the engine and tender take twice the above, *i. e.*, in pounds per T use $12 + 0.6\ (v' - 10)$.

D. K. Clark's formula. He considers part of the resistance to be a constant quantity, and the rest to vary as the square of the velocity. He gives for the resistance of the train in pounds per T $6 + \frac{v^2}{240}$. For the engine and tender take the above amount per T for them as carriages, and in addition, for the resistance of the machinery, take $2 + \frac{v^2}{600}$ pounds for each ton in the total weight of the train, engine, and tender.

Recent French experiments make the total resistance of the train, including the engine, at speeds of from 16 to 25 miles per hour, 0.003 to 0.0045 of the whole weight; from 25 to 37 miles per hour, from 0.0045 to 0.0085 of the whole weight. Excluding the engine and tender, it was, at 24 miles per hour, 0.004 ; at 31 miles per hour, 0,0066 ; and at 35 miles per hour, 0.008.

[NOTE TO PAGE 276.]

Resistance on an ascent in a straight line.

The friction of the axle and of the wheel is now reduced in the ratio of 1 : cosine of the angle of the slope with the horizon; but this difference is so small that it may be neglected. The resistance of the air is not changed.

The new resistance of gravity equals w . sin. angle of slope; or, near enough, w . tan. angle of slope $= w \cdot \frac{\text{rise}}{\text{horizontal distance}}$.

Steep grades in practice.

The Baltimore and Ohio Railroad has grades of 116 ft. per mile for 7 miles, with some curves of 600 ft. radius.

Ellet's Mountain Top Track, in Virginia, has an average of 257 ft. per mile for 2 miles, and a maximum grade of 296 ft. per mile.

Near Genoa a railroad has a grade of 147 ft. per mile for 6 miles, with a maximum of 185 ft.

The Austrian Semmering Railroad has a grade of 132 ft. for several miles, with an average of 113 ft. for 13 miles, and curves of 660 ft. radius.

The Copiapo Railroad, in Chili, has a grade of 196 ft. per mile, for 17 miles. At its chief incline it has 211 ft. per mile for 23 miles.

The Mexico and Vera Cruz Railroad ascends 7000 ft. in 55 miles.

The railroad *over* Mount Cenis has a grade of 440 ft. per mile for 1¼ miles, with one curve of 139 ft. radius. Its gauge is 3.6 ft. It has a middle line of rail, gripped between two horizontal wheels, to get more adhesion.

[NOTE TO PAGE 271.]

Resistance on curves.

There is a three-fold difficulty in determining the resistance on curves, viz., that the facts are few; that those we have are deficient in details of speed, character of engine, condition of track, etc.; and that we do not know what allowances we should make for these differences, even if they were all given. The French engineers have worked out elaborate formulas for these resistances, but they are less valuable practically than the results of observation.

It is now proposed to give some of these results, and to reduce the resistances of the curves to their equivalent grades and lbs. per ton, and finally to the equivalent increase of distance: this last being the most important point for the purpose of equating lines.

It will be assumed that the resistances of curves are inversely proportional to their radii, or directly to their "degree," which equals 5730 divided by the radius in feet. This assumption is true hypothetically, though practically the sharper curve would cause greater proportional resistance.

No. 1. Mr. Latrobe's experiments in 1844, on the Baltimore and Ohio Railroad, indicate that a curve of 400 feet radius ($14\frac{1}{3}°$) doubles the resistance as compared with a straight and level line, for an eight-wheel car at $3\frac{1}{2}$ miles per hour, the original resistance being 7.5 lbs. per ton. Then a 1° curve, or 5730 feet radius, would be equivalent to a resistance $= 7.5 \div 14\frac{1}{3} = 0.52$ lb. per ton, or to an ascent per mile $= 0.52 \times \frac{5280}{2240} = 1.23$ feet.

No. 2. Mr. Ellwood Morris considers this too much, and regards a 1° curve as equivalent to an ascent of 1 foot per mile. This corresponds to 0.42 lb. per ton.

No. 3. On the Pennsylvania Central Railroad (under Mr. Haupt) the grade was reduced on curves at the rate of 0.025 foot per 100 feet per degree of curvature. This makes a 1° curve = 1.32 feet per mile = 0.56 lb. per ton.

No. 4. Another writer says he finds a 400 feet curve = 21 feet per mile. Then a 1° curve = 1.47 feet per mile = 0.62 lb. per ton.

No. 5. Mr. W. C. Young, when superintendent of the Utica and Schenectady Railroad, found the trains to increase their speed very decidedly on passing from a 20 feet straight grade to a level curve of 700 feet radius. Then a 1° curve gives very decidedly less resistance than a grade of 2.4 feet per mile.

No. 6. On the New York and Erie Railroad, a curve of 955 feet radius causes more resistance than a 10 feet grade. Then a 1° curve would cause more than an ascent of 1.67 feet per mile, or 0.7 lb. per ton.

No. 7. On the Virginia Central Railroad, Mr. Ellet found a 300 feet curve to cause more resistance than 58 feet greater grade, or about as much when the engine flanges were oiled. Then a 1° curve would cause more than a 3 feet ascent, or more than $1\frac{1}{4}$ lb. per ton. This is excessive, but is partly accounted for by the length of the wheel-base of the engine. The exceedingly small radius also removes this case from the ordinary category.

I will now reduce the above resistances to the equivalent distances, taking the resistance on a straight level road, at the freight speed of 12 miles per hour, as $10\frac{1}{4}$ lbs. per ton, which is equivalent to 24 feet ascent per mile, and the resistance at the passenger speed of 30 miles per hour as twice this.

The different resistances of curves for different speeds will not be taken into account, for want of data. That portion of it due to friction is the same at all velocities; but that due to concussion must increase as the square of the velocity, since it consumes "Living force." With this omission, and the preceding assumptions, we make, approximately, the resistance caused by turning a complete circle, or 360° of curvature, equivalent to the following increase of distances on a straight and level line.

No. 1. This makes 360° equivalent to a grade of 1.23 feet per mile, for 360 × 100 feet = 6.8 miles, or 8.3 feet for one mile. This is equivalent to an additional distance of 8.3 ÷ 24 = 0.35 mile at freight speed; or to about half this, or 0.18 mile, at passenger speed. It would be equivalent to about half a mile at the slow speed of the experiment, since a resistance of 7.5 lb. per ton would be doubled by a grade of 17 feet per mile, and 8.3 ÷ 17 = 0.48 mile. The same result is also obtained by noticing that a complete circle of 400 feet radius is 2513 feet, or nearly half a mile long.

No. 2. This makes 360° equivalent to a grade of 1 foot per mile for 6.8 miles, or 6.8 feet for one mile: which is equivalent to 6.8 ÷ 24 = 0.28 mile at freight speed, or 0.14 at passenger speed.

No. 3. By similar reasoning, this gives 360° = 9 feet ascent for one mile = 0.38 mile at freight speed, or 0.19 mile at passenger speed.

No. 4. This makes 360° = 10 feet ascent for one mile; or 0.42 mile at freight speed, and 0.21 mile at passenger speed.

Nos. 5, 6, 7, may be analyzed in the same manner.

The average of the first four is that turning 360° of curvature is equivalent to the running an additional distance of 0.36 mile at freight speed, or 0.18 mile at passenger speed. No. 5 agrees with this; No. 6 gives more, and No. 7 much more.

The great disparity between the *proportional* resistances of curves at low and high speeds would be lessened by taking into account the increase of the *absolute* resistance of the curves at high speeds.

Perhaps, in ordinary cases, one-third of a mile per 360° would be about a fair equivalent in equating for curves, particularly taking into account the other objections to them.

[NOTE TO PAGE 146.]

Staking out the side-slopes. The "line," which has been so often spoken of, is the centre-line of the road—its axis—and the stakes which have now been set at every hundred feet, on both straight lines and curves, have marked out only this centre-line. Before the "construction" of the road is commenced, other stakes must be set to show how far on each side of the centre-line the cuttings and fillings will extend. The *data* required are the width of the road, the depth of the necessary cuttings or fillings, and the ratio of the side-slopes to unity.

Assume that the road is to be 20 feet wide, the slopes 2 to 1, and the cuttings 6 feet. Add half the bottom width to twice the depth, and the sum $(10 + 2 \times 6) = 22$, is the "distance out" from the centre stakes, at which the cutting stakes must be set. They should be marked "6.+," or "Cut 6," and be driven obliquely, so as to point in the direction of the slope. If the road had been in filling, the "distance out" would have been the same, but the stakes would have been marked "6.—," or "Fill 6."

Staking out the side-slopes is thus seen to be very easy when the ground is level in its cross-section. But when it is sidelong, farther calculations, or repeated trials with a levelling instrument, are required to find the "distance out" which will correspond to the height of the ground above or below the grade line at that precise distance out. Take the same width of road-bed, side-slopes and depth at the centre-line, as in the preceding paragraph, and suppose the work to be in excavation and the ground to have a sidelong slope. The distance out from the centre stake to the stake on the up-hill side will now be more than 22 feet, for the ground rises in that direction.

Estimate by eye the rise from the centre to where the stake is to be set, add it to the centre height, and calculate the distance out, as before, by multiplying the new depth (*i. e.*, the depth at the centre plus the estimated rise) by the side-slope, and to the product add half the width of the road-bed. Find the height at this distance out with the levelling instrument, and if it agrees with the estimated height, the point has been correctly taken; if not, try again, until the estimated height agrees nearly enough with that found by the instrument (on railroad work, usually to within one-tenth). On the down-hill side the distance out will be less than if the ground were level. It is estimated in a similar manner.

In staking out ground for an embankment the same method is used. A rise in the ground will now decrease the height of the bank, and consequently the distance out, and *vice versa.*

When the difference in heights between the upper and lower side-slope stakes is so great as to necessitate changing the instrument in setting the stakes on the same cross-section, then set the stakes on one side of the line for several stations, and then change the instrument and set those on the other side.

"Cross-section rods" are often used for this work. See Gillespie's "Levelling and Higher Surveying," Fig. 53.

A general formula for any case may be readily investigated. Examining first the up-hill side, and calling the slope of the ground m to 1; that of the side-slopes n to 1; the desired distance from

Fig. 24.

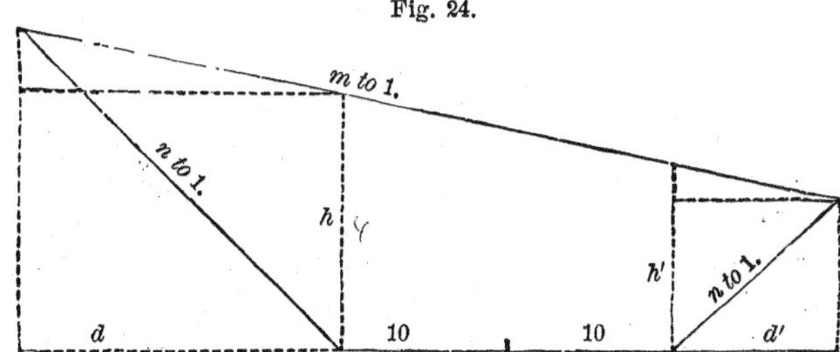

the bottom angle of the cutting, d; and the height of the ground above that bottom angle, h; we obtain (as on page 121),

$$\frac{d}{n} = h + \frac{d}{m}; \text{ whence, } d = h \, . \, \frac{mn}{m - n}$$

If $h = 6$, $n = 2$, and $m = 10$, $d = 6 \times \frac{20}{8} = 15$. Then the up-hill cutting stake will be $10 + 15 = 25$ feet from the centre stake.

Examining next the down-hill side, and using a symmetrical notation, we have, $\frac{d'}{n} = h' - \frac{d'}{m}$, whence, $d' = h' \, . \, \frac{mn}{m + n}$. Let $h' = 4$, $n = 2$, and $m = 10$, $d' = 4 \times \frac{20}{12} = 6.7$, and the "distance out" of the down-hill stake will be $10 + 6.7 = 16.7$ from centre.

Cases of embankment will be represented by the above figure inverted.

Let p and q be the reciprocals of the slope ratios $\left(i.\,e., p = \frac{1}{m}, \text{ and } q = \frac{1}{n}\right)$, or p and q = the heights divided by the bases. Then the formulas are simplified, and become,

$$d = \frac{h}{p + q}, \text{ and } d' = \frac{h'}{p - q}.$$

Formulas of this kind are seldom used in practice. Side-slope stakes can be set very rapidly by the method of repeated trials, given before.

TABLE I.—SLOPES $1\frac{1}{2}$ to 1.—BASE 20.

Feet.	1	2	3	4	5	6	7	8	9	10	Feet.
0	0.39	0.81	1.28	1.78	2.31	2.89	3.50	4.15	4.83	5.55	0
1	0.80	1.24	1.72	2.24	2.80	3.39	4.02	4.68	5.39	6.13	1
2	1.24	1.70	2.20	2.74	3.31	3.92	4.57	5.26	5.98	6.74	2
3	1.72	2.20	2.72	3.28	3.87	4.50	5.17	5.87	6.61	7.39	3
4	2.24	2.74	3.28	3.85	4.46	5.11	5.80	6.52	7.28	8.07	4
5	2.80	3.31	3.87	4.46	5.09	5.76	6.46	7.20	7.98	8.80	5
6	3.39	3.92	4.50	5.11	5.76	6.44	7.17	7.92	8.72	9.55	6
7	4.02	4.57	5.17	5.80	6.46	7.17	7.91	8.68	9.50	10.35	7
8	4.68	5.26	5.87	6.52	7.20	7.92	8.68	9.48	10.31	11.18	8
9	5.39	5.98	6.61	7.28	7.98	8.72	9.50	10.31	11.16	12.05	9
10	6.13	6.74	7.39	8.07	8.80	9.55	10.35	11.18	12.05	12.96	10
11	6.91	7.54	8.20	8.91	9.65	10.42	11.24	12.09	12.98	13.91	11
12	7.72	8.37	9.05	9.78	10.54	11.33	12.17	13.04	13.94	14.89	12
13	8.57	9.24	9.94	10.68	11.46	12.28	13.13	14.02	14.94	15.91	13
14	9.46	10.15	10.87	11.63	12.42	13.26	14.13	15.04	15.98	16.96	14
15	10.39	11.09	11.83	12.61	13.42	14.28	15.17	16.09	17.05	18.05	15
16	11.35	12.07	12 83	13.63	14.46	15.33	16.24	17.18	18.16	19.18	16
17	12.35	13.09	13.87	14.68	15.54	16.42	17.35	18.31	19.31	20.35	17
18	13.39	14.15	14.94	15.78	16.65	17.55	18.50	19.48	20.50	21.55	18
19	14.46	15.24	16.05	16.91	17.80	18.72	19.68	20.68	21.72	22.80	19
20	15.57	16.37	17.20	18.07	18.98	19.92	20.91	21.92	22.98	24.07	20
	1	2	3	4	5	6	7	8	9	10	

Feet.	11	12	13	14	15	16	17	18	19	20	Feet.
0	6.31	7.11	7.94	8.81	9.72	10.67	11.65	12.67	13.72	14.81	0
1	6.91	7.72	8.57	9.46	10.39	11.35	12.35	13.39	14.46	15.57	1
2	7.54	8.37	9.24	10.15	11.09	12.07	13.09	14.15	15.24	16.37	2
3	8.20	9.05	9.94	10.87	11.83	12.83	13.87	14.94	16.05	17.20	3
4	8.91	9.78	10.68	11.63	12.61	13.63	14.68	15.78	16.91	18.07	4
5	9.65	10.54	11.46	12.42	13.42	14.46	15.54	16.65	17.80	18.98	5
6	10.42	11.33	12.28	13.26	14.28	15.33	16.42	17.55	18.72	19.92	6
7	11.24	12.17	13.13	14.13	15.17	16.24	17.35	18.50	19.68	20.91	7
8	12.09	13.04	14.02	15.04	16.09	17.18	18.31	19.48	20.68	21.92	8
9	12.98	13.94	14.94	15.98	17.05	18.16	19.31	20.50	21.72	22.98	9
10	13.91	14.89	15.91	16.96	18.05	19.18	20.35	21.55	22.80	24.07	10
11	14.87	15.87	16.91	17.98	19.09	20.24	21.42	22.65	23.91	25.20	11
12	15.87	16.89	17.94	19.03	20.17	21.33	22.54	23.78	25.05	26.37	12
13	16.91	17.94	19.02	20.13	21.28	22.46	23.68	24.94	26.24	27.57	13
14	17.98	19.03	20.13	21.26	22.42	23.63	24.87	26.15	27.46	28.81	14
15	19.09	20.17	21.28	22.42	23.61	24.83	26.09	27.39	28.72	30.09	15
16	20.24	21.33	22.46	23.63	24.83	26.07	27.35	28.67	30.02	31.41	16
17	21.42	22.54	23.68	24.87	26.09	27.35	28.65	29.98	31.35	32.76	17
18	22.65	23.78	24.94	26.15	27.39	28.67	29.98	31.33	32.72	34.15	18
19	23.91	25.05	26.24	27.46	28.72	30.02	31.35	32.72	34.13	35.57	19
20	25.20	26.37	27.57	28.81	30.09	31.41	32.76	34.15	35.57	37.04	20
	11	12	13	14	15	16	17	18	19	20	

TABLE I.—SLOPES $1\frac{1}{2}$ to 1.—BASE 30.

Feet.	1	2	3	4	5	6	7	8	9	10	Feet.
0	0 57	1.19	1.83	2.52	3.24	4.00	4.80	5.63	6.49	7.41	0
1	1.17	1.80	2.46	3.17	3.91	4.69	5.50	6.35	7.24	8.17	1
2	1.80	2.44	3.13	3.85	4.61	5.41	6.24	7.11	8.02	8.96	2
3	2.46	3.13	3.83	4.57	5.35	6.17	7.02	7.91	8.83	9.80	3
4	3.17	3.85	4.57	5.33	6.13	6.96	7.83	8.74	9.69	10.67	4
5	3.91	4.61	5.35	6.13	6.94	7.80	8.69	9.61	10.57	11.57	5
6	4.69	5.41	6.17	6.96	7.80	8.67	9.57	10.52	11.50	12.52	6
7	5.50	6.24	7.02	7.83	8.69	9.57	10.50	11.46	12.46	13.50	7
8	6 35	7.11	7.91	8.74	9.61	10.52	11.46	12.44	13.46	14.52	8
9	7.24	8.02	8.83	9.69	10.57	11.50	12.46	13.46	14.50	15.57	9
10	8.17	8.96	9.80	10.67	11.57	12.52	13.50	14.52	15.57	16.67	10
11	9.13	9.94	10.80	11.69	12.61	13.57	14.57	15.61	16.69	17.80	11
12	10.13	10.96	11 83	12.74	13.69	14.67	15.69	16.74	17.83	18.96	12
13	11.17	12.02	12.91	13.83	14.80	15.80	16.83	17.91	19.02	20.17	13
14	12.24	13.11	14.02	14.96	15.94	16.96	18.02	19.11	20.24	21.41	14
15	13.35	14.24	15.17	16.13	17.13	18.17	19.24	20.35	21.50	22.69	15
16	14.50	15.41	16.35	17.33	18.35	19.41	20.50	21.63	22.80	24.00	16
17	15.69	16.61	17.57	18.57	19.61	20.69	21.80	22.94	24.13	25.35	17
18	16.91	17.85	18.83	19.85	20.91	22.00	23.13	24.30	25.50	26.74	18
19	18.17	19.13	20.13	21.17	22.24	23.35	24.50	25.69	26.91	28.17	19
20	19.46	20.44	21.46	22.52	23.61	24.74	25.91	27.11	28.35	29.63	20
	1	2	3	4	5	6	7	8	9	10	

Feet.	11	12	13	14	15	16	17	18	19	20	Feet.
0	8.35	9.33	10.35	11.41	12.50	13.63	14.80	16.00	17.24	18.52	0
1	9.13	10.13	11.17	12.24	13.35	14.50	15.69	16.91	18.17	19.46	1
2	9.94	10.96	12.02	13.11	14.24	15.41	16.61	17.85	19.13	20.44	2
3	10.80	11.83	12.91	14.02	15.17	16.35	17.57	18.83	20.13	21.46	3
4	11.69	12.74	13.83	14.96	16.13	17.33	18.57	19.85	21.17	22.52	4
5	12.61	13.69	14.80	15.94	17.13	18.35	19.61	20.91	22.24	23.61	5
6	13.57	14.67	15.80	16.96	18.17	19.41	20.69	22.00	23.35	24.74	6
7	14.57	15.69	16.83	18.02	19.24	20.50	21.80	23.13	24.50	25.91	7
8	15.61	16.74	17.91	19.11	20.35	21.63	22.94	24.30	25.69	27.11	8
9	16.69	17.83	19.02	20.24	21.50	22.80	24.13	25.50	26.91	28.35	9
10	17.80	18.96	20.17	21.41	22.69	24.00	25.35	26.74	28.17	29.63	10
11	18.94	20.13	21.35	22.61	23.91	25.24	26.61	28.02	29.46	30.94	11
12	20.13	21.33	22.57	23.85	25.17	26.52	27.91	29.33	30.80	32.30	12
13	21.35	22.57	23.83	25.13	26.46	27.83	29.24	30.69	32.17	33.69	13
14	22.61	23.85	25.13	26.44	27.80	29.19	30.61	32.07	33.57	35.11	14
15	23.91	25.17	26.46	27.80	29.17	30.57	32.02	33.50	35.02	36.57	15
16	25.24	26.52	27.83	29.19	30.57	32.00	33.46	34.96	36.50	38.07	16
17	26.61	27.91	29.24	30.61	32.02	33.46	34.98	36.46	38.02	39.61	17
18	28.02	29.33	30.69	32.07	33.50	34.96	36.46	38.00	39.57	41.19	18
19	29.46	30.80	32.17	33.57	35.02	36.50	38.02	39.57	41.17	42.80	19
20	30.94	32.30	33.69	35.11	36.57	38.07	39.61	41.19	42.80	44.44	20
	11	12	13	14	15	16	17	18	19	20	

TABLE III.—SLOPES 2 to 1.—BASE 20.

Feet.	1	2	3	4	5	6	7	8	9	10	Feet.
0	0.40	0.84	1.33	1.88	2.47	3.11	3.80	4.54	5.33	6.17	0
1	0.81	1.28	1.80	2.37	2.99	3.65	4.37	5.13	5.95	6.81	1
2	1.28	1.78	2.32	2.91	3.55	4.25	4.99	5.78	6.62	7.51	2
3	1.80	2.32	2.89	3.51	4.17	4.89	5.65	6.47	7.33	8.25	3
4	2.37	2.91	3.51	4.15	4.84	5.58	6.37	7.21	8.10	9.04	4
5	2.99	3.55	4.17	4.84	5.55	6.32	7.13	8.00	8.91	9.88	5
6	3.65	4.25	4.89	5.58	6.32	7.11	7.95	8.84	9.78	10.76	6
7	4.37	4.99	5.65	6.37	7.13	7.95	8.81	9.73	10.69	11.70	7
8	5.13	5.78	6.47	7.21	8.00	8.84	9.73	10.67	11.65	12.69	8
9	5.95	6.62	7.33	8.10	8.91	9.78	10.69	11.65	12.67	13.73	9
10	6.81	7.51	8.25	9.04	9.88	10.76	11.70	12.69	13.73	14.81	10
11	7.73	8.44	9.21	10.02	10.89	11.80	12.76	13.78	14.84	15.95	11
12	8.69	9.43	10.22	11.06	11.95	12.89	13.88	14.91	16.00	17.13	12
13	9.70	10.47	11.28	12.15	13.06	14.02	15.04	16.10	17.21	18.37	13
14	10.76	11.55	12.39	13.28	14.22	15.21	16.25	17.33	18.47	19.65	14
15	11.88	12.69	13.55	14.47	15.43	16.44	17.51	18.62	19.78	20.99	15
16	13.04	13.88	14.76	15.70	16.69	17.73	18.81	19.95	21.13	22.37	16
17	14.25	15.11	16.02	16.99	18.00	19.06	20.17	21.33	22.54	23.80	17
18	15.51	16.39	17.33	18.32	19.36	20.44	21.58	22.76	24.00	25.28	18
19	16.81	17.73	18.69	19.70	20.77	21.88	23.04	24.25	25.51	26.81	19
20	18.17	19.11	20.10	21.13	22.22	23.36	24.54	25.78	27.06	28.40	20
	1	2	3	4	5	6	7	8	9	10	

Feet.	11	12	13	14	15	16	17	18	19	20	Feet.
0	7.06	8.00	8.99	10.02	11.11	12.25	13.43	14.67	15.95	17.28	0
1	7.73	8.69	9.70	10.76	11.88	13.04	14.25	15.51	16.81	18.17	1
2	8.44	9.43	10.47	11.55	12.69	13.88	15.11	16.39	17.73	19.11	2
3	9.21	10.22	11.28	12.39	13.55	14.76	16.02	17.33	18.69	20.10	3
4	10.02	11.07	12.15	13.28	14.47	15.70	16.99	18.32	19.70	21.13	4
5	10.89	11.95	13.06	14.22	15.43	16.69	18.00	19.36	20.77	22.22	5
6	11.80	12.89	14.02	15.21	16.44	17.73	19.06	20.44	21.88	23.36	6
7	12.77	13.88	15.04	16.25	17.51	18.81	20.17	21.58	23.04	24.54	7
8	13.78	14.91	16.10	17.33	18.62	19.95	21.33	22.76	24.25	25.78	8
9	14.84	16.00	17.21	18.47	19.78	21.13	22.54	24.00	25.51	27.06	9
10	15.95	17.13	18.37	19.65	20.99	22.37	23.80	25.28	26.81	28.40	10
11	17.11	18.32	19.58	20.89	22.25	23.65	25.11	26.62	28.17	29.78	11
12	18.32	19.55	20.84	22.17	23.55	24.99	26.47	28.00	29.58	31.21	12
13	19.58	20.84	22.15	23.51	24.91	26.37	27.88	29.43	31.04	32.69	13
14	20.89	22.17	23.51	24.89	26.32	27.80	29.33	30.91	32.54	34.22	14
15	22.25	23.55	24.91	26.32	27.78	29.28	30.84	32.44	34.10	35.80	15
16	23.65	24.99	26.37	27.80	29.28	30.81	32.39	34.02	35.70	37.43	16
17	25.11	26.47	27.88	29.33	30.84	32.39	34.00	35.65	37.36	39.11	17
18	26.62	28.00	29.43	30.91	32.44	34.02	35.65	37.33	39.06	40.84	18
19	28.17	29.58	31.04	32.54	34.10	35.70	37.36	39.06	40.81	42.62	19
20	29.78	31.21	32.69	34.22	35.80	37.43	39.11	40.84	42.62	44.44	20
	11	12	13	14	15	16	17	18	19	20	

TABLE IV.—SLOPES 2 to 1.—BASE 30.

Feet.	1	2	3	4	5	6	7	8	9	10	Feet.
0	0.58	1.21	1.89	2.62	3.40	4.22	5.10	6.02	7.00	8.02	0
1	1.18	1.84	2.54	3.30	4.10	4.95	5.85	6.80	7.80	8.85	1
2	1.84	2.52	3.25	4.02	4.85	5.73	6.65	7.63	8.65	9.73	2
3	2.54	3.25	4.00	4.80	5.65	6.55	7.51	8.51	9.55	10.65	3
4	3.30	4.02	4.80	5.63	6.51	7.43	8.41	9.43	10.51	11.63	4
5	4.10	4.85	5.65	6.51	7.41	8.36	9.36	10.41	11.51	12.65	5
6	4.95	5.73	6.55	7.43	8.36	9.33	10.36	11.43	12.55	13.73	6
7	5.85	6.65	7.51	8.41	9.36	10.36	11.41	12.51	13.65	14.85	7
8	6.80	7.63	8.51	9.43	10.41	11.43	12.51	13.63	14.80	16.02	8
9	7.80	8.65	9.55	10.51	11.51	12.55	13.65	14.80	16.00	17.25	9
10	8.85	9.73	10.65	11.63	12.65	13.73	14.85	16.02	17.25	18.52	10
11	9.95	10.85	11.80	12.80	13.85	14.95	16.10	17.29	18.54	19.84	11
12	11.10	12.02	13.00	14.02	15.10	16.22	17.39	18.62	19.89	21.21	12
13	12.30	13.25	14.25	15.30	16.40	17.54	18.74	19.99	21.28	22.63	13
14	13.54	14.52	15.54	16.62	17.74	18.91	20.13	21.41	22.73	24.10	14
15	14.84	15.84	16.89	17.99	19.13	20.33	21.58	22.88	24.22	25.62	15
16	16.18	17.21	18.28	19.41	20.58	21.80	23.07	24.39	25.76	27.18	16
17	17.58	18.63	19.73	20.88	22.07	23.32	24.62	25.96	27.36	28.80	17
18	19.02	20.10	21.22	22.39	23.62	24.89	26.21	27.58	29.00	30.47	18
19	20.52	21.62	22.76	23.96	25.21	26.51	27.85	29.25	30.69	32.18	19
20	22.06	23.18	24.36	25.58	26.85	28.17	29.54	30.96	32.43	33.95	20
	1	2	3	4	5	6	7	8	9	10	

Feet.	11	12	13	14	15	16	17	18	19	20	Feet.
0	9.10	10.22	11.40	12.62	13.89	15.21	16.58	18.00	19.47	20.99	0
1	9.95	11.10	12.30	13.54	14.84	16.18	17.58	19.02	20.52	22.06	1
2	10.85	12.02	13.25	14.52	15.84	17.21	18.63	20.10	21.62	23.18	2
3	11.80	13.00	14.25	15.54	16.89	18.28	19.73	21.22	22.76	24.36	3
4	12.80	14.02	15.30	16.62	17.99	19.41	20.88	22.39	23.96	25.58	4
5	13.85	15.10	16.40	17.74	19.13	20.58	22.07	23.62	25.21	26.85	5
6	14.95	16.22	17.54	18.91	20.33	21.80	23.32	24.89	26.51	28.17	6
7	16.10	17.39	18.74	20.13	21.58	23.07	24.62	26.21	27.85	29.54	7
8	17.29	18.62	19.99	21.41	22.88	24.39	25.96	27.58	29.25	30.96	8
9	18.54	19.89	21.28	22.73	24.22	25.76	27.36	29.00	30.69	32.43	9
10	19.84	21.21	22.63	24.10	25.62	27.18	28.80	30.47	32.18	33.95	10
11	21.18	22.58	24.02	25.52	27.06	28.65	30.30	31.99	33.73	35.52	11
12	22.58	24.00	25.47	26.99	28.55	30.17	31.84	33.55	35.32	37.13	12
13	24.02	25.47	26.96	28.51	30.10	31.74	33.43	35.17	36.96	38.80	13
14	25.52	26.99	28.51	30.07	31.69	33.36	35.07	36.84	38.65	40.52	14
15	26.06	28.55	30.10	31.69	33.33	35.02	36.76	38.55	40.39	42.28	15
16	28.65	30.17	31.74	33.36	35.02	36.74	38.51	40.32	42.18	44.10	16
17	30.30	31.84	33.43	35.07	36.76	38.51	40.30	42.13	44.92	45.96	17
18	31.99	33.55	35.17	36.84	38.55	40.32	42.13	44.00	45.91	47.88	18
19	33.73	35.32	36.96	38.65	40.39	42.18	44.92	45.91	47.85	49.84	19
20	35.52	37.13	38.80	40.52	42.28	44.10	45.96	47.88	49.84	51.85	20
	11	12	13	14	15	16	17	18	19	20	

BOOK-KEEPING.

Folsom's Logical Book-keeping, ... $2 00
Folsom's Blanks to Book-keeping, ... 4 50

This treatise embraces the interesting and important discoveries of Prof. Folsom (of the Albany "Bryant & Stratton College"), the partial enunciation of which in lectures and otherwise has attracted so much attention in circles interested in commercial education.

After studying business phenomena for many years, he has arrived at the positive laws and principles that underlie the whole subject of Accounts; finds that the science is based in *Value* as a generic term; that value divides into *two classes* with varied species; that all the exchanges of values are reducible to nine equations; and that all the results of all these exchanges are limited to *thirteen* in number.

As accounts have been universally taught hitherto, without setting out from a radical analysis or definition of values, the science has been kept in great obscurity, and been made as difficult to impart as to acquire. On the new theory, however, these obstacles are chiefly removed. In reading over the first part of it, in which the governing laws and principles are discussed, a person with ordinary intelligence will obtain a fair conception of the *double entry* process of accounts. But when he comes to study thoroughly these laws and principles as there enunciated, and works out the examples and memoranda which elucidate the *thirteen results* of business, the student will neither fail in readily acquiring the science as it is, nor in becoming able intelligently to apply it in the interpretation of business.

Smith & Martin's Book-keeping, ... 1 25
Smith & Martin's Blanks, ... *60

This work is by a practical teacher and a practical book-keeper. It is of a thoroughly popular class, and will be welcomed by every one who loves to see theory and practice combined in an easy, concise, and methodical form.

The Single Entry portion is well adapted to supply a want felt in nearly all other treatises, which seem to be prepared mainly for the use of wholesale merchants, leaving retailers, mechanics, farmers, etc., who transact the greater portion of the business of the country, without a guide. The work is also commended, on this account, for general use in Young Ladies' Seminaries, where a thorough grounding in the simpler form of accounts will be invaluable to the future housekeepers of the nation.

The treatise on Double Entry Book-keeping combines all the advantages of the most recent methods, with the utmost simplicity of application, thus affording the pupil all the advantages of actual experience in the counting-house, and giving a clear comprehension of the entire subject through a judicious course of mercantile transactions.

The shape of the book is such that the transactions can be presented as in actual practice; and the simplified form of Blanks—three in number—adds greatly to the ease experienced in acquiring the science.

DRAWING.

Chapman's American Drawing Book, . . . *$6 00

The standard American text-book and authority in all branches of art. A compilation of art principles. A manual for the amateur, and basis of study for the professional artist. Adapted for schools and private instruction.

CONTENTS.—"Any one who can Learn to Write can Learn to Draw."—Primary Instruction in Drawing.—Rudiments of Drawing the Human Head.—Rudiments in Drawing the Human Figure.—Rudiments of Drawing.—The Elements of Geometry.—Perspective.—Of Studying and Sketching from Nature.—Of Painting.—Etching and Engraving.—Of Modeling.—Of Composition.—Advice to the American Art-Student. The work is of course magnificently illustrated with all the original designs.

Chapman's Elementary Drawing Book, . . 1 50

A Progressive Course of Practical Exercises, or a text-book for the training of the eye and hand. It contains the elements from the larger work, and a copy should be in the hands of every pupil; while a copy of the "American Drawing Book," named above, should be at hand for reference by the class.

The Little Artist's Portfolio, ... *50

25 Drawing Cards (progressive patterns), 25 Blanks, and a fine Artist's Pencil, all in one neat envelope.

Clark's Elements of Drawing, ... *1 00

A complete course in this graceful art, from the first rudiments of outline to the finished sketches of landscape and scenery.

Fowle's Linear and Perspective Drawing, . *60

For the cultivation of the eye and hand, with copious illustrations and directions for the guidance of the unskilled teacher.

Monk's Drawing Books—Six Numbers, per set, *2 25

Each book contains *eleven* large patterns, with opposing blanks. No. 1. Elementary Studies. No. 2. Studies of Foliage. No. 3. Landscapes. No. 4. Animals, I. No. 5. Animals, II. No. 6. Marine Views, etc.

Allen's Map-Drawing, . . . 25 cts.; Scale, 25

This method introduces a new era in Map-Drawing, for the following reasons:—1. It is a system. This is its greatest merit.—2. It is easily understood and taught.—3. The eye is trained to exact measurement by the use of a scale.—4. By no special effort of the memory, distance and comparative size are fixed in the mind.—5. It discards useless construction of lines.—6. It can be taught by any teacher, even though there may have been no previous practice in Map-Drawing.—7. Any pupil old enough to study Geography can learn by this System, in a short time, to draw accurate maps.—8. The System is not the result of theory, but comes directly from the school-room. It has been thoroughly and successfully tested there, with all grades of pupils.—9. It is economical, as it requires no mapping plates. It gives the pupil the ability of rapidly drawing accurate maps.

Ripley's Map-Drawing, ... 1 25

Based on the Circle. One of the most efficient aids to the acquirement of a knowledge of Geography is the practice of map-drawing. It is useful for the same reason that the best exercise in orthography is the *writing* of difficult words. Sight comes to the aid of hearing, and a double impression is produced upon the memory. Knowledge becomes less mechanical and more intuitive. The student who has sketched the outlines of a country, and dotted the important places, is little likely to forget either. The impression produced may be compared to that of a traveller who has been over the ground, while more comprehensive and accurate in detail.

NATURAL SCIENCE.

FAMILIAR SCIENCE.

Norton & Porter's First Book of Science, . $1 75

By eminent Professors of Yale College. Contains the principles of Natural Philosophy, Astronomy, Chemistry, Physiology, and Geology. Arranged on the Catechetical plan for primary classes and beginners.

Chambers' Treasury of Knowledge, ... 1 25

Progressive lessons upon—*first*, common things which lie most immediately around us, and first attract the attention of the young mind; *second*, common objects from the Mineral, Animal, and Vegetable kingdoms, manufactured articles, and miscellaneous substances; *third*, a systematic view of Nature under the various sciences. May be used as a Reader or Text-book.

NATURAL PHILOSOPHY.

Norton's First Book in Natural Philosophy, 1 00

By Prof. NORTON, of Yale College. Designed for beginners. Profusely illustrated and arranged on the Catechetical plan.

Peck's Ganot's Course of Nat. Philosophy, . 1 75

The standard text-book of France, Americanized and popularized by Prof. PECK, of Columbia College. The most magnificent system of illustration ever adopted in an American school-book is here found. For intermediate classes.

Peck's Elements of Mechanics, ... 2 00

A suitable introduction to Bartlett's higher treatises on Mechanical Philosophy, and adequate in itself for a complete academical course.

Bartlett's SYNTHETIC AND ANALYTIC **Mechanics,** . each 5 00

Bartlett's Acoustics and Optics, ... 3 50

A system of Collegiate Philosophy, by Prof. BARTLETT, of West Point Military Academy.

Steele's 14 Weeks Course in Philos. (see p. 34) 1 50

Steele's Philosophical Apparatus, ... *125 00

Adequate to performing the experiments in the ordinary text-books. The articles will be sold separately, if desired. See special circular for details.

GEOLOGY.

Page's Elements of Geology, ... 1 25

A volume of Chambers' Educational Course. Practical, simple, and eminently calculated to make the study interesting.

Emmons' Manual of Geology, ... 1 25

The first Geologist of the country has here produced a work worthy of his reputation.

Steele's 14 Weeks Course (see p. 34) 1 50

Steele's Geological Cabinet, ... *40 00

Containing 125 carefully selected specimens. In four parts. Sold separately, if desired. See circular for details.

CHEMISTRY.

Porter's First Book of Chemistry, ... $1 00

Porter's Principles of Chemistry, ... 2 00

The above are widely known as the productions of one of the most eminent scientific men of America. The extreme simplicity in the method of presenting the science, while exhaustively treated, has excited universal commendation.

Darby's Text-Book of Chemistry, ... 1 75

Purely a Chemistry, divesting the subject of matters comparatively foreign to it (such as heat, light, electricity, etc.), but usually allowed to engross too much attention in ordinary school-books.

Gregory's Organic Chemistry, ... 2 50

Gregory's Inorganic Chemistry, ... 2 50

The science exhaustively treated. For colleges and medical students.

Steele's Fourteen Weeks Course, ... 1 50

A successful effort to reduce the study to the limits of a *single term*, thereby making feasible its general introduction in institutions of every character. The author's felicity of style and success in making the science pre-eminently *interesting* are peculiarly noticeable features. (See page 34.)

Steele's Chemical Apparatus, ... *20 00

Adequate to the performance of all the important experiments in the ordinary text-book.

Steele's New Chemistry, (see p. 34) 1 50

Contains the new nomenclature.

BOTANY.

Thinker's First Lessons in Botany, ... 40

For children. The technical terms are largely dispensed with in favor of an easy and familiar style adapted to the smallest learner.

Wood's Object-Lessons in Botany, ... 1 50

Wood's American Botanist and Florist, . . 2 50

Wood's New Class-Book of Botany, ... 3 50

The standard text-books of the United States in this department. In style they are simple, popular, and lively; in arrangement, easy and natural; in description, graphic and strictly exact. The Tables for Analysis are reduced to a perfect system. More are annually sold than of all others combined.

Wood's Plant Record, ... *75

A simple form of Blanks for recording observations in the field.

Wood's Botanical Apparatus, ... *8 00

A portable Trunk, containing Drying Press, Knife, Trowel, Microscope, and Tweezers, and a copy of Wood's Plant Record—composing a complete outfit for the collector.

Young's Familiar Lessons, ... 2 00

Darby's Southern Botany, ... 2 00

Embracing general Structural and Physiological Botany, with vegetable products, and descriptions of Southern plants, and a complete Flora of the Southern States.

Jarvis' Physiology and Laws of Health.

TESTIMONIALS.

From SAMUEL B. McLANE, *Superintendent Public Schools, Keokuk, Iowa.*

I am glad to see a really good text-book on this much neglected branch. This is clear, concise, accurate, and eminently adapted to the *class-room.*

From WILLIAM F. WYERS, *Principal of Academy, West Chester, Pennsylvania*

A thorough examination has satisfied me of its superior claims as a text-book to the attention of teacher and taught. I shall introduce it at once.

From H. R. SANFORD, *Principal of East Genesee Conference Seminary, N. Y.*

"Jarvis' Physiology" is received, and fully met our expectations. We immediately adopted it.

From ISAAC T. GOODNOW, *State Superintendent of Kansas—published in connection with the "School Law."*

"Jarvis' Physiology," a common-sense, practical work, with just enough of anatomy to understand the physiological portions. The last six pages, on Man's Responsibility for his own health, are worth the price of the book.

From D. W. STEVENS, *Superintendent Public Schools, Fall River, Mass.*

I have examined Jarvis' "Physiology and Laws of Health," which you had the kindness to send to me a short time ago. In my judgment it is far the best work of the kind within my knowledge. It has been adopted as a text-book in our public schools.

From HENRY G. DENNY, *Chairman Book Committee, Boston, Mass.*

The very excellent "Physiology" of D. Jarvis I had introduced into our High School, where the study had been temporarily dropped, believing it to be by far the best work of the kind that had come under my observation; indeed, the reintroduction of the study was delayed for some months, because Dr. Jarvis' book could not be had, and we were unwilling to take any other.

From PROF. A. P. PEABODY, D.D., LL.D., *Harvard University.*

* * I have been in the habit of examining school-books with great care, and I hesitate not to say that, of all the text-books on Physiology which have been given to the public, Dr. Jarvis' deserves the first place on the score of accuracy, thoroughness, method, simplicity of statement, and constant reference to topics of practical interest and utility.

From JAMES N. TOWNSEND, *Superintendent Public Schools, Hudson, N. Y.*

Every human being is appointed to take charge of his own body; and of all books written upon this subject, I know of none which will so well prepare one to do this as "Jarvis' Physiology"—that is, in so small a compass of matter. It considers the pure, simple *laws of health* paramount to science; and though the work is thoroughly scientific, it is divested of all cumbrous technicalities, and presents the subject of physical life in a manner and style really charming. It is unquestionably the best text-book on physiology I have ever seen. It is giving great satisfaction in the schools of this city, where it has been adopted as the standard.

From L. J. SANFORD, M.D., *Prof. Anatomy and Physiology in Yale College*

Books on human physiology, designed for the use of schools, are more generally a failure perhaps than are school-books on most other subjects.

The great want in this department is met, we think, in the well-written treatise of Dr. Jarvis, entitled "Physiology and Laws of Health." * * The work is not too detailed nor too expansive in any department, and is clear and concise in all. It is not burdened with an excess of anatomical description, nor rendered discursive by many zoological references. Anatomical statements are made to the extent of qualifying the student to attend, understandingly, to an exposition of those functional processes which, collectively, make up health; thus the laws of health are enunciated, and many suggestions are given which, if heeded, will tend to its preservation.

☞ **For further testimony of similar character, see current numbers of the Illustrated Educational Bulletin.**

NATURAL SCIENCE.

"FOURTEEN WEEKS" IN EACH BRANCH.

By J. DORMAN STEELE, A. M.

Steele's 14 Weeks Course in Chemistry NEW ED., $1 50

Steele's 14 Weeks Course in Astronomy . 1 50

Steele's 14 Weeks Course in Philosophy . 1 50

Steele's 14 Weeks Course in Geology. . 1 50

Steele's 14 Weeks Course in Physiology . 1 50

Our Text-Books in these studies are, as a general thing, dull and uninteresting. They contain from 400 to 600 pages of dry facts and unconnected details. They abound in that which the student cannot learn, much less remember. The pupil commences the study, is confused by the fine print and coarse print, and neither knowing exactly what to learn nor what to hasten over, is crowded through the single term generally assigned to each branch, and frequently comes to the close without a definite and exact idea of a single scientific principle.

Steele's Fourteen Weeks Courses contain only that which every well-informed person should know, while all that which concerns only the professional scientist is omitted. The language is clear, simple, and interesting, and the illustrations bring the subject within the range of home life and daily experience. They give such of the general principles and the prominent facts as a pupil can make familiar as household words within a single term. The type is large and open; there is no fine print to annoy; the cuts are copies of genuine experiments or natural phenomena, and are of fine execution.

In fine, by a system of condensation peculiarly his own, the author reduces each branch to the limits of a single term of study, while sacrificing nothing that is essential, and nothing that is usually retained from the study of the larger manuals in common use. Thus the student has rare opportunity to *economize his time*, or rather to employ that which he has to the best advantage.

A notable feature is the author's charming "style," fortified by an enthusiasm over his subject in which the student will not fail to partake. Believing that Natural Science is full of fascination, he has moulded it into a form that attracts the attention and kindles the enthusiasm of the pupil.

The recent editions contain the author's "Practical Questions" on a plan never before attempted in scientific text-books. These are questions as to the nature and cause of common phenomena, and are not directly answered in the text, the design being to test and promote an intelligent use of the student's knowledge of the foregoing principles.

Steele's General Key to his Works *1 50

This work is mainly composed of Answers to the Practical Questions and Solutions of the Problems in the author's celebrated "Fourteen Weeks Courses" in the several sciences, with many hints to teachers, minor Tables, &c. Should be on every teacher's desk.

Steele's 14 Weeks in each Science.

TESTIMONIALS.

From L. A. BIKLE, *President N. C. College.*

I have not been disappointed. Shall take pleasure in introducing this series.

From J. F. COX, *Prest. Southern Female College, Ga.*

I am much pleased with these books, and expect to introduce them.

From J. R. BRANHAM, *Prin. Brownsville Female College, Tenn.*

They are capital little books, and are now in use in our institution.

From W. H. GOODALE, *Professor Readville Seminary, La.*

We are using your 14 Weeks Course, and are much pleased with them.

From W. A. BOLES, *Supt. Shelbyville Graded School, Ind.*

They are as entertaining as a story book, and much more improving to the mind.

From S. A. SNOW, *Principal of High School, Uxbridge, Mass.*

Steele's 14 Weeks Courses in the Sciences are a perfect success.

From JOHN W. DOUGHTY, *Newburg Free Academy, N. Y.*

I was prepared to find Prof. Steele's Course both attractive and instructive. My highest expectations have been fully realized.

From J. S. BLACKWELL, *Prest. Ghent College, Ky.*

Prof. Steele's unexampled success in providing for the wants of academic classes, has led me to look forward with high anticipations to his forthcoming issue.

From J. F. COOK, *Prest. La Grange College, Mo.*

I am pleased with the neatness of these books and the delightful diction. I have been teaching for years, and have never seen a lovelier little volume than the Astronomy.

From M. W. SMITH, *Prin. of High School, Morrison, Ill.*

They seem to me to be admirably adapted to the wants of a public school, containing, as they do, a sufficiently comprehensive arrangement of elementary principles to excite a healthy thirst for a more thorough knowledge of those sciences.

From J. D. BARTLEY, *Prin. of High School, Concord, N. H.*

They are just such books as I have looked for, viz., those of interesting style, not cumbersome and filled up with things to be omitted by the pupil, and yet sufficiently full of facts for the purpose of most scholars in these sciences in our high schools; there is nothing but what a pupil of average ability can thoroughly master.

From ALONZO NORTON LEWIS, *Principal of Parker Academy, Conn.*

I consider Steele's Fourteen Weeks Courses in Philosophy, Chemistry, &c., the *best* school-books that have been issued in this country.

As an introduction to the various branches of which they treat, and especially for that numerous class of pupils who have not the time for a more extended course, I consider them *invaluable.*

From EDWARD BROOKS, *Prin. State Normal School, Millersville, Pa.*

At the meeting of Normal School Principals, I presented the following resolution, which was unanimously adopted: "*Resolved,* That Steele's 14 Weeks Courses in Natural Philosophy and Astronomy, or an amount equivalent to what is contained in them, be adopted for use in the State Normal Schools of Pennsylvania." The works themselves will be adopted by at least three of the schools, and, I presume, by them all.

LITERATURE.

Cleveland's Compendiums . . . . each, $*2 50

ENGLISH LITERATURE. AMERICAN LITERATURE.
ENGLISH LITERATURE OF THE XIXTH CENTURY.

In these volumes are gathered the cream of the literature of the English speaking people for the school-room and the general reader. Their reputation is national. More than 125,000 copies have been sold.

Boyd's English Classics . . . . . each, *1 25

MILTON'S PARADISE LOST. THOMSON'S SEASONS.
YOUNG'S NIGHT THOUGHTS. POLLOK'S COURSE OF TIME.
COWPER'S TASK, TABLE TALK, &c. LORD BACON'S ESSAYS.

This series of annotated editions of great English writers, in prose and poetry, is designed for critical reading and parsing in schools. Prof. J. R. Boyd proves himself an editor of high capacity, and the works themselves need no encomium. As auxiliary to the study of Belles Lettres, etc., these works have no equal.

Pope's Essay on Man *20

Pope's Homer's Iliad *80

The metrical translation of the great poet of antiquity, and the matchless "Essay on the Nature and State of Man," by ALEXANDER POPE, afford superior exercise in literature and parsing.

Steele's Brief History of Literature, . . . 1 50

AESTHETICS.

Huntington's Manual of the Fine Arts . .*1 75

A view of the rise and progress of Art in different countries, a brief account of the most eminent masters of Art, and an analysis of the principles of Art. It is complete in itself, or may precede to advantage the critical work of Lord Kames.

Boyd's Kames' Elements of Criticism . .*1 75

The best edition of this standard work; without the study of which none may be considered proficient in the science of the Perceptions. No other study can be pursued with so marked an effect upon the taste and refinement of the pupil.

POLITICAL ECONOMY.

Champlin's Lessons on Political Economy 1 25

An improvement on previous treatises, being shorter, yet containing every thing essential, with a view of recent questions in finance, etc., which is not elsewhere found.

CLEVELAND'S COMPENDIUMS.

TESTIMONIALS.

From the New Englander.

This is the very best book of the kind we have ever examined.

From GEORGE B. EMERSON, Esq., *Boston.*

The Biographical Sketches are just and discriminating; the selections are admirable, and I have adopted the work as a text-book for my first class.

From PROF. MOSES COIT TYLER, *of the Michigan University.*

I have given your book a thorough examination, and am greatly delighted with it; and shall have great pleasure in directing the attention of my classes to a work which affords so admirable a bird's-eye view of recent "English Literature."

From the Saturday Review.

It acquaints the reader with the characteristic method, tone, and quality of all the chief notabilities of the period, and will give the careful student a better idea of the recent history of English Literature than nine educated Englishmen in ten possess.

From the Methodist Quarterly Review, New York.

This work is a transcript of the best American mind; a vehicle of the noblest American spirit. No parent who would introduce his child to a knowledge of our country's literature, and at the same time indoctrinate his heart in the purest principles, need fear to put this manual in the youthful hand.

From REV. C. PEIRCE, *Principal, West Newton, Mass.*

I do not believe the work is to be found from which, within the same limits, so much interesting and valuable information in regard to English writers and English literature of every age, can be obtained; and it deserves to find a place in all our high schools and academies, as well as in every private library.

From the Independent.

The work of selection and compilation—requiring a perfect familiarity with the whole range of English literature, a judgment clear and impartial, a taste at once delicate and severe, and a most sensitive regard to purity of thought or feeling—has been better accomplished in this than in any kindred volume with which we are acquainted.

From the Christian Examiner.

To form such a Compendium, good taste, fine scholarship, familiar acquaintance with English literature, unwearied industry, tact acquired by practice, an interest in the culture of the young, a regard for truth, purity, philanthropy, religion, as the highest attainment and the highest beauty,—all these were needed, and they are united in Mr. Cleveland.

CHAMPLIN'S POLITICAL ECONOMY.

From J. L. BOTHWELL, *Prin. Public School No. 14, Albany, N.Y.*

I have examined Champlin's Political Economy with much pleasure, and shall be pleased to put it into the hands of my pupils. In quantity and quality I think it superior to anything that I have examined.

From PRES. N. E. COBLEIGH, *East Tennessee Wesleyan University.*

An examination of Champlin's Political Economy has satisfied me that it is the book I want. For brevity and compactness, division of the subject, and clear statement, and for appropriateness of treatment, I consider it a better text-book than any other in the market.

From the Evening Mail, New York.

A new interest has been imparted to the science of political economy since we have been necessitated to raise such vast sums of money for the support of the government. The time, therefore, is favorable for the introduction of works like the above. This little volume of two hundred pages is intended for beginners, for the common school and academy. It is intended as a basis upon which to rear a more elaborate superstructure. There is nothing in the principles of political economy above the comprehension of average scholars, when they are clearly set forth. This seems to have been done by President Champlin in an easy and graceful manner.

MENTAL PHILOSOPHY.

Mahan's Intellectual Philosophy $1 75

The subject exhaustively considered. The author has evinced learning, candor, and independent thinking.

Mahan's Science of Logic 2 00

A profound analysis of the laws of thought. The system possesses the merit of being intelligible and self consistent. In addition to the author's carefully elaborated views, it embraces results attained by the ablest minds of Great Britain, Germany, and France, in this department.

Boyd's Elements of Logic 1 25

A systematic and philosophic condensation of the subject, fortified with additions from Watts, Abercrombie, Whately, &c.

Watts on the Mind 50

The Improvement of the Mind, by Isaac Watts, is designed as a guide for the attainment of useful knowledge. As a text-book it is unparalleled; and the discipline it affords cannot be too highly esteemed by the educator.

MORALS.

Peabody's Moral Philosophy, ... 1 25

For Colleges and High Schools.

Willard's Morals for the Young *75

Lessons in conversational tyle to inculcate the elements of moral philosophy. The study is made attractive by narratives and engravings.

GOVERNMENT.

Howe's Young Citizen's Catechism 75

Explaining the duties of District, Town, City, County, State, and United States Officers, with rules for parliamentary and commercial business—that which every future "sovereign" ought to know, and so few are taught.

Young's Lessons in Civil Government . . 1 25

A comprehensive view of Government, and abstract of the laws show ing the rights, duties, and responsibilities of citizens.

Mansfield's Political Manual 1 25

This is a complete view of the theory and practice of the General and State Governments of the United States, designed as a text-book. The author is an esteemed and able professor of constitutional law, widely known for his sagacious utterances in matters of statecraft through the public press. Recent events teach with emphasis the vital necessity that the rising generation should comprehend the noble polity of the American government, that they may act intelligently when endowed with a voice in it.

MODERN LANGUAGE.

French and English Primer, ... $ 10
German and English Primer, ... 10
Spanish and English Primer, ... 10

The names of common objects properly illustrated and arranged in easy lessons.

Ledru's French Fables, ... 75
Ledru's French Grammar, ... 1 00
Ledru's French Reader, ... 1 00

The author's long experience has enabled him to present the most thoroughly practical text-books extant, in this branch. The system of pronunciation (by phonetic illustration) is original with this author, and will commend itself to all American teachers, as it enables their pupils to secure an absolutely correct pronunciation without the assistance of a native master. This feature is peculiarly valuable also to "self-taught" students. The directions for ascertaining the gender of French nouns—also a great stumbling-block—are peculiar to this work, and will be found remarkably competent to the end proposed. The criticism of teachers and the test of the school-room is invited to this excellent series, with confidence.

Worman's French Echo, ... 1 25

To teach conversational French by actual practice, on an entirely new plan, which recognizes the importance of the student learning *to think* in the language which he speaks. It furnishes an extensive vocabulary of words and expressions in common use, and suffices to free the learner from the embarrassments which the peculiarities of his own tongue are likely to be to him, and to make him thoroughly familiar with the use of proper idioms.

Worman's German Echo, ... 1 25

On the same plan. See Worman's German Series, page 29.

Pujol's Complete French Class-Book, . . . 2 25

Offers, in one volume, methodically arranged, a complete French course—usually embraced in series of from five to twelve books, including the bulky and expensive Lexicon. Here are Grammar, Conversation, and choice Literature—selected from the best French authors. Each branch is thoroughly handled; and the student, having diligently completed the course as prescribed, may consider himself, without further application, *au fait* in the most polite and elegant language of modern times.

Maurice-Poitevin's Grammaire Francaise, . 1 00

American schools are at last supplied with an American edition of this famous text-book. Many of our best institutions have for years been procuring it from abroad rather than forego the advantages it offers. The policy of putting students who have acquired some proficiency from the ordinary text-books, into a Grammar written in the vernacular, can not be too highly commended. It affords an opportunity for finish and review at once; while embodying abundant practice of its own rules.

Joynes' French Pronunciation, ... 30

Willard's Historia de los Estados Unidos, . 2 00

The History of the United States, translated by Professors Tolon and De Tornos, will be found a valuable, instructive, and entertaining reading-book for Spanish classes.

Pujol's Complete French Class-Book.

TESTIMONIALS.

From PROF. ELIAS PEISSNER, *Union College.*

I take great pleasure in recommending Pujol and Van Norman's French Class-Book, as there is no French grammar or class-book which can be compared with it in completeness, system, clearness, and general utility.

From EDWARD NORTH, *President of Hamilton College.*

I have carefully examined Pujol and Van Norman's French Class-Book, and am satisfied of its superiority, for college purposes, over any other heretofore used. We shall not fail to use it with our next class in French.

From A. CURTIS, *Pres't of Cincinnati Literary and Scientific Institute.*

I am confident that it may be made an instrument in conveying to the student, in from six months to a year, the art of speaking and writing the French with almost native fluency and propriety.

From HIRAM ORCUTT, A. M., *Prin. Glenwood and Tilden Ladies' Seminaries.*

I have used Pujol's French Grammar in my two seminaries, exclusively, for more than a year, and have no hesitation in saying that I regard it the best text-book in this department extant. And my opinion is confirmed by the testimony of Prof. F. De Launay and Mademoiselle Marindin. They assure me that the book is eminently accurate and practical, as tested in the school-room.

From PROF. THEO. F. DE FUMAT, *Hebrew Educational Institute, Memphis, Tenn.*

M. Pujol's French Grammar is one of the best and most practical works. The French language is chosen and elegant in style—modern and easy. It is far superior to the other French class-books in this country. The selection of the conversational part is very good, and will interest pupils; and being all completed in only one volume, it is especially desirable to have it introduced in our schools.

From PROF. JAMES H. WORMAN, *Bordentown Female College, N. J.*

The work is upon the same plan as the text-books for the study of French and English published in Berlin, for the study of those who have not the aid of a teacher, and these books are considered, by the first authorities, the best books. In most of our institutions, Americans teach the modern languages, and heretofore the trouble has been to give them a text-book that would dispose of the difficulties of the French pronunciation. This difficulty is successfully removed by P. and Van N., and I have every reason to believe it will soon make its way into most of our best schools.

From PROF. CHARLES S. DOD, *Ann Smith Academy, Lexington, Va.*

I cannot do better than to recommend "Pujol and Van Norman." For comprehensive and systematic arrangement, progressive and thorough development of all grammatical principles and idioms, with a due admixture of theoretical knowledge and practical exercise, I regard it as superior to any (other) book of the kind.

From A. A. FORSTER, *Prin. Pinehurst School, Toronto, C. W.*

I have great satisfaction in bearing testimony to M. Pujol's System of French Instruction, as given in his complete class-book. For clearness and comprehensiveness, adapted for all classes of pupils, I have found it superior to any other work of the kind, and have now used it for some years in my establishment with great success.

From PROF. OTTO FEDDER, *Maplewood Institute, Pittsfield, Mass.*

The conversational exercises will prove an immense saving of the hardest kind of labor to teachers. There is scarcely any thing more trying in the way of teaching language, than to rack your brain for short and easily intelligible bits of conversation, and to repeat them time and again with no better result than extorting at long intervals a doubting "oui," or a hesitating "non, monsieur"

☞ For further testimony of a similar character, see special circular, and current numbers of the Educational Bulletin.

GERMAN.

A COMPLETE COURSE IN THE GERMAN.

By JAMES H. WORMAN, A. M.

Worman's Elementary German Grammar . $1 50
Worman's Complete German Grammar . 2 00

These volumes are designed for intermediate and advanced classes respectively. The bitterness with which they have been attacked, and their extraordinary success in the face of an unprincipled opposition, are facts which have stamped them as possessing unparalleled merit.

Though following the same general method with "Otto" (that of 'Gaspey'). our author differs essentially in its application. He is more practical, more systematic, more accurate, and besides introduces a number of invaluable features which have never before been combined in a German grammar.

Among other things, it may be claimed for Prof. Worman that he has been *the first* to introduce in an American text-book for learning German, a system of analogy and comparison with other languages. Our best teachers are also enthusiastic about his methods of inculcating the art of speaking, of understanding the spoken language, of correct pronunciation; the sensible and convenient original classification of nouns (in four declensions), and of irregular verbs, also deserves much praise. We also note the use of heavy type to indicate etymological changes in the paradigms and, in the exercises, the parts which specially illustrate preceding rules.

Worman's Elementary German Reader, . 1 00
Worman's German Reader 1 75

The finest compilation of classical and standard German Literature ever offered to American students. It embraces, progressively arranged, selections from the masterpieces of Goethe, Schiller, Korner, Seume, Uhland, Freiligrath, Heine, Schlegel, Holty, Lenau, Wieland, Herder, Lessing, Kant, Fichte, Schelling, Winkelmann, Humboldt, Ranke, Raumer, Menzel, Gervinus, &c., and contains complete Goethe's "Iphigenie," Schiller's "Jungfrau;" also, for instruction in modern conversational German, Benedix's "Eigensinn."

There are besides, Biographical Sketches of each author contributing, Notes, explanatory and philological (after the text), Grammatical References to all leading grammars, as well as the editor's own, and an adequate Vocabulary.

Worman's German Echo 1 25

Consists of exercises in colloquial style entirely in the German, with an adequate vocabulary, not only of words but of idioms. The object of the system developed in this work (and its companion volume in the French) is to break up the laborious and tedious habit of *translating the thoughts*, which is the student's most effectual bar to fluent conversation, and to lead him to *think in the language in which he speaks*. As the exercises illustrate scenes in actual life, a considerable knowledge of the manners and customs of the German people is also acquired from the use of this manual.

www.ingramcontent.com/pod-product-compliance
Lightning Source LLC
LaVergne TN
LVHW020925110826
845150LV00004B/772